GED

PREPARATION FOR THE
HIGH SCHOOL EQUIVALENCY EXAMINATION

MATHEMATICS

NEW GED TEST 5

JERRY HOWETT

Consultant
M. Michael Dean
Bair Community School
VTAE Broward County
Florida

Project Editor
Ellen Carley Frechette

CONTEMPORARY
BOOKS, INC.
CHICAGO • NEW YORK

Library of Congress Cataloging-in-Publication Data

Howett, Jerry.
Contemporary's new GED test 5, mathematics.

1. Mathematics—Problems, exercises, etc.
2. General educational development tests.
I. Contemporary Books, inc. II. Title. III. Title:
New GED test 5, mathematics. IV. Title: Contemporary's
new GED test five, mathematics. V. Title: New GED
test five, mathematics.
QA43.H73 1987 510'.76 87-5366
ISBN 0-8092-5228-7

Published by Contemporary Books, Inc.
180 North Michigan Avenue, Chicago, Illinois 60601
Manufactured in the United States of America
International Standard Book Number: 0-8092-5228-7

Published simultaneously in Canada by
Beaverbooks, Ltd.
195 Allstate Parkway
Valleywood Business Park
Markham, Ontario L3R 4T8
Canada

Editorial Director
Caren Van Slyke

Editorial
Sarah Schmidt
Chris Benton
Karen Schenkenfelder

Production Editor
Patricia Reid

Art Director and Cover Design
Georgene G. Sainati

Illustrators
Miguel Ibarra
Ellen Sherwood

Art and Production
Princess Louise El
Sue Springston
Lois Koehler

Typography
J-B Typesetting

Cover photo © Image Bank

GED

PREPARATION FOR THE
HIGH SCHOOL EQUIVALENCY EXAMINATION

MATHEMATICS

NEW GED TEST 5

CONTENTS

TO THE STUDENT — vii
HOW TO USE THIS BOOK — xiii
MATHEMATICS PRETEST — 1

1 WHOLE NUMBER REVIEW — 10
Number Values and Facts 10
The Basic Operations 14

2 PROBLEM SOLVING AND SPECIAL TOPICS — 19
The Five-Step Approach to Solving Word Problems 19
Special Topics with Whole Numbers 35
▶ PROBLEM SOLVING: Introducing Set-up Questions 42

3 INTRODUCING NUMBERS SMALLER THAN 1 — 47

4 DECIMALS — 50
What Are Decimals? 50
▶ PROBLEM SOLVING: Comparing and Ordering 56
Adding and Subtracting Decimals 58
Multiplying Decimals 62
Dividing Decimals 64
▶ PROBLEM SOLVING: Introducing Item Sets 70

5 FRACTIONS — 76
Forms of Fractions 76
Adding Like Fractions and Mixed Numbers 85
Subtracting Like Fractions 87
Unlike Fractions 89
Adding and Subtracting Unlike Fractions and Mixed Numbers 92
Multiplying Fractions 96
Dividing Fractions 101
▶ PROBLEM SOLVING: Multiplying and Dividing with Fractions 103
Fractions and Decimals 107
▶ PROBLEM SOLVING: When Both Decimals and Fractions Appear in Word Problems 109

6 PROBABILITY, RATIO, AND PROPORTION — 115

Probability 115
Ratio and Proportion 119
Proportion 121
▶ PROBLEM SOLVING: Using Proportion to Solve Word Problems 124

7 PERCENTS — 128

Understanding Percents 128
▶ PROBLEM SOLVING: Working with Item Sets 148
▶ PROBLEM SOLVING: When Not Enough Information Is Given 152

8 MEASUREMENT — 158

Standard or Imperial Measurements 158
Metric Measurements 162
▶ PROBLEM SOLVING: Comparing and Ordering Measurements 165
Simplifying Measurements 166
Arithmetic Operations 168
▶ PROBLEM SOLVING: What More Do You Need to Know? 170
Scales and Meters 174

9 GRAPHS AND TABLES — 180

What Is A Graph? 180
▶ PROBLEM SOLVING: Using Approximation With Graphs 190
▶ PROBLEM SOLVING: Reading a Graph Carefully 194
Tables 197
Graphs and Tables Summary 199

10 THE BASICS OF ALGEBRA — 202

Writing Algebraic Expressions 202
Equations 205
Solving Longer Equations 210
Writing and Solving Longer Equations 214
▶ PROBLEM SOLVING: Using Algebra to Solve Word Problems 216
Formulas 219

11 GEOMETRY — **224**
Common Geometric Figures 224
Perimeter and Circumference 227
▶ PROBLEM SOLVING: Drawing Pictures 230
Powers and Roots 234
Area 238
Volume 244
Angles 251
Triangles 257
Similarity 260
Congruence 264
Pythagorean Relationship 267
▶ PROBLEM SOLVING: Using the Formulas Page 269

12 SPECIAL TOPICS IN ALGEBRA AND GEOMETRY — **276**
The Number Line 276
Inequalities 286
Multiplying and Dividing Monomials 288
Factoring 289
Using Algebra to Solve Geometry Problems 293
Rectangular Coordinates 295
Linear Equations 300
Slope and Intercepts 302
Quadratic Equations 305
Scientific Notation 307

MATHEMATICS POSTTEST — **311**
ANSWER KEY — **322**
INDEX — **381**

TO THE STUDENT

Introducing the Mathematics Test

Part of preparing for a test is working through the doubts and questions that you have. Perhaps it has been a long time since you took an important test. Maybe you are not sure what is required to pass the GED Mathematics Test, or you are nervous about the test-taking situation.

This book has been designed so that you can succeed on the test. It will provide you with instruction in the skills you need to pass the test, background information on key mathematics concepts, and plenty of GED-type practice. If you work carefully through this book, you should do well. The material on pages xiii–xiv will give you advice on how to use this book.

What Kind of Test Is This?

The GED Mathematics Test consists of multiple-choice questions. Almost all of the questions will be word problems—a short story or situation that asks a question. Each of these questions is followed by five possible answer choices, only one of which is correct.

These math items will test not only your ability to perform arithmetic and algebraic operations but also your ability to use problem-solving skills. Some problem-solving skills that you will need to exercise are: selecting necessary information in a problem, knowing when you have not been given enough information to solve a problem, and estimating an approximate answer. This book will give you lots of practice in these and other problem-solving areas.

What Does the Test Look Like?

There are 56 multiple-choice questions, and you will be given 90 minutes to complete the test. About $\frac{1}{3}$ of the items will be based on diagrams, charts, or graphs. Some of the questions will appear in what are called "item sets." An item set is a short passage or situation that is followed by two or more questions based on that passage.

Some of the questions on the GED Mathematics Test will not require you to do any computation. Instead, you will have to decide the best way to "set up" the problem in order to solve it, or you will have to decide what information is necessary to solve a given problem. In some problems, the answer choice *not enough information is provided* will appear as one of the five possible answers, and it may be the correct answer. Again, this book will give you plenty of opportunities to practice with these types of questions.

To get an idea of what the test is like, look at the practice test at the end of this book. This practice test is based on the real GED Test.

What's on the Test?

The GED Mathematics Test can be broken down into these content areas:

Arithmetic	50%
Algebra	30%
Geometry	20%

The twelve chapters of this book will provide you with explanations, examples, and exercises in all three of these content areas.

The Five GED Tests

The following section answers some of the questions asked most frequently about the GED Tests.

Q: WHAT DOES *GED* STAND FOR?

A: *GED* stands for the Tests of General Educational Development. The GED is a national examination developed by the GED Testing Service of the American Council on Education. The credential received for passing the test is widely recognized by colleges, training schools, and employers as equivalent to a high school diploma.

While the GED measures skills and knowledge normally acquired in four years of high school, much that you have learned informally or through other types of training can help you pass the test.

The GED Test is available in English, French, and Spanish, on audiocassette, in Braille, and in large print.

Q: WHAT SHOULD I KNOW TO PASS THE TEST?

A: The test consists of five examinations in the areas of writing skills, social studies, science, literature and the arts, and mathematics. The chart on the next page outlines the main content areas, the breakdown of questions, and the time allowed per test.

THE GED TESTS

Test	Minutes	Questions	Percentage
Writing Skills			
Part I	75	55	Sentence Structure 35% Usage 35% Mechanics 30%
Part II	45	1 topic	
Social Studies	85	64	History 25% Economics 20% Political Science 20% Geography 15%* Behavioral Science 20%
Science	95	66	Biology 50% Physical Sciences 50%
Literature and the Arts	65	45	Popular Literature 50% Classical Literature 25% Commentary 25%
Mathematics	90	56	Arithmetic 50% Algebra 30% Geometry 20%

* *In Canada, 20% of the test is based on geography and 15% on behavioral sciences.*

On all five tests, you are expected to demonstrate the ability to think about many issues. You are also tested on knowledge and skills you have acquired from life experiences, television, radio, books and newspapers, consumer products, and advertising. In addition to the above information, keep these facts in mind:

1. Three of the five tests—Literature and the Arts, Science, and Social Studies—require that you answer questions based on reading or interpreting cartoons, diagrams, maps, charts, and graphs in these content areas. Developing strong reading and thinking skills is the key to succeeding on these tests.

2. The Writing Skills Test requires you to be able to detect and correct errors in sentence structure, grammar, punctuation, and spelling. You will also have to write a composition of approximately 200 words on a topic familiar to most adults.
3. The Math Test consists mainly of word problems to be solved. Therefore, you must be able to combine your ability to perform computations with problem-solving skills.

Someone once said that an education is what remains after you've forgotten everything else. In many ways, this is what the GED measures.

Q: CAN I TAKE THE TEST?

A: Each year, more than 700,000 people take the GED Test. In the United States, Canada, and many territories, people who have not graduated from high school and who meet specific eligibility requirements (age, residency, etc.) may take the test. Since eligibility requirements vary, it would be useful to contact your local GED testing center or the director of adult education in your state, province, or territory for specific information.

Q: WHAT IS A PASSING SCORE ON THE GED?

A: Again, this varies from area to area. To find out what you need to pass the test, contact your local GED testing center. However, you must keep two scores in mind. One score represents the minimum score you must get on each test. For example, if your state requires minimum scores of 40, you must get at least 40 points on every test. Additionally, you must meet the requirements of a minimum average score on all five tests. For example, if your state requires a minimum average score of 45, you must get a total of 225 points to pass. The two scores together, the minimum score and the minimum average score, determine whether you pass or fail the GED.

To understand this better, look at the scores of three people who took the test in a state that requires a minimum score of 40 and a minimum average score of 45 (225 total). Tom and Sarah did not pass, but Ana did. See if you can tell why.

	Tom	Sarah	Ana
Test 1	44	42	43
Test 2	43	43	48
Test 3	38	42	47
Test 4	50	40	52
Test 5	50	40	49
	225	207	239

Tom made the total of 225 points but fell below the minimum score on Test 3. Sarah passed each test but failed to get the 225 points needed; just passing the individual tests was not enough. Ana passed all the tests and exceeded the minimum average score. Generally, to receive a GED credential, you must correctly answer half or a little more than half of the questions on each test.

Q: WHAT HAPPENS IF I DON'T PASS THE TEST?

A: You are allowed to retake some or all of the tests. Again, the regulations governing the number of times that you may retake the tests and the time you must wait before retaking them are set by your state, province, or territory. Some states require you to take a review class or to study on your own for a certain amount of time before taking the test again.

Q: HOW CAN I BEST PREPARE FOR THE TEST?

A: Many libraries, community colleges, adult education centers, churches, and other institutions offer GED preparation classes. Some television stations broadcast classes to prepare people for the test. If you cannot find a GED preparation class locally, contact the director of adult education in your state, province, or territory.

Q: I NEED TO STUDY FOR THE OTHER TESTS. ARE THERE OTHER MATERIALS AVAILABLE?

A: Contemporary Books publishes a wide range of materials to help you prepare for the tests. These books are designed for home study or class use. Contemporary's GED preparation books are available through schools and bookstores and directly from the publisher, at Contemporary Books, 180 N. Michigan Ave., Chicago, IL 60601.

Now let's focus on some useful test-taking tips. As you read this section, you should feel more confident about your ability to succeed on the Mathematics Test.

Test-Taking Tips

1. **Prepare physically.** Get plenty of rest and eat a well-balanced meal before the test so that you will have energy and will be able to think clearly. Last-minute cramming will probably not help as much as a relaxed and rested mind.
2. **Arrive early.** Be at the testing center at least fifteen to twenty minutes before the starting time. Make sure you have time to find the room and to get situated. Keep in mind that many testing centers refuse to admit latecomers.
3. **Think positively.** Tell yourself you will do well. If you have studied and prepared for the test, you should succeed.
4. **Relax during the test.** Take half a minute several times during the test to stretch and breathe deeply, especially if you are feeling anxious or confused.
5. **Read the test directions carefully.** Be sure you understand how to answer the questions. If you have any questions about the test or about filling in the answer form, ask before the test begins.

6. **Know the time limit for each test.** The math test has a time limit of 90 minutes.

   Some testing centers allow extra time, while others do not. You may be able to find out the policy of your testing center before you take the test, but always work according to the official time limit. If you have extra time, go back and check your answers.

   For this 56-question test, you should allow about one-and-a-half minutes per question. However, this is not a hard and fast rule. Use it only as a guide to keep yourself within the time limit.

7. **Have a strategy for answering questions.** Read each question carefully; reread it if you are having trouble understanding what is being asked of you. You will probably want to use the five-step approach to solving word problems discussed on pages 19–28 of this book.

8. **Don't spend a lot of time on difficult questions.** If you're not sure of an answer, go on to the next question. Answer easier questions first and then go back to the harder questions. However, when you skip a question, be sure that you have skipped the same number on your answer sheet. Although skipping difficult questions is a good strategy for making the most of your time, it is very easy to get confused and throw off your whole answer key.

   Lightly mark the margin of your answer sheet next to the numbers of the questions you did not answer so that you know what to go back to. To prevent confusion when your test is graded, be sure to erase these marks completely after you answer the questions.

9. **Answer every question on the test.** If you're not sure of an answer, take an educated guess. When you leave a question unanswered, you will *always* lose points, but you can gain points if you make a correct guess.

   If you must guess, try to eliminate one or more answers that you are sure are not correct. Then choose from the remaining answers. Remember, you greatly increase your chances if you can eliminate one or two answers before guessing. Of course, guessing should be used only when all else has failed.

10. **Clearly fill in the circle for each answer choice.** If you erase something, erase it completely. Be sure that you give only one answer per question; otherwise, no answer will count.

11. **Practice test-taking.** Use the exercises, reviews, and especially the posttest in this book to better understand your test-taking habits and weaknesses. Use them to practice different strategies such as skimming questions first or skipping hard questions until the end. Knowing your own personal test-taking style is important to success on the GED.

HOW TO USE THIS BOOK

If you are a student about to prepare for the GED Tests, you are to be admired. You have decided to resume an education that has been cut short. It is never easy to get back on track after you have been derailed. While it may not be easy, it will not be impossible. It will require determination and a lot of hard work.

Contemporary's GED Mathematics book has been designed to give you the necessary skills to succeed on the test. Take a moment to look down the table of contents, and you will get an idea of the content areas you will be working on. Don't be afraid if some areas sound a little overwhelming; each chapter builds upon the skills acquired in previous chapters.

Here are some things to note before you begin working in this book:

1. Before beginning in Chapter 1, you should take the three pretests (pages 1–9). These tests will give you a preview of what skills will be necessary for success on the GED Math Test, but more importantly, they will help you to identify which areas you need to concentrate on most. Use the chart for each test to pinpoint the types of questions you answered incorrectly and to determine what skills you need special work in. You may take each of the three tests individually, or take all three in one sitting.

2. As you work through this book, pay special attention to the "Problem Solving for the GED" feature. These sections focus on problem solving skills that are important for success on the GED Math Test. It is a good idea to work carefully through the example and explanation in these sections, even if you find that the computation issues covered in the chapter are easy for you. The problem-solving strategies described can be helpful in many different situations.

3. The math tips found throughout the book offer some useful hints in areas that can sometimes be troublesome to students.

4. Included on page 270 is a copy of the Formulas Page that you will be given when you take the GED Mathematics Test. When you get to Chapter 11 of this book, be sure to familiarize yourself with the contents of this page so that you will be ready to use it efficiently when you get to the testing situation.

5. The answers and solutions are a special feature of this book. If you get a problem wrong in an exercise, the information provided in the answer key will help you to figure out where you went wrong. Work carefully through the solutions provided and compare them to the math that you did to arrive at your answer.

6. After you have worked through the contents of this book, take the posttest (page 311). The posttest is a simulated GED Test that presents questions in the format, level of difficulty, and percentages found on the actual test. The posttest will help you to determine whether you are ready to take the GED Mathematics Test, and if not, what areas you need to do more work in. The evaluation chart at the end of the posttest will be especially helpful in making this decision.

7. An index appears on page 381 of the book. Use it as a handy reference tool when you need to find a specific concept or definition.

8. The twelve chapters of this book cover all of the content areas that will appear on the GED Math Test.

   - The symbol 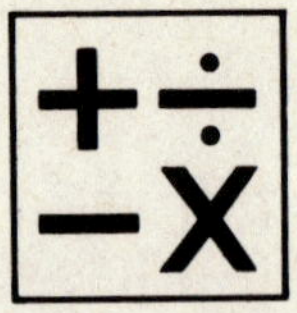 appears at the beginning of some chapters. This symbol indicates that you will be working in arithmetic.

   - The symbol 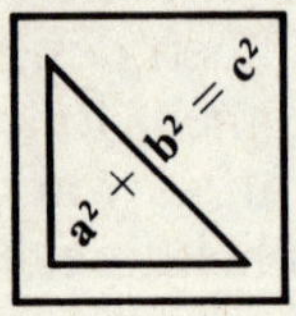

 appears at the beginning of other chapters. This symbol indicates that you will be working in algebra or geometry.

MATHEMATICS PRETEST

The three tests that follow will help you to evaluate your strengths and weaknesses in mathematics. You can take all three tests at once, or you can come back to these tests before you start each major section of the book.

Read each problem carefully. If you find some problems too difficult, don't give up on the whole test. There may be problems toward the end of each section that you don't know how to do. Skip those that are too hard, and come back to them if you have time later.

When you finish, check the answers and solutions that follow the tests. Then look at the charts at the end of the tests. Use these charts as a guide to tell you the areas where you need the most work.

Even if you do very well on these pretests, work through the entire book. Every section includes problems and hints that will help you on the GED Test.

Pretest Formulas

Following is a list of formulas that you may need on the pretests. You can refer back to this list as you work through the tests.

Area of a square	$A = s^2$; where A = area, s = side.
Area of a rectangle	$A = lw$; where A = area, l = length, w = width.
Circumference of a circle	$C = \pi d$; where C = circumference, $\pi = 3.14$, d = diameter.
Volume of a rectangular container	$V = lwh$; where V = volume, l = length, w = width, h = height.
Pythagorean relationship	$c^2 = a^2 + b^2$; where c = hypotenuse, a and b are the legs of the right triangle.
Slope of a line	$m = \frac{y_2 - y_1}{x_2 - x_1}$; where (x_1, y_1) and (x_2, y_2) represent two points in the plane.
Simple interest	$i = prt$; where i = interest, p = principal, r = rate, t = time.
Distance as function of rate and time	$d = rt$; where d = distance, r = rate, t = time
Total cost	$c = nr$; where c = total cost, n = number of units, r = rate (cost per unit).

Pretest 1: Arithmetic

Directions: Solve each problem.

1. Find the difference between 7,986 and 20,500.

2. What is the product of 709 and 68?

3. Find the sum of 16, 385, 4012, and 856.

4. What is the quotient of 27,438 divided by 34?

5. A plane flew for six hours. For two hours the average speed was 265 mph. For four hours the average speed was 490 mph. How far did the plane travel in six hours?

6. Abe worked three hours overtime on Thursday and four hours overtime on Friday. He makes $12.50 an hour for overtime work. Which of the following expresses the amount Abe made for overtime work those two days?
(1) $(12.50) \times 3 + 4$
(2) $12.50(3 + 4)$
(3) $3(12.50 + 4)$
(4) $4 \times 3 \times 12.50$
(5) $12.50(3 \times 4)$

7. Subtract 7.42 from 300.

8. Find the sum of 2.91, 46, and 1.085.

9. Simplify $\frac{35.36}{17}$.

10. Find the product of 1.84 and 2.5.

11. Divide 156 by 6.5.

12. Round off 214.0863 to the nearest hundredth.

13. Sam drove for 262 miles on 18 gallons of gasoline. To the nearest tenth, find the average number of miles he drove on one gallon of gasoline.

14. In 1960 the population of Central County was 1.75 million; in 1970, it was 2.3 million; and in 1980, 2.6 million. By how much did the population increase from 1960 to 1980?

15. Find the sum of $9\frac{1}{2}$, $3\frac{2}{3}$, and $6\frac{5}{6}$.

16. How much greater is $7\frac{1}{3}$ than $5\frac{7}{8}$?

17. Find the quotient of 8 divided by $2\frac{2}{3}$.

18. What is the product of $2\frac{4}{5}$ and $3\frac{3}{4}$?

19. Find the quotient of $3\frac{1}{2}$ divided by $\frac{3}{8}$.

20. Arrange the following in order from smallest to largest: $\frac{3}{4}$, $\frac{2}{3}$, $\frac{5}{9}$, and $\frac{5}{6}$.

21. From a pipe 4 meters long, Nick cut a piece that was $1\frac{3}{4}$ meters long. Assuming no waste, which of the following expresses the length of the remaining piece?
(1) $4 - 1.75$
(2) $4 - \frac{3}{4}$
(3) $4 - 1.25$
(4) $\frac{3}{4} - 4$
(5) $4 - 1.34$

22. The Chungs spend $\frac{1}{4}$ of their income on rent and $\frac{1}{3}$ on food. They take home $1200 a month. How much do the Chungs have left each month after paying for rent and food?

23. Simplify the ratio 36:48.

24. Solve for x in the proportion: $9{:}x = 12{:}20$.

25. In a GED class of 18 students, there are 10 women. What is the ratio of the number of men to the number of women?

26. Sam Stroud bought a raffle ticket at the Uptown Community Center. His wife bought three tickets, and his son bought two. Altogether there were 300 tickets. What is the probability that someone in the Stroud family will win?

Directions: Use the table below to answer questions 27 and 28.

Starting Yearly Salaries		
	Midvale	*Central County*
Police	$18,500	$22,200
Firefighters	16,800	18,400
Teachers	11,100	13,500

27. What is the ratio of the starting salary of a firefighter in Midvale to the starting salary of a firefighter in Central County?

28. What is the ratio of the starting salary of a teacher in Midvale to the starting salary of a police officer in Central County?

---

29. Change $83\frac{1}{3}\%$ to a fraction.

30. Find 6.5% of 200.

31. 24 is what percent of 40?

32. 35% of what number is 91?

33. Find the interest on $1000 at 18% annual interest for 1 year and 4 months.

Directions: Use the passage below to answer questions 34–36.

Carla started a new job with a gross salary of $1650 a month. Her employer will deduct 20% of her salary for taxes and social security. After a year, Carla will be eligible for a raise of 8% of her gross salary.

34. Which of the following tells the amount Carla's employer withholds each month for social security?
(1) $330
(2) $290
(3) $240
(4) $40
(5) Not enough information is provided.

35. What will be Carla's net pay for her first year of employment?
(1) $19,800
(2) $16,500
(3) $15,840
(4) $13,200
(5) Not enough information is provided.

36. What will Carla's monthly gross salary be during her second year of employment?
(1) $1815
(2) $1782
(3) $1730
(4) $1650
(5) Not enough information is provided.

ANSWERS AND SOLUTIONS START ON PAGE 6.

Pretest 2: Geometry

Directions: Solve each problem.

1. Find the circumference of the circle shown at the right.

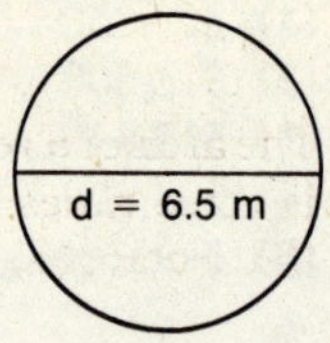

2. What is the volume of a rectangular container with a length of 10 feet, a height of $3\frac{1}{2}$ feet, and a width of 4 feet?

3. Find the area of the rectangle pictured at the right.

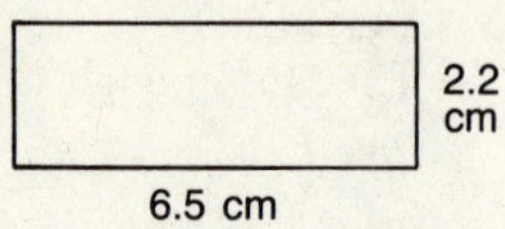

4. The shaded part of the diagram at the right shows the walkway around a swimming pool. Find the number of square feet of space on the walkway.

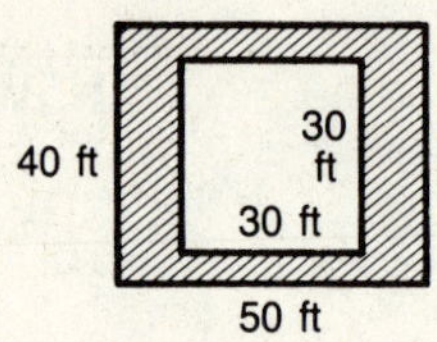

5. What is the value of 80^2?

6. Simplify $16^2 + 5^3 - 8^1$.

7. Find the value of $\sqrt{8464}$.

8. In the diagram at the right, $\angle a = 102.5°$. Find the measurement of $\angle b$.

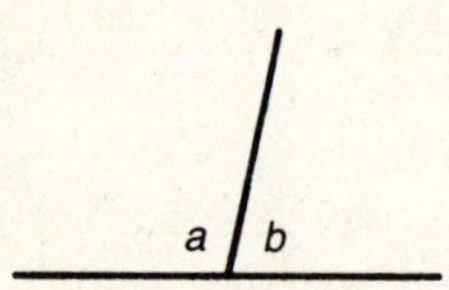

9. In the diagram at the right, $\angle a = 72°$. Find the measurement of $\angle d$.

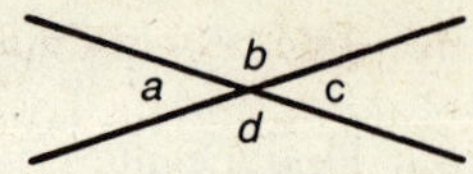

10. In $\triangle XYZ$ at the right, side $XY = 12$ inches and side $XZ = 10$ inches. The perimeter of the triangle is 34 inches. What kind of triangle is $\triangle XYZ$?

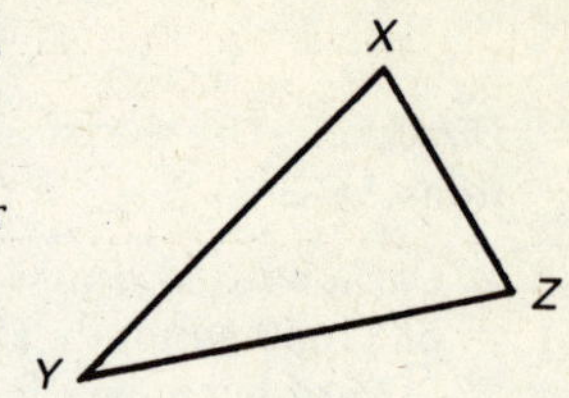

11. In $\triangle ABC$, $\angle A = 38°$ and $\angle B = 52°$. What kind of triangle is $\triangle ABC$?

12. A 5-foot-high vertical stick casts a shadow 3 feet long at the same time that a vertical pole casts a shadow 36 feet long. How tall is the pole?

13. In the picture below, $\angle C = \angle F$, and $\angle A = \angle D$. $AC = 24$, $DF = 15$, and $EF = 10$. Find BC.

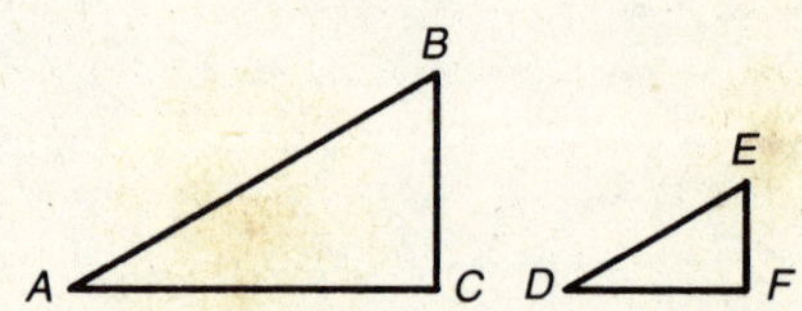

14. In $\triangle MNO$ at the right, $MN = 9$ and $MO = 12$. Find the length of NO.

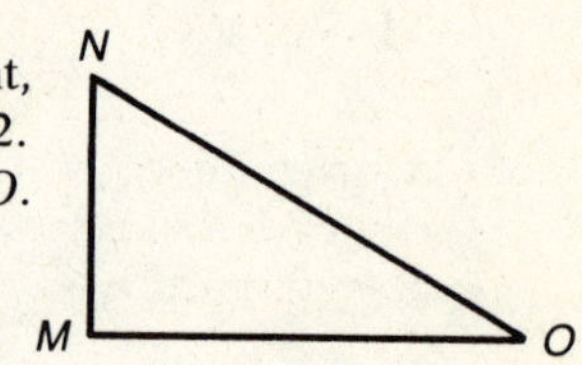

ANSWERS AND SOLUTIONS START ON PAGE 7.

Pretest 3: Algebra

Directions: Solve each problem.

1. Which point on the number line below corresponds to the value $\frac{3}{2}$?

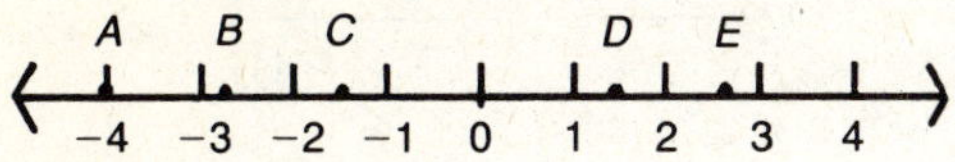

2. Simplify $+9 - (-4) + (-6) - 8$.

3. Simplify $-\frac{30}{6} + 4(-10)$.

4. Janet weighs w pounds. Which of the following expresses her weight after she loses 12 pounds?
   (1) $12w$
   (2) $12 - w$
   (3) $w - 12$
   (4) $\frac{4}{12}$
   (5) $w + 12$

5. Jack makes $6 an hour more than his wife. Which expression tells their combined income for a 40-hour week?
   (1) $40x + 6$
   (2) $40(2x + 6)$
   (3) $x(6 + 40)$
   (4) $x + x + 6$
   (5) $40 + 6x$

6. Which of the following expresses the sum of a number and three times the same number, all multiplied by five?
   (1) $5x + 3$
   (2) $x(5 + 3)$
   (3) $5\ (x + 3)$
   (4) $5(3x + 3)$
   (5) $5(x + 3x)$

7. Solve for x in $12x - 4 = 3x + 17$.

8. The ratio of Anton's age to his son's age is 9:2. Together their ages add up to 55. How old is Anton?

9. The area of a rectangle is 147 in^2. The length is three times the width. Find the length of the rectangle.

10. Solve for m in $6(m - 2) > 5(m + 5)$.

11. Factor the expression $n^2 - 6n$.

12. Which of the following is equal to $\sqrt{75}$?
    (1) 15
    (2) 10
    (3) 5
    (4) $3\sqrt{5}$
    (5) $5\sqrt{3}$

Directions: Use the graph below to answer questions 13 and 14.

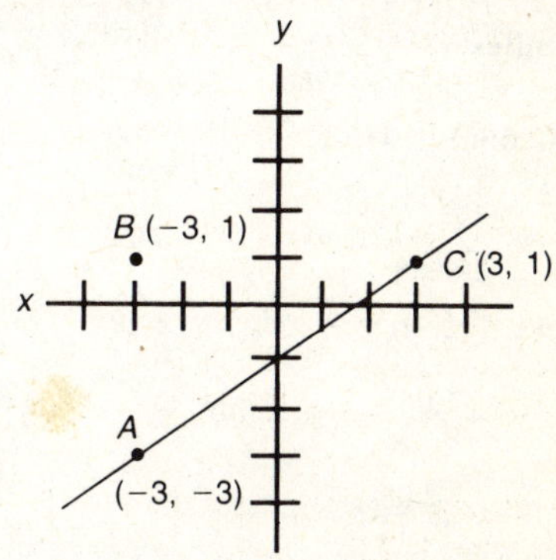

13. What is the slope of the line that passes through points A and C?

14. What is the distance from B to C on the graph?

---

15. Is the point (2, 5) on the graph of the equation $y = 4x - 3$?

ANSWERS AND SOLUTIONS START ON PAGE 7.

PRETEST ANSWERS AND SOLUTIONS

PRETEST 1: ARITHMETIC

1. **12,514** $\begin{array}{r} 20{,}500 \\ -\ 7{,}986 \\ \hline 12{,}514 \end{array}$

2. **48,212** $\begin{array}{r} 709 \\ \times\ 68 \\ \hline 5\,672 \\ 42\,54\ \\ \hline 48{,}212 \end{array}$

3. **5269** $\begin{array}{r} 16 \\ 385 \\ 4012 \\ +\ 856 \\ \hline 5269 \end{array}$

4. **807** $\begin{array}{r} 807 \\ 34\overline{)27{,}438} \\ \underline{27\,2} \\ 23 \\ \underline{0} \\ 238 \\ \underline{238} \\ 0 \end{array}$

5. **2490 miles** $(2 \times 265) + (4 \times 490) =$
$530 + 1960 = 2490$ miles

6. **(2) 12.50(3 + 4)**

7. **292.58** $\begin{array}{r} 300.00 \\ -\ 7.42 \\ \hline 292.58 \end{array}$

8. **49.995** $\begin{array}{r} 2.91 \\ 46. \\ +\ 1.085 \\ \hline 49.995 \end{array}$

9. **2.08** $\begin{array}{r} 2.08 \\ 17\overline{)35.36} \\ \underline{34} \\ 1\,3 \\ \underline{0} \\ 1\,36 \\ \underline{1\,36} \end{array}$

10. **4.6** $\begin{array}{r} 1.8\,4 \\ \times\ \ 2.5 \\ \hline 9\,2\,0 \\ 3\,6\,8 \\ \hline 4.6\,0\,0 \end{array} = 4.6$

11. **24** $\begin{array}{r} 2\,4. \\ 6.5\overline{)156.0} \\ \underline{130} \\ 26\,0 \\ \underline{26\,0} \\ 0 \end{array}$

12. **214.09**

13. **14.6 mpg** $\begin{array}{r} 14.55 \\ 18\overline{)262.00} \end{array}$ to the nearest tenth = 14.6 mi/gal

14. **.85 million** $\begin{array}{r} 2.60 \\ -\ 1.75 \\ \hline 0.85 \end{array}$ million

15. **20** $\begin{array}{rcl} 9\frac{1}{2} & = & 9\frac{3}{6} \\ 3\frac{2}{3} & = & 3\frac{4}{6} \\ +\ 6\frac{5}{6} & = & 6\frac{5}{6} \\ \hline & & 18\frac{12}{6} = 20 \end{array}$

16. **$1\frac{11}{24}$** $\begin{array}{rcl} 7\frac{1}{3} = 7\frac{8}{24} = 6\frac{8}{24} + \frac{24}{24} & = & 6\frac{32}{24} \\ -\ 5\frac{7}{8} = 5\frac{21}{24} = & & 5\frac{21}{24} \\ \hline & & 1\frac{11}{24} \end{array}$

17. **3** $8 \div 2\frac{2}{3} =$
$\frac{8}{1} \div \frac{8}{3} = \frac{8}{1} \times \frac{3}{8} = \frac{3}{1} = 3$

18. **$10\frac{1}{2}$** $2\frac{4}{5} \times 3\frac{3}{4} =$
$\frac{14}{5} \times \frac{15}{4} = \frac{21}{2} = 10\frac{1}{2}$

19. **$9\frac{1}{3}$** $3\frac{1}{2} \div \frac{3}{8} =$
$\frac{7}{2} \times \frac{8}{3} = \frac{28}{3} = 9\frac{1}{3}$

20. In order: **$\frac{5}{9}, \frac{2}{3}, \frac{3}{4}, \frac{5}{6}$**
$\frac{3}{4} = \frac{27}{36}, \frac{2}{3} = \frac{24}{36}, \frac{5}{9} = \frac{20}{36}, \frac{5}{6} = \frac{30}{36}$

21. **(1) 4 − 1.75**

22. **\$500** $\begin{array}{r} \frac{1}{4} = \frac{3}{12} \\ +\frac{1}{3} = \frac{4}{12} \\ \hline \frac{7}{12} \end{array}$ $\frac{7}{12} \times \frac{1200}{1} = 700$ $\begin{array}{r} \$1200 \\ -\ 700 \\ \hline \$\ 500 \end{array}$

23. 36:48 = **3:4**

24. **15** $\frac{9}{x} = \frac{12}{20}$ $\begin{array}{r} 20 \\ \times\ 9 \\ \hline 180 \end{array}$ $\begin{array}{r} 15 \\ 12\overline{)180} \end{array}$ $x = 15$

25. **4:5** $\begin{array}{lr} \text{total} & 18 \\ \text{women} & -\ 10 \\ \text{men} & \overline{8} \end{array}$ men:women = 8:10 = 4:5

26. **$\frac{1}{50}$** $1 + 3 + 2 = 6$ $\frac{6}{300} = \frac{1}{50}$

27. \$16,800:\$18,400 = **21:23**

28. \$11,100:\$22,200 = **1:2**

29. **$\frac{5}{6}$** $83\frac{1}{3} = \frac{83\frac{1}{3}}{100}$
$83\frac{1}{3} \div \frac{100}{1} = \frac{250}{3} \times \frac{1}{100} = \frac{5}{6}$

30. **13** $\frac{x}{200} = \frac{6.5}{100}$ $\begin{array}{r} 6.5 \\ \times\ 20\,0 \\ \hline 1300.0 \end{array}$ $\begin{array}{r} 13 \\ 100\overline{)1300} \end{array}$

31. **60%** $\frac{24}{40} = \frac{x}{100}$ $\begin{array}{r} 24 \\ \times\ 100 \\ \hline 2400 \end{array}$ $\begin{array}{r} 60 \\ 40\overline{)2400} \end{array}$

32. **260** $\frac{91}{x} = \frac{35}{100}$ $\begin{array}{r} 91 \\ \times\ 100 \\ \hline 9100 \end{array}$ $\begin{array}{r} 260 \\ 35\overline{)9100} \end{array}$

33. **\$240** 1 yr. 4 mos. $= 1\frac{4}{12}$ yr. $= 1\frac{1}{3}$ yr.

$i = prt$

$i = \frac{1000}{1} \times \frac{18}{100} \times \frac{4}{3} = \$240$

34. **(5)** Not enough information is provided. You do not know what percent is withheld for social security.

35. **(3)** **\$15,840**

First find the amount of deductions.

$\frac{p}{1650} = \frac{20}{100}$ 1650 × 20 = 33,000 $33{,}000 \div 100 = 330$

Subtract deductions from monthly salary and multiply by 12 to get yearly net pay.

\$1650 − 330 = \$1320

\$1 320 × 12: 2 640 + 13 20 = \$15,840

36. **(2)** **\$1782**

First find the amount of her raise.

$\frac{p}{1650} = \frac{8}{100}$ 1650 × 8 = 13,200 $13{,}200 \div 100 = 132$

Add the raise to her current salary.

\$1650 + 132 = \$1782

PRETEST 2: GEOMETRY

1. **20.41 m** $C = \pi d$
$C = 3.14 \times 6.5$
$C = 20.41$

2. **140 ft³** $V = lwh$
$V = 10 \times 3\frac{1}{2} \times 4$
$V = \frac{10}{1} \times \frac{7}{2} \times \frac{4}{1} = 140 \text{ ft}^3$

3. **14.3 cm²** $A = lw$
$A = 6.5 \times 2.2$
$A = 14.3 \text{ cm}^2$

4. **1100 ft²** Area of rectangle: $A = lw$
$A = 50 \times 40$
$A = 2000 \text{ ft}^2$
Area of square: $A = s^2$
$A = 30^2$
$A = 900 \text{ ft}^2$
Subtract the areas: $2000 - 900 = 1100 \text{ ft}^2$

5. $80^2 = 80 \times 80 =$ **6400**

6. **373** $16^2 = 16 \times 16 = 256$
$5^3 = 5 \times 5 \times 5 = 125$
$8^1 = 8$
$256 + 125 - 8 = 373$

7. **92**
Guess 90 because $90 \times 90 = 8100$
$8464 \div 90 = 94$ (810, 364, 360, remainder 4) 94 + 90 = 184 $184 \div 2 = 92$ 92 × 92 = 8464

8. **77.5°** 180.0° − 102.5° = 77.5°

9. **108°** 180° − 72° = 108°

10. **isosceles** 12 in. + 10 = 22 in. 34 in. − 22 = 12 in.
Since two sides are the same, $\triangle XYZ$ is isosceles.

11. **right triangle** 38° + 52° = 90° 180° − 90° = 90°
Since $\angle C = 90°$, $\triangle ABC$ is a right triangle.

12. **60 ft** $\frac{\text{height}}{\text{shadow}}$ $\frac{5}{3} = \frac{x}{36}$ 36 × 5 = 180 $180 \div 3 = 60$ ft

13. **16** $\frac{\text{base}}{\text{height}}$ $\frac{24}{x} = \frac{15}{10}$
24 × 10 = 240 $240 \div 15 = 16$ (15, 90, 90, 0)

14. **15** $c^2 = a^2 + b^2$
$c^2 = 9^2 + 12^2$
$c^2 = 81 + 144$
$c^2 = 225$
$c = \sqrt{225}$
$c = 15$

PRETEST 3: ALGEBRA

1. **D**

2. **−1** $+9 - (-4) + (-6) - 8$
$= +9 + 4 - 6 - 8$
$= +13 - 14 = -1$

3. **−45** $-\frac{30}{6} + 4(-10)$
$= -5 - 40 = -45$

4. **(3)** $\boldsymbol{w - 12}$

5. **(2)** $\boldsymbol{40(2x + 6)}$
$40(x + x + 6) = 40(2x + 6)$

6. **(5)** $\boldsymbol{5(x + 3x)}$

7. $\boldsymbol{2\frac{1}{3}}$
$12x - 4 = 3x + 17$
$-3x \qquad -3x$
$9x - 4 = 17$
$+4 \qquad +4$
$\frac{9x}{9} = \frac{21}{9}$
$x = 2\frac{3}{9} = 2\frac{1}{3}$

8. **45** Son's age $= 2x$
Anton's age $= 9x$
$2x + 9x = 55$
$\frac{11x}{11} = \frac{55}{11}$
$x = 5$
Anton's age $= 9(5) = 45$

9. **21 in** width $= x$ length $= 3x$
$A = lw$
$147 = 3x \cdot x$
$147 = 3x^2$
$\frac{147}{3} = \frac{3x^2}{3}$
$49 = x^2$
$\sqrt{49} = x$
$7 = x$
$21 = 3x$
The length is 21 inches.

10. **$m > 37$**

$$\begin{array}{rcl} 6(m-2) & > & 5(m+5) \\ 6m-12 & > & 5m+25 \\ -5m & & -5m \\ m-12 & > & 25 \\ +12 & & +12 \\ m & > & 37 \end{array}$$

11. **$n(n-6)$**

12. **(5) $5\sqrt{3}$**

$$\sqrt{75} = \sqrt{25}\cdot\sqrt{3} = 5\sqrt{3}$$

13. **$\frac{2}{3}$** $m = \dfrac{y_2 - y_1}{x_2 - x_1}$

$$m = \frac{1-(-3)}{3-(-3)} = \frac{1+3}{3+3} = \frac{4}{6} = \frac{2}{3}$$

14. **6** B is 3 units to the left of the y-axis.
C is 3 units to the right of the y-axis.
The distance between them is $3 + 3 = 6$.

15. **Yes** Substitute 2 for x in $y = 4x - 3$

$$y = 4(2) - 3 = 8 - 3 = 5$$

The point (2, 5) is on the graph.

Evaluation Charts

On the following charts, circle the number of any problem you got wrong. After each problem you will see the name of the section where you can find the skill (or skills) you need to solve the problem. When a problem involves more than one skill, the sections are separated by a slash (/).

These charts should help you decide which areas you need to work on most in this book.

Pretest 1 Evaluation

Problem	Section	Page
1, 2, 3, 4	The Basic Operations	14
5	Problems of Two or More Parts	31
6	Set-up Questions	42
7, 8	Adding and Subtracting Decimals	58
9	Dividing Decimals by Whole Numbers	64
10	Multiplying Decimals	62
11	Dividing by Decimals	65
12	Rounding Off Decimals	68
13	Multiplying and Dividing Decimals: Word Problems	67
14	Adding and Subtracting Decimals: Word Problems	60
15, 16	Adding and Subtracting Unlike Fractions and Mixed Numbers	92
17	Dividing Fractions by Whole or Mixed Numbers	102
18	Multiplying Mixed Numbers	99
19	Dividing Fractions	101
20	Comparing Fractions	111
21	Changing Fractions to Decimals	108
22	Multiplying and Dividing Fractions: Word Problems	104
23, 25	Ratio	119
24	Proportion	121
26	Probability	115
27, 28	Ratio/Tables	119/197
29	Changing a Percent to a Fraction	133
30, 31, 32	Proportion and Percent	136
33	Interest	149
34, 35, 36	Proportion and Percent/Item Sets	136/148

Pretest 2 Evaluation Chart

Problem	Section	Page
1	Perimeter and Circumference	227
2	Volume	244
3	Area	238
4	More Perimeter, Area, and Volume Word Problems	247
5, 6	Powers	234
7	Square Roots	236
8, 9	Angle Relationships	254
10, 11	Triangles	257
12, 13	Similarity	260
14	Pythagorean Relationship	267

Pretest 3 Evaluation Chart

Problem	Section	Page
1	The Number Line	276
2	Subtraction of Signed Numbers	280
3	Signed Numbers Review	285
4	Writing Algebraic Expressions	202
5, 6	Writing Longer Algebraic Expressions	210
7	Solving Equations with Separated Unknowns	215
8	Algebra Word Problems/Solving Equations with Separated Unknowns	216/215
9	Using the Formulas Page	269
10	Solving Longer Inequalities	287
11	Factoring	289
12	Simplifying Square Roots	291
13	Slope and Intercepts	302
14	Finding Distances on the Rectangular Coordinate System	297
15	Linear Equations	300

1
WHOLE NUMBER REVIEW

Number Values and Facts

Place Value

Digits (0, 1, 2, 3, 4, 5, 6, 7, 8, 9) are used to write whole numbers. 506 is a three-digit number. 5600 is a four-digit number even though the last two digits (00) are the same.

Place value means that the position of each digit in a whole number determines its value. The diagram below shows the names of the first 10 whole number places.

Place Values of Whole Numbers

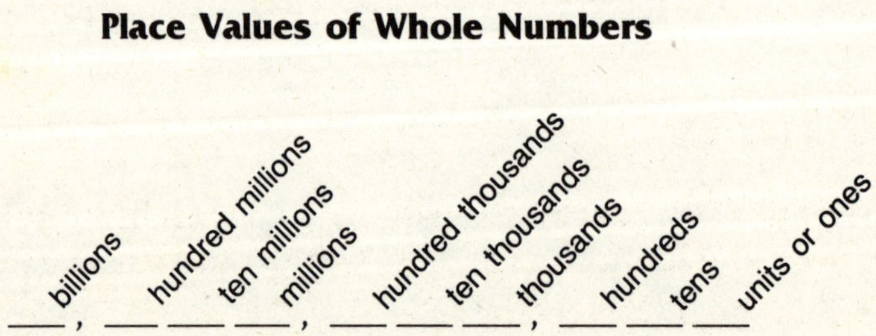

Let's compare the digit 5 as it appears in two different numbers. The 5 in the number 506 is in the hundreds place. The value of the 5 in 506 is $5 \times 100 = 500$. The digit 5 in the number 5600 is in the thousands place. The value of the 5 in 5600 is $5 \times 1000 = 5000$.

EXAMPLE 1: What is the value of each digit in 7403?

STEP 1. 7 is in the thousands place.
$7 \times 1000 =$ **7000**

STEP 2. 4 is in the hundreds place.
$4 \times 100 =$ **400**

STEP 3. 0 is in the tens place.
$0 \times 10 =$ **0**

STEP 4. 3 is in the units (ones) place.
$3 \times 1 =$ **3**

EXAMPLE 2: What is the value of 6 in 5600?

STEP 1. Tell what place the 6 is in. 6 is in the third place from the right, the hundreds place.
5$\underline{6}$00

STEP 2. Multiply 6 by 100.
$6 \times 100 =$ **600**

Whole Number Review Exercise 1

Directions: Write the value of each underlined digit.

1. 1$\underline{4}$6 $\underline{2}$0,019 167,28$\underline{3}$

2. 9,$\underline{3}$60,280 480,$\underline{9}$23 8$\underline{1}$,240

3. $\underline{2}$7 $\underline{6}$7,209,488 8$\underline{4}$43

4. 6,$\underline{4}$09,255,108 3$\underline{4}$2,774 21,$\underline{6}$40,456

5. What is the value of each digit in 10,496?

value of 1 =

value of 0 =

value of 4 =

value of 9 =

value of 6 =

ANSWERS AND SOLUTIONS START ON PAGE 322.

Number Facts

The Mathematics Test of the GED measures your ability to solve practical problems. To solve most problems, you need to know the arithmetic facts. If you cannot recall these facts quickly, you will waste time.

The Addition Facts

Exercise 2 is a short review of the addition facts that you use in both addition and subtraction. Write down the answers that you know. Do not count to get any answers. Time yourself. This exercise should not take more than three minutes.

Whole Number Review Exercise 2

Directions: In each box, write the sum of the number on the top and the number on the side. Several have been done as examples.

	1	3	8	4	6	9	7	10	2	5
8	9									
1		4								
3										
6										
4					10					
9										
5										
7								17		
2										
10										

ANSWERS START ON PAGE 322.

The Multiplication Facts

Knowing the multiplication tables is also essential for your work in mathematics. Take the time now to make sure that you can remember them accurately and quickly.

Whole Number Review Exercise 3

Directions: For each box on the table below, write the product (the answer to a multiplication problem) of the number on the top and the number on the side. A few answers have been filled in as examples.

	12	3	1	4	10	9	7	5	6	8	11	0	2
9													
7													
4		12											
5													
2							14						
1													
8													
10													

	12	3	1	4	10	9	7	5	6	8	11	0	2
12			12										
0													
11										88			
3													
6													

ANSWERS START ON PAGE 322.

You will also use these facts in division problems. For example,

$$7 \times 3 = 21 \text{ and } 21 \div 3 = 7.$$

The following exercise will give you a chance to recall the multiplication facts in both multiplication and division problems.

Whole Number Review Exercise 4

Directions: Without looking at the multiplication table that you filled in, do the exercise below. Time yourself. This exercise should not take more than three minutes.

1. $5 \times 4 =$

2. $7 \times 8 =$

3. $3 \times 8 =$

4. $5 \times 9 =$

5. $6 \times 5 =$

6. $7 \times 6 =$

7. $9 \times 2 =$

8. $11 \times 11 =$

9. $12 \times 4 =$

10. $8 \times 5 =$

11. $6 \times 3 =$

12. $9 \times 6 =$

13. $8 \times 9 =$

14. $7 \times 4 =$

15. $9 \times 4 =$

16. $64 \div 8 =$

17. $42 \div 6 =$

18. $25 \div 5 =$

19. $24 \div 8 =$

20. $54 \div 9 =$

21. $36 \div 6 =$

22. $42 \div 7 =$

23. $81 \div 9 =$

24. $56 \div 8 =$

25. $49 \div 7 =$

26. $45 \div 5 =$

27. $32 \div 4 =$

28. $63 \div 9 =$

29. $48 \div 6 =$

30. $21 \div 3 =$

ANSWERS START ON PAGE 323.

Now is the time to review any addition or multiplication facts that you got wrong. Learn any fact that you missed before continuing to work with whole numbers.

The Basic Operations

This section gives you a chance to review the four basic whole number operations: addition, subtraction, multiplication, and division.

The following examples illustrate methods commonly used in the United States. You may use any method that works best for you as long as you are careful and accurate.

Addition

Addition is the process of combining two or more numbers to find a total. The answer to an addition problem is called a ***sum***.

EXAMPLE: 2723 + 8 + 700 + 925 = ?

$$\textbf{(1)}\quad \begin{array}{r} 2723 \\ 8 \\ 700 \\ \underline{+\ \ 925} \end{array} \qquad \textbf{(2)}\quad \begin{array}{r} {\scriptstyle 2\ 1\ \ } \\ 2723 \\ 8 \\ 700 \\ \underline{+\ \ 925} \\ 4356 \end{array}$$

STEP 1. Line up the problem with units under units, tens under tens, and hundreds under hundreds.

STEP 2. Start with the units and add down each column. If the sum of any column is a two-digit number, put the digit on the right under the column you are adding and *carry* the number on the left to the next column. The total of the units column here is 16, so you carry the 1. The hundreds column totals 23, so you carry the 2.

To check an addition problem, add the numbers from the bottom to the top.

$$\textbf{EXAMPLE:}\quad \begin{array}{r} 34 \\ \underline{+28} \\ 62 \end{array} \qquad \textbf{CHECK:}\quad \begin{array}{r} 28 \\ \underline{+34} \\ 62 \end{array}$$

Subtraction

When you subtract, you take one number away from another. The answer to a subtraction problem is called the ***difference***. When you cannot subtract one digit from another in a subtraction problem, you must *borrow*.

Borrowing doesn't change the value of a number. It allows you to take from a higher place value and add to a lower place value.

EXAMPLE 1: Find the difference between 1982 and 6039.

$$\textbf{(1)}\ \begin{array}{r} 6039 \\ \underline{-\ 1982} \end{array} \quad \textbf{(2)}\ \begin{array}{r} 6039 \\ \underline{-\ 1982} \\ 7 \end{array} \quad \textbf{(3)}\ \begin{array}{r} {\scriptstyle 5\,1\ \ \ \ } \\ \not{6}039 \\ \underline{-\ 1982} \\ 7 \end{array} \quad \textbf{(4)}\ \begin{array}{r} {\scriptstyle 5\,9\ \ \ \ } \\ \not{6}\not{0}39 \\ \underline{-\ 1982} \\ 4057 \end{array}$$

STEP 1. Put the larger number (6039) on top. Line up the digits with units under units, tens under tens, etc.

STEP 2. Start subtracting with the units. Since you cannot take 8 from 3 in the tens column, you must borrow.

STEP 3. You cannot borrow from 0. Therefore, you must go one more place to the left. Borrow 1 from 6 to make 5 in the thousands place. Put 1 next to the 0 in the hundreds place to make 10.

STEP 4. This still doesn't help the 3 in the tens place. Borrow again from the 10. Put a 9 in the hundreds place and put a 1 next to the 3 in the tens place. Subtract each column.

To check a subtraction problem, add the answer to the number being subtracted. The result should be the top number of the original problem.

EXAMPLE:
$$\begin{array}{r} 49 \\ -28 \\ \hline 21 \end{array}$$

CHECK:
$$\begin{array}{r} 28 \\ +21 \\ \hline 49 \end{array}$$

Sometimes you will need to subtract from a number that has more than one zero in a row, and borrowing can get tricky. If you do your work neatly and carefully, you should have no problem. Look at the following example:

EXAMPLE 2: What is the difference between 10,000 and 1450?

(1)
$$\begin{array}{r} 10{,}000 \\ -\ 1{,}450 \\ \hline \end{array}$$

(2)
$$\begin{array}{r} {\scriptstyle 0\,1}\quad\;\; \\ \cancel{1}\cancel{0}{,}000 \\ -\ 1{,}450 \\ \hline 0 \end{array}$$

(3)
$$\begin{array}{r} {\scriptstyle 0\,9\;1}\quad \\ \cancel{1}\cancel{0}{,}000 \\ -\ 1{,}450 \\ \hline 0 \end{array}$$

(4 & 5)
$$\begin{array}{r} {\scriptstyle 0\,9\;9\,1}\;\; \\ \cancel{1}\cancel{0}{,}\cancel{0}\cancel{0}0 \\ -\ 1{,}450 \\ \hline 8{,}550 \end{array}$$

STEP 1. Put the larger number (10,000) on top. Line up the smaller number underneath, with units under units, tens under tens, and so on.

STEP 2. Start subtracting from the units and proceed to the tens column. Since you cannot take 5 from 0, you will need to borrow. You cannot borrow from the hundreds column or the thousands column because there are zeros in these places. You will need to borrow from a whole number. Cross out the 1 in the ten thousands column and put a 1 next to the zero in the thousands place to make 10.

STEP 3. Borrow from this 10 in the thousands place. Cross it out and write a 9 in its place. Put the borrowed 1 next to the 0 in the hundreds place.

STEP 4. Now borrow from this 10 in the hundreds place to help you subtract 5 from 0 in the tens place.

STEP 5. Subtract each column.

Multiplication

The answer to a multiplication problem is called the ***product***. When you multiply two numbers, put the number with more digits on top. Start by multiplying the units.

EXAMPLE: Find the product of 28 and 407.

$$
\text{(1)}\;\begin{array}{r} 407 \\ \times\ 28 \\ \hline \end{array}
\qquad
\text{(2)}\;\begin{array}{r} {\scriptstyle 5} \\ 407 \\ \times\ 28 \\ \hline 3256 \end{array}
\qquad
\text{(3)}\;\begin{array}{r} {\scriptstyle 5} \\ 407 \\ \times\ 28 \\ \hline 3256 \\ 814 \end{array}
\qquad
\text{(4)}\;\begin{array}{r} {\scriptstyle 5} \\ 407 \\ \times\ 28 \\ \hline 3\,256 \\ 8\,14 \\ \hline 11{,}396 \end{array}
$$

STEP 1. Put 407 on top.
STEP 2. Multiply 407 by 8. Start this part of the answer under the units column. Note that $8 \times 0 = 0$, so you just bring down the carried digit, 5.
STEP 3. Multiply 407 by 2. Start this part of the answer under the tens column.
STEP 4. Add the results from steps 2 and 3.

To check a multiplication problem, divide the answer by one of the numbers you multiplied. The result should be the other number you multiplied.

EXAMPLE: $\begin{array}{r} 18 \\ \times\ 6 \\ \hline 108 \end{array}$ **CHECK:** $\begin{array}{r} 18 \\ 6\overline{)108} \end{array}$

Division

The symbols $\div$ and $\overline{)}$ both mean to divide. 56 divided by 8 is written as $56 \div 8$ or $8\overline{)56}$. The answer to a division problem is called the ***quotient***.

EXAMPLE: $798 \div 38 = ?$

$$
\text{(1)}\; 38\overline{)798}
\qquad
\text{(2)}\;\begin{array}{r} 2 \\ 38\overline{)798} \\ \underline{76} \\ 3 \end{array}
\qquad
\text{(3)}\;\begin{array}{r} 21 \\ 38\overline{)798} \\ \underline{76} \\ 38 \\ \underline{38} \\ 0 \end{array}
$$

STEP 1. Set up your problem carefully. Put the number being divided inside the frame. Here 798 is being divided.
STEP 2. How many times does 38 divide into 7? None. Since 38 is too large to go into 7, look at the first two digits together — 79. Ask yourself, "How many times does 38 go into 79?" Two times. Put a 2 above the tens place. Multiply 2 by 38; then subtract.
STEP 3. Bring down the 8. How many times does 38 go into 38? One time. Put the 1 above the units place. Multiply 1 by 38. Subtract. There is no remainder. 38 divides evenly into 798.

In division problems with a ***remainder***, indicate the remainder as part of the answer. Later in the book, you will learn to express the remainder in different ways.

To check a division problem, multiply the answer by the number you divided by. Then add the remainder if there is one. The result should be the number you divided into.

EXAMPLE:

$$\begin{array}{r} 17\text{r}3 \\ 9\overline{)156} \\ \underline{9} \\ 66 \\ \underline{63} \\ 3 \end{array}$$

CHECK:

$$\begin{array}{r} 17 \\ \underline{\times\ 9} \\ 153 \\ \underline{+\ 3} \\ 156 \end{array}$$

Division is the most difficult whole number operation. Often you have to guess and test possible answers. Zeros in division problems can create extra trouble. Look at the following example.

EXAMPLE:

$$\begin{array}{r} 207 \\ 13\overline{)2691} \\ \underline{26} \\ 09 \\ \underline{0} \\ 91 \\ \underline{91} \\ 0 \end{array}$$

CHECK:

$$\begin{array}{r} 207 \\ \underline{\times 13} \\ 621 \\ \underline{207} \\ 2691 \end{array}$$

The zero in the answer above holds the tens place. This zero means that 13 did not divide into 9.

Zeros sometimes cause trouble. The exercise below has several examples of common problems with zeros in all four of the basic operations. If you get any of the following problems wrong, study the solution carefully. Then try the exercise again until you get every problem right.

Whole Number Review Exercise 5

Directions: Solve and check each problem.

1. 25,624 + 92,183 =
2. 60,845 − 2,926 =
3. Find the difference between 48,005 and 6,774.
4. 83 + 2096 + 194 =
5. 6 × 5708 =
6. What is the product of 349 and 74?
7. 872 × 409 =
8. Find the product of 65 and 50,000.
9. 446 ÷ 17 =
10. What is the quotient of 2464 divided by 8?
11. Find the quotient of 54,036 divided by 6.
12. 30,045 − 15,586 =

13. Find the sum of 194, 8, 2366, and 850.
14. 500 × 96 =
15. What is the difference between 3,000,000 and 816,000?
16. 7 × 29,058 =
17. 37,600 ÷ 800 =
18. What is the product of 8000 and 74?
19. What is the difference between 5,040,000 and 264,500?
20. What is the quotient of 10,710 divided by 35?
21. Find the product of 8 and 5090.
22. How much is 5076 divided by 12?
23. What is the difference between 88 and 7000?
24. Find the difference between 400,300 and 9,216.
25. Find the quotient of 2820 divided by 4.
26. What is the product of 346 and 450?
27. Find the sum of 2,900, 857, 11,630, and 405.
28. 56,423 ÷ 7 =
29. Find the difference between 807,631 and 1,300,500.
30. Find the quotient of 14,060 divided by 19.

ANSWERS AND SOLUTIONS START ON PAGE 323.

You should have at least 24 problems right on the last exercise. If you did not get 24 problems right, review your whole numbers skills before you go on. If you got 24 or more right, correct any problem you got wrong. Then go on to the next chapter.

2
PROBLEM SOLVING AND SPECIAL TOPICS

The Five-Step Approach to Solving Word Problems

Most problems on the Mathematics Test of the GED will be word problems. One goal of this book is to help you master such problems.

> The key to solving a word problem is to organize your thinking about the problem.

Take the time to do the following:

STEP 1. Find the question — what is being asked for?
STEP 2. Decide what information you need to answer the question.
STEP 3. Decide what arithmetic operation to use (addition, subtraction, multiplication, or division).
STEP 4. Do the arithmetic carefully and check your work.
STEP 5. Make sure that you answered the question asked *and* that your answer is sensible.

Study the following example closely.

EXAMPLE: Laura works at a part-time job 20 hours a week. She makes \$5 an hour. How much money does she make a week?

STEP 1. *Question:* How much money does she make a week?

STEP 2. *Information:* 20 hours at \$5 an hour.

STEP 3. *Arithmetic:* Since she makes \$5 in one hour, and she works 20 hours, multiply.

STEP 4. 20 hours × \$5 = \$100 per week.
Check: 100 ÷ 5 = 20.

STEP 5. **\$100** is a *sensible* amount of money to make in one week for a part-time job. You have answered the question asked.

On the next few pages, you will get a chance to practice each of these five steps.

Step 1: Understanding the Question

After you have read the problem, the first step in solving a word problem is to get a clear understanding of what the question is asking you to find. You will use this knowledge to select necessary information and later to see if your answer makes sense.

> **TIP:** One good strategy is to try to put the question in your own words. Sometimes this can help you visualize the problem more clearly.

Which of the answer choices below accurately restates what you are being asked to find?

Sarah bought 3 pounds of meat, 5 pounds of fruit, and 7 pounds of vegetables. What was the total weight of her purchases?

(1) Find the difference in weight between the fruit and the vegetables.
(2) Find how much Sarah's purchases weighed altogether.
(3) Find the combined weight of the meat and the vegetables Sarah bought.

Answer choice (2) means the same thing as "What was the total weight of her purchases?"

In the exercise on the next page, practice deciding what you are being asked to find in each problem. Restate the problem in your own words if you think it will help.

Problem-Solving Exercise 1

Directions: Which choice best expresses the question asked in each problem? **Do not solve these problems now.** You will have a chance to solve them later in the chapter.

1. In a recent election Mr. Sanders got 67,576 votes, and his opponent got 62,881 votes. How many more votes did Mr. Sanders get than his opponent?
(1) Find how many people voted.
(2) Find how many registered voters turned up at the polls.
(3) Find the difference between the number of votes Mr. Sanders received and the number of votes his opponent received.

2. A television was on sale for $198. This was $60 less than the original price. Find the original price.
(1) What was the sale price of the TV?
(2) How much more did the TV cost before it was on sale?
(3) How much did the TV cost before it was on sale?

3. Manny makes monthly payments of $190 for a new car. How much does Manny pay in one year for the car?
(1) Find how much more Manny owes on his car.
(2) Find Manny's total car payments for twelve months.
(3) Find the down payment Manny made for his car.

4. Pancho drove 460 miles and used 20 gallons of gasoline. On the average, how far did he drive on one gallon of gasoline?
(1) Find how much 20 gallons of gasoline cost.
(2) Find out how far Pancho can drive on 20 gallons of gas.
(3) Find out how far Pancho drove on one gallon of gasoline.

5. In September after a long diet, Gordon weighed 178 pounds. He went off his diet, and by the end of December he had gained 33 pounds. Find Gordon's weight at the end of December.
(1) How much did Gordon weigh before his diet?
(2) How much did Gordon weigh at the end of December?
(3) How much did Gordon weigh in early September?

6. Together Marilyn and Calvin Howard take home $22,200 a year. Find their monthly take-home pay.
(1) How much does Marilyn Howard take home in a month?
(2) How much do both Marilyn and Calvin take home in a month?
(3) How much do both Marilyn and Calvin make in a year?

7. On the open road Carmen's car gets 24 miles to the gallon. How far can she drive with 13 gallons of gas in her tank?
(1) Find how many miles Carmen gets on one gallon of gas.
(2) Find the number of gallons Carmen's gas tank holds.
(3) Find the distance Carmen can drive with 13 gallons of gas.

8. Gloria is saving for a color TV listed at $385. So far she has saved $290. Find the additional amount Gloria has to save.
(1) How much more does Gloria need to save for the TV?
(2) How much has Gloria saved already?
(3) What is the list price of the color TV?

ANSWERS AND SOLUTIONS START ON PAGE 324.

Step 2: Selecting the Necessary Information

Step 2 of the five steps to solving word problems is: Decide what information you need to answer the question. Some word problems can be confusing because they contain information you do not need. Think about the following problem:

EXAMPLE: The Chung family made \$17,400 last year. They paid \$2,580 for rent and \$3,160 in taxes. After paying taxes, how much money did the Chungs have left for all their expenses?

INFORMATION NEEDED. Yearly income of \$17,400 and taxes of \$3,160.

DO NOT NEED. \$2,580 rent.

The question asks: How much money did the Chungs have left after paying taxes? You do not need to know the amount of their rent to answer the question. The amount of rent they paid is *unnecessary* information. To solve this problem, simply subtract the amount the Chungs paid in taxes from the amount they made in a year.

$$\begin{array}{r} \$17{,}400 \\ -\ 3{,}160 \\ \hline \$14{,}240 \end{array}$$

One good strategy in working with word problems is to take a minute to decide what information you need to get an answer. Then cross out those numbers you do *not* need so that they will not confuse you.

Problem-Solving Exercise 2

Directions: For each problem, tell what information you need to solve each problem. Also tell what information is given that you do not need. (You can lightly cross this information out, if you wish.) Look back at the example above if you need help. **Do not solve these problems now.**

1. Of the 943 cars sold in Midvale last month, 387 of them were made in the U.S. Of the foreign cars sold, 198 were made in Japan. How many of the cars sold in Midvale last month were made outside the U.S.?

Need:

Do not need:

2. In March 1985, 10,549 people in Center City were officially unemployed. In December 1984, 9,307 were unemployed. In March 1984, 8,248 were unemployed. By how many people did the unemployment count rise from March 1984 to March 1985?

Need:

Do not need:

3. Huey drove 132 miles in 3 hours. He gets 22 miles on one gallon of gas. If he drove at the same rate, how far did he drive in one hour?

Need:

Do not need:

4. Huey drove 132 miles in 3 hours. He gets 22 miles on one gallon of gas. How many gallons of gas did he use?

Need:

Do not need:

5. A child's ticket to the circus cost \$6.50. An adult's ticket cost \$10.50. Altogether, 1230 children attended the circus. How much money was paid for children's tickets to the circus?

Need:

Do not need:

6. Of the 2500 adults in Cripple Creek, 1250 are high school graduates and 750 finished college. How many adults did not finish high school?

Need:

Do not need:

7. California has 23,668,000 people, while New York has 17,558,000 and Texas has 14,229,000. How many more people live in the most populous state than in the second most populous?

Need:

Do not need:

8. Paul paid \$76.80 for 64 feet of floor joists that weigh 4 pounds a foot. What was the price per foot of the joists?

Need:

Do not need:

9. Adrienne makes \$1100 a month. She pays \$285 a month for rent and \$400 a month for food. How much more does she pay for food than for rent in a month?

Need:

Do not need:

10. Guadalupe paid \$2.49 a pound for 3 pounds of lamb chops and \$1.89 a pound for 2 pounds of beef roast. How much did she pay for the lamb chops?

Need:

Do not need:

ANSWERS START ON PAGE 324.

Step 3: Choosing the Operation

Step 3 of the five steps to solving a word problem is: Decide what arithmetic operation to use. In this section we will talk about some of the key words and the situations that give you a clue as to what arithmetic operation to use in a word problem.

Addition and Subtraction

When you have to find a *total amount*, consider adding. Words such as *sum, total, altogether,* and *combined* in a question often indicate addition.

EXAMPLE 1: Mr. Swenson makes $260 a week, and Mrs. Swenson makes $240. What is their combined weekly income?
QUESTION. Find combined income.
OPERATION. Add their weekly incomes.

The following phrases usually mean to subtract: *How much more? How much less? How much greater? How much smaller? How much farther? Find the balance. Find the difference.*

EXAMPLE 2: Mary weighs 120 pounds. Her daughter weighs 50 pounds. How much more does Mary weigh than her daughter?
QUESTION. Find how much more mother weighs than daughter.
OPERATION. Subtract the daughter's weight from Mary's weight.

The words *net* and *gross* may also suggest subtraction. ***Gross*** refers to an amount before something has been subtracted from it. ***Net*** refers to the remaining amount after something has been subtracted from the gross.

EXAMPLE 3: Mark's gross monthly salary is $1000. His employer withholds $200 for taxes and social security. What is Mark's net salary for a month?
QUESTION. Find net salary.
OPERATION. Subtract the amount withheld from the gross amount.

Multiplication and Division

Watch for situations that suggest multiplication. A problem may give you information for one thing. Then you will be asked to find information for several things.

EXAMPLE 1: There are 3 feet in one yard. How many feet are there in 20 yards?
QUESTION. Find number of feet in 20 yards. (Here you have information for one thing and you want to find information for several things.)
OPERATION. Multiply the number of feet in one yard by 20.

EXAMPLE 2: On the open road, Jack can drive 24 miles on a gallon of gasoline. How far can he drive using 10 gallons of gas?

QUESTION. Find the distance Jack can drive with 10 gallons of gas. (Here, again, you have information for one thing and you want to find information for several things.)

OPERATION. Multiply the number of miles Jack can drive on one gallon of gas by 10.

The following situations suggest division. A problem may give you information about several things. Then you will be asked to find information about one thing. Or, you may be given information about a *whole* and asked to find a *part*.

EXAMPLE 3: Last year, Ruth and Gordon together paid $3000 for rent. How much rent did they pay each month?

QUESTION. Find monthly rent. [Here you have information about a *whole* (year) and you want to find information about a *part* (month).]

OPERATION. Divide the amount of rent they paid in a year by the number of months in a year.

A problem may also ask you to find how many of a certain item are contained in some bigger item. This situation also means to divide.

EXAMPLE 4: Soup is packed with 24 cans in a box. How many boxes are needed to hold 1200 cans of soup?

QUESTION. Find the number of boxes needed to hold 1200 cans. (Here you want to find how many times 24 goes into 1200 or how many 24s there are in 1200.)

OPERATION. Divide the total number of cans by the number of cans one box can hold.

The ideas of sharing, cutting, or splitting also indicate division.

EXAMPLE 5: Fatima won $200 in a lottery. She decided to share the money equally among three friends and herself. How much did each one get?

QUESTION. Find what part of $200 each of the 4 people received.

OPERATION. Divide the amount Fatima won by the number of people who are going to share the money.

Problem-Solving Exercise 3

Directions: In the next problems, first restate the question in your own words, as shown in the examples above. Then tell what arithmetic operation you need to solve each problem. **Do not solve these problems now.**

1. Pieter is 37 years old. His son Christopher is 8. How much older is Pieter than his son?

Question:

Operation:

2. Caren works 40 hours a week for $260. How much does she make in one hour?

Question:

Operation:

3. In November the Johnsons' utility bills were $12 for electricity, $25 for the telephone, and $53 for heat. Find the total for their November utilities.

   Question:

   Operation:

4. At the corner grocery all but 5 dozen of the 20 dozen eggs in the cooler are white. How many dozen eggs are white?

   Question:

   Operation:

5. Cheryl's gross income for the year was $18,296. Her deductions totaled $4680. What was her net income for the year?

   Question:

   Operation:

6. Ezra Hollis died in 1985 at the age of 92. In what year was he born?

   Question:

   Operation:

7. Pat works 8 hours a day for $7.50 an hour. How much does she make in a day?

   Question:

   Operation:

8. Alfredo's yearly salary is $16,380. How much does he make in a month?

   Question:

   Operation:

9. Kim types 85 words per minute. How long does it take her to type a letter that contains 1700 words?

   Question:

   Operation:

10. In June the Sports Center Fund Drive had $85,000. By the end of August contributors sent another $27,000. How much was in the fund at the end of August?

    Question:

    Operation:

ANSWERS START ON PAGE 324.

Step 4: Solving Word Problems

Step 4 of the five steps to solving word problems is: Do the arithmetic carefully and check your work. This is where your skills in the basic arithmetic operations play an important part. Later, you will see how algebra and geometry can also help you solve word problems.

When you get to this point in solving a problem, be sure to do your math neatly and accurately. All the good planning in the world will not get you a correct answer if you make mistakes in your computation.

Problem-Solving Exercise 4

Directions: It's now time to solve some of the word problems you have been working on. Go back to Exercises 1 (page 21), 2 (page 22), and 3 (page 25), and solve each problem. Don't forget to use the steps you have learned so far. Check your answers.

ANSWERS AND SOLUTIONS START ON PAGE 324.

Step 5: Being Sure That Your Answer Makes Sense

Is the Answer Sensible?

You may have thought that once you have computed an answer to a problem, you are finished. However, on tests like the GED, you want to take one more step to avoid making unnecessary mistakes. Even if you have already checked your computation, you should always go back to make sure your answer is sensible. To show you how important this is, take a look at how Carol, a GED student, solved the following problem:

Anthony walked a total of 77 miles over the last seven days. On the average, how many miles did he walk each day?

(1) 539
(2) 84
(3) 70
(4) 11
(5) 7

Carol looked quickly at the problem and decided she would multiply 77 by 7 to find the number of miles. Doing her computation carefully, and checking her math over twice, she came up with 539 total miles. She was happy to see that answer listed as one of the choices, and she answered (1).

What's wrong with Carol's work? You may have already figured out that she should have *divided* instead of multiplied. If Carol had taken an extra minute to decide *whether her answer was sensible*, she probably would have found her error. The question asks how many miles Anthony walked in *one* day. It is not sensible to think that he could walk 539 miles in a day, especially when the problem tells you he walked only 77 miles in 7 days! Even though Carol's computation was accurate, she did not use her common sense. If she had, she would have seen that she needed to rethink her solution to the problem.

Have I Answered the Question Asked?

Another part of step 5 is making sure you answered the question that was asked. Sometimes, in your haste to get a problem finished, you may skip an important step. Take a look at this example:

A tailor needs 14 inches of blue thread, 11 inches of green, and 11 inches of white to finish his project. How many *feet* of thread does he need in all?
(1) 2
(2) 3
(3) 4
(4) 36
(5) 432

Laticia, a GED student, knew that she needed to add 14, 11 and 11 inches together to find the total length of string. She did this, checked her addition carefully, and chose answer (4) 36. Then she went on to the next problem.

If Laticia had taken a minute to *see if she had answered the question asked*, she would have seen that the question asks for *feet*, whereas she has given her answer in *inches*. Answer choice (2) is the correct response. Once Laticia got her answer in inches, she should have divided by 12 to find the number of feet.

You may want to take some time now to go over your work from Exercises 1–3. If you got a wrong answer, try to figure out why this happened. Did you choose the wrong information? Did you fail to answer the question asked? Taking time to check your answer is an important step.

Rounding and Estimating

Rounding Off

Think about the number 362. Is 362 closer to 300 or to 400? You could subtract to find out.

$$\begin{array}{r} 362 \\ -300 \\ \hline 62 \end{array} \qquad \begin{array}{r} 400 \\ -362 \\ \hline 38 \end{array}$$

362 is closer to 400. 400 is called a ***round number*** because it ends with zeros. 400 is 362 **rounded off** to the nearest hundred. Rounding off numbers can sometimes help you decide whether or not your answer is sensible. Here are some steps to help you learn to round off.

Rules for Rounding Off Whole Numbers

1. Underline the digit in the place you are rounding off to.
2. a. If the digit to the right of the underlined digit is 5 or more, add 1 to the underlined digit.
   b. If the digit to the right of the underlined digit is less than 5, leave the underlined digit as it is.
3. Put zeros in all the places to the right of the underlined digit.

EXAMPLE 1: Round off 26,489 to the nearest thousand.

2<u>6</u>,489 → 26,000

STEP 1. Underline the digit in the thousands place, 6. Look at the digit to the right of 6. The digit is 4.

STEP 2. Because this number is less than 5, leave the 6 and the digits to the left of 6. Put 0s to the right of 6.

EXAMPLE 2: Round off 196,274 to the nearest ten thousand.

1<u>9</u>6,274 → 200,000

STEP 1. Underline the digit in the ten thousands place, 9. Look at the digit to the right of 9. The digit is 6.

STEP 2. Because this number is more than 5, add 1 to the underlined digit. 9 + 1 = 10. When you add 1 to 9, carry 1 to the next column to the left. Add 1 to the 1 already there. Then put 0s to the right.

Problem-Solving Exercise 5

Directions: Round off each number to the place indicated.

1. 63 to the nearest ten.

2. 228 to the nearest ten.

3. 439 to the nearest hundred.

4. 5620 to the nearest thousand.

5. 8098 to the nearest thousand.

6. 6982 to the nearest hundred.

7. 24,507 to the nearest thousand.

8. 38,496 to the nearest ten thousand.

9. 1063 to the nearest ten.

10. 28,092 to the nearest hundred.

11. 16,236 to the nearest hundred.

12. 1,475,290 to the nearest ten thousand.

13. 2951 to the nearest hundred.

14. 312 to the nearest ten.

15. 8059 to the nearest hundred.

ANSWERS AND SOLUTIONS START ON PAGE 325.

Estimating

As you know, the last step of the five steps to solving a word problem is: Make sure that your answer is sensible. How can rounding off numbers help you do this?

A good way to see if your answer is sensible is to estimate an answer. Round off at least one number in the problem and solve the new problem using the round number. This "rounded" problem should be easier to solve.

Now compare the answer to the "rounded" problem and the answer to the original problem. The answers should be close.

EXAMPLE: When Mr. MacDonald died, he left his 588-acre farm to be divided up equally by his three children. How many acres did each child inherit?

STEP 1. *Question:* How many acres did each child inherit?
STEP 2. *Information:* 588 acres altogether and 3 children.
STEP 3. *Arithmetic:* Sharing equally means to divide.
STEP 4. Divide 588 by 3. 588 ÷ 3 = 196.
STEP 5. Estimate an answer. Round off 588 to the nearest 100. Then divide 600 by the number of children. The approximate number of acres each child inherits is 200. The answer, **196**, is close to 200.

Estimation is a useful tool, but it is no substitute for your own judgment. Think about every answer you get. Be sure it makes sense to you.

Problem-Solving Exercise 6

Directions: Find an approximate answer for each problem below by rounding off at least one number in each problem. You are given a suggestion of which numbers to round off. Then compute the actual answer and compare.

1. Fred works 9 hours a day. He earns $8.25 an hour. How much does Fred earn in a day?

Round off wage to nearest dollar:

Approximate answer:

Actual answer:

2. Mrs. Meida drove for 6 hours at a speed of 52 miles per hour. How far did she drive?

Round off speed to nearest ten:

Approximate answer:

Actual answer:

3. On Friday 487 people attended the Central School Carnival. On Saturday 662 people attended, and on Sunday 523 people attended. Find the total attendance.

Round off attendance to nearest 100:

Approximate answer:

Actual answer:

4. Marcello makes $24,960 a year. How much does he make in a month?

Round off salary to a number easily divided by 12:

Approximate answer:

Actual answer:

5. The Central County courthouse was built in 1897. How old was the courthouse in 1988?

   Round off year of construction to nearest ten:

   Round off 1988 to nearest ten:

   Approximate answer:

   Actual answer:

6. The television Magda bought was on sale for $198. She saved $58 by buying the TV on sale. What was the original price?

   Round off sale price and amount of savings to nearest ten:

   Approximate answer:

   Actual answer:

7. A half-gallon carton of orange juice costs $1.89. How much do 5 cartons cost?

   Round off price of one carton to nearest dollar:

   Approximate answer:

   Actual answer:

8. From a board 96 inches long, Paul cut a piece 27 inches long. How long was the leftover piece?

   Round off each length to nearest ten:

   Approximate answer:

   Actual answer:

ANSWERS AND SOLUTIONS START ON PAGE 326.

Problems of Two or More Parts

To solve many word problems on the GED test you will have to perform more than one operation in Step 4 (the computation step) in order to answer the question asked. Study the following example carefully.

EXAMPLE 1: Celia bought two pounds of apples that cost $.60 a pound and five pounds of potatoes that cost $.40 a pound. What was the total cost of her purchases?

STEP 1. *Question:* What is the total cost of the apples and the potatoes?

STEP 2. *Information:* 2 pounds of apples for $.60 a pound and 5 pounds of potatoes for $.40 a pound.

STEP 3. *Arithmetic:* Here it is useful to outline each step.
a. Multiply the price of a pound of apples by 2.
b. Multiply the price of a pound of potatoes by 5.
c. Add the results of steps a and b.

STEP 4. *Solution:*
a. $.60 × 2 = $1.20.
b. $.40 × 5 = $2.00.
c. $1.20 + $2.00 = **$3.20.**

STEP 5. *Make sure the answer is sensible:* $3.20 is a reasonable amount to pay for the apples and potatoes.

EXAMPLE 2: Andrea finished cutting out 42 paper dolls in 3 hours. Her daughter cut out 12 dolls in 3 hours. Together, how many dolls did the two of them cut out per hour?

STEP 1. *Question:* Approximately how many dolls did they cut per hour?

STEP 2. *Information:* 42 dolls in 3 hours and 12 dolls in 3 hours

STEP 3. *Arithmetic:* Outline each step.
a. Divide 42 by 3.
b. Divide 12 by 3.
c. Add results of steps a and b.

STEP 4. *Solution:*
a. $42 \div 3 = 14$
b. $12 \div 3 = 4$
c. $14 + 4 = \mathbf{18}$

STEP 5. *Make sure the answer is sensible:* 18 is a sensible answer, given the information provided.

You may have chosen a different way to set up the problem above. For example, in Step 3, you may have decided to add 42 and 12, then divide by 3. Whatever strategy you find most helpful is the one you should use.

Problem-Solving Exercise 7

Directions: In this exercise, first make an outline for each problem as in Step 3 in the examples above. Then solve each problem. The first one is done for you.

1. The Upperville school district spends $7800 a year to educate a child. The Central City school district spends $2100 per child. Find the difference between the costs of educating 30 children in Upperville and 30 children in Central City.

a. Find cost for 30 children at Upperville.

b. Find cost for 30 children at Central City.

c. Subtract the result of step b. from the result of step a.

Solution: a. $7800 × 30 = $234,000   b. $2100 × 30 = $63,000   c. $234,000 − $63,000 = $171,000

2. Each week Everett makes $240 and his wife makes $200. Together, how much do they make in a year?

a.

b.

Solution:

3. For her store, Mrs. Rivera ordered 4 cases of green beans, 6 cases of canned corn, and 10 cases of beets. If there are 20 cans to a case, what is the total number of cans in Mrs. Rivera's order?

a.

b.

Solution:

4. Jermaine drove for 5 hours at a rate of 55 mph and then for 3 hours at a rate of 35 mph. How far did she drive altogether?

a.

b.

c.

Solution:

5. Alfonso works 35 hours a week for $8.50 an hour and 6 hours a week for $12.75 an hour. How much does he make in a week?

a.

b.

c.

Solution:

6. Mrs. Seltzer paid $150 down and $45 a month for 18 months for new living room furniture. What total amount did she pay for the furniture?

a.

b.

Solution:

7. A charity drive raised $30,400. This was $3,000 less than last year's total. Last year's money was shared equally by five agencies. How much did each agency get last year?

a.

b.

Solution:

8. Grace can type 75 words per minute. How long will it take her to type both a 1000-word letter and a 500-word memo?

a.

b.

Solution:

9. Jose makes $1300 a month. He pays $345 for a month's rent and $90 per week for food. How much does he have left each month after paying for rent and food?

a.

b.

Solution:

10. One month Jack of Jack's Autos sold two station wagons for $11,500 each, three sedans for $6,500 each, and a truck for $24,800. What were his total sales for the month?

a.

b.

c.

Solution:

ANSWERS AND SOLUTIONS START ON PAGE 326.

Special Topics with Whole Numbers

In the next few sections you will learn to apply your computation skills and word-problem skills to special types of problems.

Topic 1: Mean and Median

A problem may ask you to find the mean or the median for a group of numbers. ***Mean*** is another name for *average*. To find the average of a group of numbers, add the numbers and then divide the sum by how many numbers you have added.

EXAMPLE 1: Find the average of 10, 15, 23, and 28.

(1)
```
   10
   15
   23
 + 28
 ----
   76
```

(2)
```
    19
 4)76
   4
   --
   36
   36
   --
    0
```

STEP 1. Find the sum by adding the numbers.

STEP 2. Since you have added 4 numbers, divide the sum by 4. The average of the four numbers is **19.**

Finding an average is a common application of arithmetic skills. Look at the example below.

EXAMPLE 2: Houses on Fred's block sold for $40,000, $43,000, and $52,000. What was the average selling price?

(1)
```
 $ 40,000
   43,000
 + 52,000
 --------
 $135,000
```

(2)
```
   $45,000
 3)135,000
   12
   --
    15
    15
    --
     0 000
```

STEP 1. Find the sum by adding the cost of each house.
STEP 2. Divide by the number of houses, 3. The average selling price is **$45,000**.

Remember that another name for *average* is *mean*. The mean selling price for the three houses in example 2 is $45,000.

A ***median*** is a number in the middle of a whole group of numbers. In this group, half of the numbers are above the median and half of the numbers are below the median. In Example 2, $43,000 is the median price of the three houses. $40,000 is less, and $52,000 is more.

To find a median for a group of numbers, arrange the numbers in order from smallest to largest. The number in the middle is the median.

EXAMPLE 1: Lorenzo got scores of 90, 72, 88, 85, and 93 on 5 math tests. What was his median score?

72 85 <u>88</u> 90 93

STEP 1. Arrange the scores in order from smallest to largest.
STEP 2. Underline the number in the middle, 88. The median score on the five tests was **88**.

Sometimes there are two numbers in the middle of a group of numbers. The median is the average of these two numbers.

EXAMPLE 2: Ruby got scores of 78, 97, 72, and 84 on 4 Spanish quizzes. What was her median score?

(1) 72 <u>78</u> <u>84</u> 97 **(2)** $\begin{array}{r} 78 \\ +\,84 \\ \hline 162 \end{array}$ **(3)** $2\overline{)162}$ = 81

STEP 1. Arrange the scores in order from smallest to largest.
STEP 2. Underline and add the two numbers in the middle, 78 and 84.
STEP 3. Find the average (mean) of 78 and 84. **81** is the median score on the four tests.

Problem-Solving Exercise 8

Directions: Solve each problem. In some problems there may be extra information given. Make sure you use only necessary information.

1. Find both the mean and the median for 353, 19, and 207.
2. Find both the mean and the median for 2046, 971, 3113, and 1850.
3. Find both the mean and the median for 240, 313, 189, and 270.
4. Ramon weighs 187 pounds. His brother Manny weighs 159 pounds. What is their average weight?
5. The noon temperature Monday was 69°F; Tuesday the noon temperature was 71°F; Wednesday it was 56°F; Thursday it was 63°F; and Friday it was 66°F. What was the mean noon temperature for those days?

6. In 1985 Fran made $24,700. In 1986 she made $23,900. In 1987 she made $25,800. What was her average salary for those years?

7. The Johnsons' phone bills were $25.66 in January, $33.27 in February, $19.28 in March, and $32.51 in April. What was their average phone bill for the first three months of the year?

8. The total weekly payroll of the Central Electric Company is $108,000. 450 people work at Central Electric, and 255 of them are women. What is the mean pay for the workers?

9. A car salesman sold three used cars for $1650 each, four used cars for $1875 each, and two used cars for $2100 each. What was the average price for the cars he sold?

10. Sylvia received the GED scores shown at right. What was her average score for social studies, science, and literature?

GED Scores

Writing Skills	42
Social Studies	47
Science	44
Literature	50
Mathematics	45

ANSWERS AND SOLUTIONS START ON PAGE 327.

Topic 2: Number Series

A ***number series*** is a list of numbers in a special order or pattern. The numbers 1, 2, 3, 4, 5, 6. . . . form the most familiar series. The number 7 is the next *term* in the series.

To find a missing number of a number series, first find the pattern that changes the numbers from left to right.

EXAMPLE 1: Find the next term in the series 5, 8, 11, 14 . . .

STEP 1. Find how the series is changing. This series is changing by adding 3.

5 \ +3 / 8 \ +3 / 11 \ +3 / 14

STEP 2. Add 3 to the last term.

14 + 3 = 17

EXAMPLE 2: Find the next term in the series 2, 7, 4, 9, 6, 11 . . .

STEP 1. Find how the series is changing. This series first increases by 5 and then decreases by 3. At 11, the subtraction step is next.

2 \ +5 / 7 \ −3 / 4 \ +5 / 9 \ −3 / 6 \ +5 / 11

STEP 2. Subtract 3 from the last term.

11 − 3 = 8

TIP: Pay attention to which term you are asked to find. You will not always be asked to find the "next" term.

EXAMPLE 3: Find the seventh term in the series 1, 3, 9, 27, 81 . . .

STEP 1. Find how this series is changing. This series is changing by multiplying by 3.

1 3 9 27 81

×3 ×3 ×3 ×3

STEP 2. Find the sixth term. Multiply 81 by 3.
$81 \times 3 = 243$

STEP 3. Find the seventh term. Multiply 243 by 3.
$243 \times 3 = 729$

The number series problems on the GED Test will be word problems. You will have to figure out how a series of numbers is changing and then use that knowledge to answer the question. Look at Example 4 below.

EXAMPLE 4: One day in June the 9:00 A.M. temperature was 62°F. The 10:00 A.M. temperature was 67°F. The 11:00 A.M. temperature was 72°F. If the temperature continued in the same pattern, what was the 12:00 noon temperature that day?

STEP 1. Make a diagram that shows the temperature for each hour. Then find how the series is changing. This series changes by adding 5.

62 67 72

+5 +5

STEP 2. Add 5 to the last term.
$72 + 5 = 77$

Problem-Solving Exercise 9

Directions: Solve each problem.

1. Find the next term in the series 6, 12, 24, 48 . . .
2. What is the sixth term in the series 81, 84, 87, 90 . . . ?
3. What is the next term in the series 32, 29, 26, 23 . . .?
4. Find the seventh term in the series 13, 17, 21, 25 . . .
5. What is the next term in the series 10, 5, 15, 10, 20, 15 . . .?
6. Find the eighth term in the series 100, 81, 64, 49, 36, 25 . . .
7. Find the next term in the series 5, 9, 17, 33 . . .
8. What is the eighth term in the series 2, 3, 5, 8, 12, 17 . . .?
9. Find the next term in the series 320, 160, 80, 40, 20 . . .
10. Find the ninth term in the series 4, 8, 7, 14, 13, 26, 25 . . .
11. If the temperature in the diagram below continues in the same pattern, what will be the 7:00 P.M. temperature?

Temperature	72°F	68°F	64°F	——
Time	4:00 P.M.	5:00 P.M.	6:00 P.M.	7:00 P.M.

12. Every month, a committee is withdrawing money from a building fund at the same rate. How much will be in the account in *August*?

Balance	Month
\$30,000	March
\$24,000	April
\$18,000	May
\$12,000	June

13. This table shows the population of Bridge Creek. If the population continues in the same pattern, what will be the population of Bridge Creek in the year 2000?

Year	Pop.
1900	1,500
1920	3,000
1940	6,000
1960	12,000
1980	24,000

ANSWERS AND SOLUTIONS START ON PAGE 327.

Topic 3: Properties of Numbers

There are three properties of numbers that you should know for the GED exam. You do not need to memorize the names of these properties, but you should be able to understand them and apply them to problems.

The Commutative Property

This property applies to both addition and multiplication. For addition, it means that you can add the same two numbers in any order. The sums will be the same. For example, $6 + 3$ and $3 + 6$ both add up to 9.

We can write these properties with letters instead of numbers. You will get more practice in using letters for numbers when you work with algebra. The letters *a* and *b* stand for any two numbers.

> The commutative property for addition is:
> $a + b = b + a$

The commutative property for multiplication means that you can multiply the same two numbers in any order. The products will be the same. For example, 5×7 and 7×5 both multiply to make 35.

> We can write the commutative property for multiplication two different ways:
> $a \times b = b \times a$ OR $ab = ba$

The first method uses the multiplication sign between the two letters. The second method uses no signs. When two letters stand next to each other with no signs between them, this indicates multiplication. Again, remember that *a* and *b* stand for any two numbers.

EXAMPLE: Which of the following is the same as 9 + 8?
(1) 9×8
(2) 8×9
(3) 8 + 9
(4) 9 − 8

SOLUTION. Answer **(3)** is the correct choice. 9 + 8 and 8 + 9 both add up to 17. This is an example of the commutative property of addition.

The Associative Property

The associative property works for addition and multiplication. When you add three or more numbers, you can first add any two numbers and then add the third. The results will be the same.

> The associative property for addition is:
> $(a + b) + c = a + (b + c)$

The parentheses group numbers together. Suppose we want to add 2 + 5 + 6. We could first add 2 and 5 to get 7 and then add 7 and 6 to get 13. Or, we could first add 5 and 6 to get 11 and then add 2 and 11 to get 13. The results are the same. If we plug in these numbers to the property above, it looks like this:

$$(2 + 5) + 6 = 2 + (5 + 6)$$
$$7 + 6 = 2 + 11$$
$$13 = 13$$

> The associative property for multiplication is:
> $(ab)c = a(bc)$

Again the parentheses group numbers together. Notice that there are no multiplication signs. When letters stand next to each other, this indicates multiplication. Suppose we want to solve this problem: $2 \times 3 \times 4$. We could first multiply 2×3 to get 6 and then multiply 6×4 to get 24. Or, we could first multiply 3×4 to get 12 and then multiply 2×12 to get 24. For example:

$$(2 \times 3) \times 4 = 2 \times (3 \times 4)$$
$$6 \times 4 = 2 \times 12$$
$$24 = 24$$

EXAMPLE: Which of the following is the same as (10 + 12) + 15?
(1) 10 (12 + 15)
(2) (10 + 12) 15
(3) 10 + 15 + 12 + 15
(4) 10 + (12 + 15)

SOLUTION. Answer **(4)** is the correct choice. We could first add 10 and 12 to get 22 and then add 22 and 15 to get 37. Or, we could first add 12 and 15 to get 27 and then add 10 and 27 to get 37. This is an example of the associative property for addition.

The Distributive Property

The distributive property means that a number outside parentheses is multiplied by each number inside the parentheses. Their products are then combined according to the sign within the parentheses. You can say that the number outside parentheses is being "distributed" to both numbers on the inside.

> The distributive property applies to both addition and subtraction:
>
> $$a(b + c) = ab + ac$$
> $$a(b - c) = ab - ac$$

Suppose we want to solve this problem: 2 (6 + 3). The distributive property tells us that we can multiply the numbers inside the parentheses, 6 and 3, by the number outside the parentheses, 2:

$$\begin{aligned} 2\,(6 + 3) &= \\ (2 \times 6) + (2 \times 3) &= \\ 12 + 6 &= 18 \end{aligned}$$

To check the answer, add the numbers inside the parentheses, 6 and 3, to get 9. Then multiply 9 by the number outside the parentheses. 2 × 9 = 18.

Problem-Solving Exercise 10

Directions: Apply what you have learned about properties to choose the correct expression. Remember, a number written outside parentheses with no operation sign means multiplication.

1. Which of the following is equal to 20 + 30?
(1) 30 × 20
(2) 30 − 20
(3) 30 + 30
(4) 30 + 20
(5) 20 + 20

2. Which of the following is equal to (5 × 3) 4?
(1) (5 + 3) 4
(2) 5 (3 × 4)
(3) 5 (3 + 4)
(4) 5 + 3 × 4
(5) 5 + 4 × 3 + 4

3. Which of the following is the same as (9 + 7) + 4?
(1) 4 (9 + 7)
(2) 9 + (7 + 4)
(3) 9 (7 + 4)
(4) 7 (9 + 4)
(5) 4 (9 + 7)

4. Which of the following is equal to 16 (14 + 27)?
(1) 16 + 14 + 27
(2) 14 (16 + 27)
(3) 27 (16 + 14)
(4) 16 × 14 + 27
(5) (16 × 14) + (16 × 27)

5. Which of the following is the same as (3 × 15) − (3 × 6)?
   (1) 3 − (15 × 6)
   (2) 3 × 15 × 6
   (3) 3 × 15 − 6
   (4) 3 (15 − 6)
   (5) 15 (6 − 3)

ANSWERS START ON PAGE 328.

Topic 4: Order of Operations

In many math problems, you will have to perform more than one operation in order to get the answer. Mathematicians have agreed on a correct order to perform these operations, and it is very important that you know these rules. To see what might happen if you used an incorrect order, look at the example below:

$$10 + 4 \times 2$$

Suppose you added 10 and 4, then multiplied by 2. Your answer would be 28. However, this is an *incorrect* answer. The correct answer is found by performing the multiplication first, then adding, as shown below.

$$10 + 4 \times 2$$
$$10 + 8$$
$$18$$

Study the following **order of operations** and memorize it. Whenever you come across a problem with more than one operation, you will need to know the correct order in which to perform them.

Order of Operations

1. Do operations within parentheses.
2. Do multiplication or division.
3. Do addition or subtraction.

EXAMPLE: What is the value of the following expression?
(4 + 8) ÷ 6 + 3

$$(4 + 8) \div 6 + 3$$
$$12 \div 6 + 3$$
$$2 + 3$$
$$5$$

STEP 1. You know from the order of operations that you need to perform the operation inside the parentheses before you do anything else. Add 4 and 8.

STEP 2. Division should be performed before the addition. Divide 12 by 6 to get 2.

STEP 3. Add 2 to 3. The answer is **5**.

Problem-Solving Exercise 11

Directions: Find the values of the following expressions.

1. $(8 \times 3) \div 4 - 6$

2. $9 + (14 - 7)$

3. $80 \div (22 - 2)$

4. $17 - (16 \div 4)$

5. $(14 \times 2) \div (5 + 2)$

ANSWERS START ON PAGE 328.

Problem Solving for the GED

Introducing Set-up Questions

During a performance test, a car was driven 220 miles on the highway and 88 miles on back roads. This car averaged 22 miles per gallon of gasoline. Which of the following expressions gives the number of gallons of gas the car used in this trip?

(1) $\dfrac{220 \times 22}{88}$

(2) $\dfrac{220 + 88}{22}$

(3) $220 + 88 - 22$

(4) $22\,(220 + 88)$

(5) $\dfrac{88 + 22}{220}$

This problem is typical of some you will see on the GED Test. Notice that you *do not* have to do any computation in this problem. You need to know *what type* of computation is necessary, however, in order to get the problem right. You also need to understand and be able to use the properties of numbers you have just learned. Basically, you will have to *set up* a problem without actually solving it.

Here are some hints to help you succeed with problems like this one:

1. Remember the problem-solving steps that you have learned in this chapter. Understanding the question and knowing what information you need are extremely important skills in this type of problem.

   Question: How many gallons?
   Information needed: 220 miles, 88 miles, 22 miles per gallon

2. Think about how you go about answering the question asked and try to put it into words.

   To find the number of gallons used, you need to divide the number of miles driven by the rate of miles per gallon (mpg).

3. Once you have figured out in words how to solve the problem, plug in the appropriate numbers.

   gallons used = miles driven divided by mpg
   miles driven = 220 + 88
   miles per gallon = 22

4. Choose the answer that expresses the correct way to set the problem up. The horizontal fraction bar means the same as *divided by*.

   (2) $\frac{220 + 88}{22}$

★ GED PRACTICE ★

Set-up Questions

1. A machine at Gaston & Company can mold 45 cylinders in one minute. A slower machine molds 30 cylinders in a minute. With both machines working at once, which of the following expressions gives the number of cylinders Gaston produces in five minutes?

(1) $\frac{45 - 30}{5}$
(2) $\frac{45 + 30}{5}$
(3) $5\ (45 + 30)$
(4) $(5 \times 45) - (5 \times 30)$
(5) $(45 \div 5) + (30 \div 5)$

2. On Saturday afternoon, 265 people were at the Uptown Movie Theatre. Saturday night, 304 people were there. Everyone paid $4 for a ticket. Which of the following represents the total value of the ticket sales that day?

(1) $4 \times 265 \times 304$
(2) $4\ (265 + 304)$
(3) $265\ (4 + 304)$
(4) $4 + 265 + 304$
(5) $265 + 4\ (304)$

ANSWERS START ON PAGE 328.

Problem Solving Exercise 12

Directions: Choose the correct solution to each problem.

1. In February Herb's Furniture sold 5 dining table sets at $200 each. In March Herb sold 10 of the table sets. Which expression represents the total value of the table sets sold those two months?

(1) $200\ (5 + 10)$
(2) $5\ (200 + 10)$
(3) $10\ (200 + 5)$
(4) $200\ (5 \times 10)$
(5) $200\ (10 - 5)$

2. One week Mark worked overtime for 3 hours on Wednesday and 4 hours on Thursday. For overtime Mark makes $9 an hour. Which expression tells how much Mark made in overtime for that week?

(1) $4\ (3 + 9)$
(2) $3\ (4 + 9)$
(3) $9\ (3 + 4)$
(4) $(4 \times 3) + 9$
(5) $(4 \times 3) + (9 \times 3)$

3. Sharon drove for 4 hours at an average speed of 55 mph and then for 3 hours at an average speed of 40 mph. Which of the following expressions represents the total distance she drove?
(1) 7 (55 + 40)
(2) 55 (3 + 4)
(3) (3 × 55) + (4 × 40)
(4) (4 × 55) + (3 × 40)
(5) 40 (3 + 4)

4. All but 10 dozen out of the 50 dozen eggs were sold at Marone's grocery store. Which expression represents the number of eggs that were sold?
(1) 12 (50 − 10)
(2) 50 (12 − 10)
(3) 10 (50 − 12)
(4) (12 × 50) − (10 × 50)
(5) 12 (50 + 10)

5. Sam bought 3 shirts for $16 each and 2 pairs of pants for $24 each. Which expression represents the total price he paid for his purchases?
(1) (2 + 16) × (3 + 24)
(2) (3 × 16) + (2 × 24)
(3) (2 × 16) + (3 × 24)
(4) (22 + 24) + (2 + 3)
(5) (2 × 3) + (22 × 24)

6. On Monday, Marge's children collected $3.40 for charity. Tuesday they collected $8.00, and on Wednesday and Thursday they collected $6.50 each day. Which expression represents the average amount collected on the 4 days?
(1) $\frac{\$3.40 + \$8.00 + \$6.50 + \$6.50}{4}$
(2) $\frac{\$3.40 + \$8.00 + \$6.50}{4}$
(3) $\frac{\$3.40 + \$8.00 + \$6.50}{3}$
(4) $3.40 + $8.00 + $6.50 + $6.50
(5) 4 − ($3.40 + $8.00 + $6.50 + $6.50)

ANSWERS START ON PAGE 328.

★ GED PRACTICE ★

Problem-Solving Exercise 13: Review

Directions: Choose the best answer to each problem.

1. A grant of $12,500,000 is to be shared equally among 5 programs. How much will each program receive?

   Which of the following best expresses the question asked in this problem?
   (1) Find the amount the five programs get altogether.
   (2) Find the amount the five programs have to share.
   (3) Find the amount each program gets in a year.
   (4) Find the amount each program gets from the grant.
   (5) Find the amount each program gets in a month.

2. Of the 420 employees of Apex, Inc., 63 regularly walk to work, and 187 take public transportation. How many employees of Apex do not regularly walk to work?
(1) 124 **(2)** 233 **(3)** 250 **(4)** 357 **(5)** 483

3. In one month the leading U.S. car manufacturer sold 259,056 cars. The second largest manufacturer sold 120,947 cars. How many more cars did the leading manufacturer sell than the second largest manufacturer?

(1) 128,109
(2) 138,009
(3) 138,109
(4) 139,009
(5) 139,109

4. A year's profit of $81,480 was split evenly among the 84 employees of the Central Electric Cooperative. In addition, each employee received a bonus of $200. Which operations should you use to find how much money each employee received?
(1) Add 81,480 and 200, then divide by 84.
(2) Multiply 81,480 by 84, then add 200.
(3) Divide 81,480 by 84, then add 200.
(4) Divide 81,480 by 84, then divide 200 by 84.
(5) Add 84 and 200, then subtract it from 81,480.

5. Round off 36,498 to the nearest thousand.
(1) 36,500
(2) 37,000
(3) 36,000
(4) 37,500
(5) 38,000

6. Craig got an oil change for $12 and bought 10 gallons of gasoline at $1.29 a gallon. To the nearest dollar, how much did Craig spend for oil and gasoline?
(1) $12
(2) $13
(3) $24
(4) $25
(5) $26

7. Selma paid $200 down and $56 a month for 15 months for new furniture. What total amount did she pay for the furniture?
(1) $256
(2) $840
(3) $872
(4) $940
(5) $1040

8. For the first seven months of the year, Janet made $1100 a month. She got a raise and made $1300 a month for the rest of the year. Find how much she made altogether that year.
(1) $14,200
(2) $13,200
(3) $15,600
(4) $15,200
(5) $14,400

9. Rafael drove for 6 hours at an average speed of 55 mph. Then after he got to a big city he drove for 2 hours at an average speed of 9 mph. How many miles did he drive altogether?
(1) 318
(2) 418
(3) 385
(4) 440
(5) 348

10. Every day for a week, Sam timed his trip home. His trips took 27, 36, 28, 39, and 40 minutes. Find the average number of minutes for these trips.
(1) 28
(2) 30
(3) 32
(4) 34
(5) 36

11. Following are the attendance figures for each of four nights at a basketball tournament: 1024, 1263, 1137, and 1440. Find the median attendance at the games.
(1) 1216
(2) 1250
(3) 1024
(4) 1200
(5) 1440

12. Which of the following represents the value of this expression: $30 \div 6 + (3 + 1)$
    (1) 3
    (2) 9
    (3) 10
    (4) 13
    (5) 20

13. The chart below tells the number of homes in Central County with televisions. If the pattern continued, how many thousands of homes had televisions in 1985?

Year	1960	1965	1970	1975	1980	1985
No. of homes with TVs (in thousands)	1	3	7	15	31	—

    (1) 63
    (2) 62
    (3) 47
    (4) 32
    (5) 31

14. Which of the following is the same as $(4 \times 9) + (4 \times 15)$?
    (1) $4 \times 9 \times 15$
    (2) $4 (9 + 15)$
    (3) $4 + 9 + 15$
    (4) $9 (4 + 15)$
    (5) $15 (4 + 9)$

15. Fred worked 9 hours on Monday and 6 hours on Wednesday. He makes $12 an hour. Which expression tells how much money Fred made those two days?
    (1) $12 (9 + 6)$
    (2) $9 (12 + 6)$
    (3) $6 (12 + 9)$
    (4) $(12 \times 9) + (6 \times 9)$
    (5) $(12 \times 6) + 9$

ANSWERS AND SOLUTIONS START ON PAGE 328.

Problem Solving Review Evaluation

Problem	Section	Starting Page
1	Finding the Question	20
2	Necessary and Unnecessary Information	22
3-4	Choosing the Operation	24
5	Rounding Off	28
6	Estimation	29
7-9	Problems of Two or More Parts	31
10-11	Mean and Median	34
12	Order of Operations	41
13	Number Series	36
14	Properties of Numbers	38
15	Set-up	42

Passing Score: 12 right out of 15 problems.

Your score: ______ right out of 15 problems.

If you had less than a passing score, review the sections for those problems you missed. Then redo these questions before going on.

If you had a passing score, correct any problem you got wrong. Then go on to the next chapter.

3 INTRODUCING NUMBERS SMALLER THAN 1

Up until now, the only number smaller than 1 that you have worked with in this book is 0. However, there are many other numbers less than 1 that you will need to work with. In fact, you may already be familiar with some of them without realizing it.

As you saw in your review of whole numbers, a number or a group of objects can be divided into smaller groups. In this chapter you will see that you can also divide *one* object, or the number 1, into smaller parts. Each of these parts will be more than 0 but less than 1 in value. Look at the examples below:

Each picture at the right shows *one* object divided into smaller parts. The first one represents 3 equal parts, all less than 1 whole in value. The second picture shows you 100 equal parts of a whole, each one smaller than 1 whole in value.

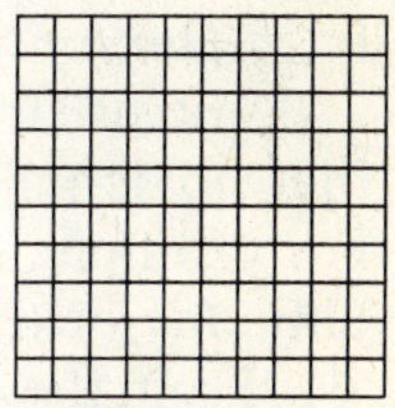

In the next few chapters, you will learn three different ways to express the values of numbers smaller than 1. You may already be familiar with one or more of them. They are:

- decimals
- fractions
- percents

It is important to remember for the GED Test and in everyday life that decimals, fractions, and percents are just different methods of expressing the same value. The more easily you can change from one form to another, the more successful you will be in working with numbers smaller than 1. Let's take a quick look at each of these methods of expression.

Decimals

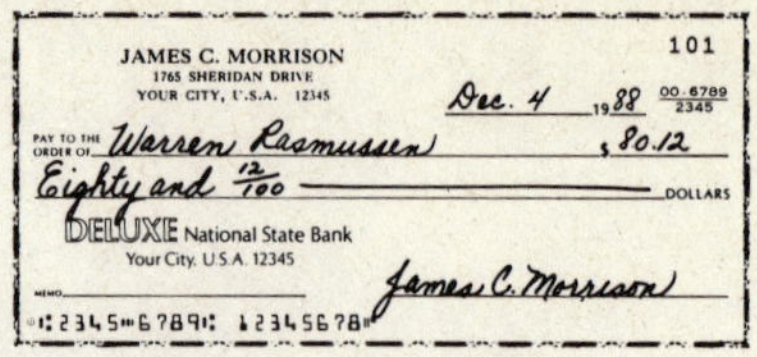

You are probably very familiar with decimals even if you don't realize it. Every time you talk about or write down amounts of money, you are using our decimal system. This system is based on the fact that *one* dollar can be divided into 100 smaller parts, or "hundredths." Each of these hundredths of a dollar is a penny, or one cent. When you have 35 cents, for example, you have *35 hundredths* of a dollar.

Parts of a whole dollar (or "cents") can be expressed as numbers smaller than one dollar by writing them after a decimal point, as shown below.

$4.13

You know this amount to be "four dollars and 13 cents." You can also think of it as 4 *whole* dollars and 13 *hundredths* of a dollar. Anything written after a decimal point is a number smaller than 1.

Decimals are used in places other than our money system, and in Chapter 4 you will have a chance to practice with many of them.

Fractions

A fraction is another way to show a whole divided into smaller parts. In a fraction, one number is written over another number, with a bar in between, as shown below:

$$\frac{1}{3}$$

In a fraction, the *bottom* number tells you how many parts the whole is divided into. The *top* number tells you how many parts of the whole you have.

You use fractions often if you use recipes to do any cooking. A measuring cup that holds one *whole* cup is divided into smaller parts that are *less* than one cup. For example, $\frac{1}{2}$ cup tells you that the cup is divided into two parts, and you need only one of those parts.

From the discussion of decimals above, you know that dollars are divided into *one hundred parts*. If you had 9 cents, or 9 hundredths of a dollar (9 parts out of 100 parts), how would you write this amount as a fraction? Remember that in a fraction the number of parts in the whole goes on the bottom, and the number of parts you have goes on top. You're right if you thought that the fraction is written as $\frac{9}{100}$.

You will learn much more about fractions in Chapter 5 of this book.

Percents

A percent is the third way to express a part of a whole, or a number smaller than 1. Percents can be easy to remember because they are *always* based on a whole divided into 100 equal parts. A percent is written as a number with the percent symbol (%) following it.

For example, 20 percent is written as 20%, and it means "twenty *hundredths* of a whole." In other words, if you took a whole and divided it into 100 equal parts, 20% would be equal to 20 of those 100 parts.

Stores often use percents to tell customers what kind of sale they are having. "50% off!" and "Only 10% down!" are ways to tell you how much money you will save or how much money you will need for a purchase.

You will learn much more about percents in Chapter 6 of this book.

Summary

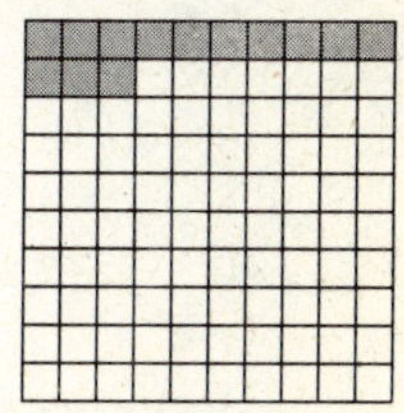

Before we go on to study each of these three ways to express parts and wholes, let's take one last look at how they are all related. It's important to remember that, although we generally use decimals when we talk about money, fractions when we talk about recipes, and percents when we talk about store sales, all three could be interchanged. For example, look at the figure at right. As you can see, it is one whole divided into 100 equal parts. How much of it is shaded?

- You learned that "hundredths" can be written after a decimal point, as we do in our money system. Therefore, you could say that **.13** is shaded.
- You also learned that to write a number smaller than 1 you can put the total number of parts on the bottom and the number of parts you have on top, in fraction form. Therefore, you could say that $\mathbf{\frac{13}{100}}$ is shaded.
- Finally, you learned that % means *hundredths*. Therefore, you could say that **13%** is shaded.

The expressions .13, $\frac{13}{100}$, and 13% all mean the same thing — the part of the figure that is shaded. They are all equal to each other.

4 DECIMALS

What Are Decimals?

Decimals are parts of a whole that are expressed in tenths or multiples of tenths. As you saw in the previous chapter, our money system is based on decimals, as is the metric system, a form of measurement we will discuss later in the book.

In our money system, dollars are represented by whole numbers. The cents, or the parts of a dollar, are written as decimals. Let's review.

dollars dimes pennies
$4.35

The 4 in $4.35 represents 4 whole dollars. The .35 represents 35 of 100 equal parts of a dollar (pennies) or 35 hundredths of a dollar. The digit 3, one place to the right of the decimal point, is in the dimes place and has a value of 30 cents. The digit 5 is in the pennies place, two places to the right of the decimal point, and has a value of 5 cents.

You already know something about decimals from handling money every day.

EXAMPLE 1: What is the value of the 2 in $7.82?

SOLUTION. Tell what place 2 is in. 2 is in the pennies place. The 2 has a value of **2¢**.

EXAMPLE 2: What is the value of the 8 in $7.82?

SOLUTION. Tell what place the 8 is in. 8 is in the dimes place. The 8 has a value of **80¢**.

Decimals Exercise 1

Directions: Find the value of each underlined digit.

1. $<u>4</u>.56 $.<u>2</u>8 $1.<u>0</u>9

2. $18.0<u>5</u> $1922.<u>4</u>3 $87.3<u>1</u>

ANSWERS START ON PAGE 329.

Place Value in the Decimal System

As you saw with dollars and cents, the first place to the right of the decimal point is the dimes place. The second place to the right of the point is the pennies place. *Without* the dollar sign, the first place to the right of the point is called the ***tenths place***. (A dime is one of 10 equal parts of a dollar.) The second place is called the ***hundredths place***. (A penny is one of the 100 equal parts of a dollar.)

Notice that the names of the places to the right of the decimal point end with the letters *ths*. Below are the place names of the digits in 4.35 without a dollar sign.

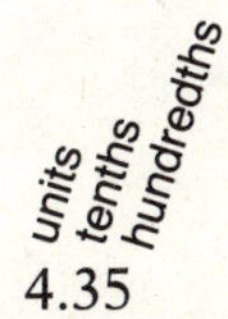

Below is a diagram of the first six whole number places and the first six decimal places. The decimal point separates whole numbers from decimal places.

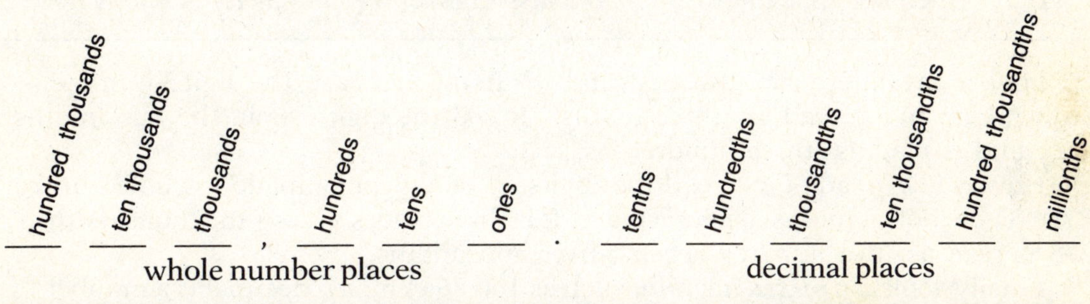

> **TIP:** To learn the decimal place names, remember that 10 has one zero and tenths has one place, 100 has two zeros and hundredths has two places, 1000 has three zeros and thousandths has three places.

EXAMPLE 1: Tell how many decimal places the number 18.406 has.
SOLUTION. Count the number of digits to the right of the decimal point. 18.406 has three digits to the right of the point. 18.406 has **three** decimal places.

EXAMPLE 2: Tell what place the 6 in 18.456 is in.
SOLUTION. Count how many places to the right of the decimal point the 6 is in. 6 is three places to the right of the point. The third place to the right of the point is called **thousandths**.

Decimals Exercise 2

Directions: For items 1 to 4, tell how many decimal places each number has.

1. a. 4.907 **b.** 3806.2 **c.** 314.26

2. a. 1,976,500 **b.** 18.04 **c.** 2.87654

3. a. 59 **b.** 9.3 **c.** 4.0035

4. a. 80.538 **b.** 1264.7 **c.** 0.004502

Directions: For items 5 to 8, tell what place each underlined digit is in.

5. a. $\$8.4\underline{7}$ **b.** $1.09\underline{3}6$ **c.** $\$267.\underline{8}1$

6. a. $0.073\underline{8}$ **b.** $\$256.3\underline{0}$ **c.** $0.0004\underline{5}$

7. a. $\$198.1\underline{3}$ **b.** $6\underline{0}.482$ **c.** $1\underline{2}53.4$

8. a. $0.00057\underline{6}$ **b.** $\$93.\underline{4}0$ **c.** $7.81\underline{5}$

ANSWERS START ON PAGE 329.

Reading Decimals

You have seen how place value works in the decimal system. Now you will learn how to read a decimal. As you work ahead, remember this:

> **TIP:** Remember that the place of the *last* digit of a decimal names its value.

Look at the difference between .3 and .03. In the first case, the digit 3 is in the tenths place and is read as "three tenths." In .03 the digit 3 is in the hundredths place and is read as "three hundredths."

How is .275 read? First read the digits as though they made a whole number (275). Then look at the last decimal place. It is three places over — the thousandths. .275 is read as "two hundred seventy-five thousandths."

Finally, look at a ***mixed decimal*** such as 100.26. A mixed decimal is a number that includes both a whole number and a decimal. Read the whole number part first (100). Read the decimal point as the word *and*. Then read the decimal part (26 with the last digit in the hundredths place). 100.26 is read as "one hundred and twenty-six hundredths."

Decimals Exercise 3

Directions: Circle the number that is equal to the written-out number.

1. thirty-one thousandths 0.31 *or* .031

2. two and twelve hundredths 2.12 *or* .212

3. seven hundred eighty ten thousandths .780 *or* .0780

4. one hundred and three hundredths 0.103 *or* 100.03

5. seventy one and five tenths 71.5 *or* .7150

6. four and seven thousandths 4.007 *or* .407

7. three and forty-nine ten thousandths .00349 *or* 3.0049

8. thirteen and thirteen hundredths 13.13 *or* 0.1313

9. four hundred three and one tenth .4031 *or* 403.1

10. ten and twenty hundredths 10.20 *or* 1020.0

ANSWERS START ON PAGE 329.

Zeros

The digit 0 has no value, but some 0s hold other digits in their places. Look at the number below:

08.0060

The digit to the left of 8 has no use. The digit to the right of the 6 also has no use. However, the two 0s to the right of the decimal point keep the digit 6 in the thousandths place.

You can rewrite the number 08.0060 and drop two of the zeros without changing its value.

08.0060 = 8.006

The 0s in the tenths place and the hundredths place mean that there are no tenths and no hundredths. Remember, you read .6, .06, and .006 differently because they have different values.

One exception to the rule of dropping unnecessary zeros is in our money system. We write $4.50 (four dollars and fifty cents) because the system is based on hundredths of a dollar, not tenths of a dollar.

TIP: Understanding the difference between necessary and unnecessary zeros will help you throughout your work with decimals.

Decimals Exercise 4

Directions: Rewrite each number keeping only the necessary zeros.

	a.	b.	c.
1.	**a.** 0.06700	**b.** 03.405	**c.** 8.0906
2.	**a.** 80.0250	**b.** 0124.0090	**c.** 007.50
3.	**a.** 5.0	**b.** 0.37080	**c.** 029.300
4.	**a.** 06.3	**b.** 0.002300	**c.** 060.0502

ANSWERS AND SOLUTIONS START ON PAGE 329.

Writing Decimals

When you begin to write a decimal, first think about how many places you need:

EXAMPLE 1: Write fifty-two thousandths as a decimal.

SOLUTION. 52 is a two-digit number, and the last word, thousandths, tells you the place value. Thousandths take three places, but 52 uses only two places. *Hold* the first place (tenths) with a zero. Fifty-two thousandths is written as **.052** or **0.052**. The zero in the units place is optional. It is often used to show that there is no whole number.

Watch for the word *and*. Remember that *and* separates whole numbers from decimals. As you recall, whole numbers with a decimal are called **mixed decimals**.

EXAMPLE 2: Write twenty-three and nine tenths as a decimal.

SOLUTION. First write the whole number (23) and the decimal point. Then think about how many decimal places you need. Tenths are in the first decimal place to the right of the point. Twenty-three and nine tenths is written as **23.9**.

Decimals Exercise 5

Directions: Write each group of words as a decimal or mixed decimal.

1. four tenths
2. six and three tenths
3. eighteen hundredths
4. nine thousandths
5. two ten thousandths
6. thirteen thousandths
7. ninety-six and four tenths
8. five and sixteen thousandths
9. seven thousand five hundred and eight tenths
10. one hundred twenty-five thousandths
11. eighty-four and nine thousandths
12. five thousand six hundred two and twenty-eight thousandths
13. three hundred twelve ten-thousandths
14. two hundred four and three hundredths
15. seventy and three hundred forty-five thousandths.

ANSWERS START ON PAGE 329.

Comparing Decimals

Which is larger, $.30 or $.04? You know that 30 cents is more than 4 cents. The decimals .30 and .04 have the same number of places. This makes them easy to compare. You know that 30 hundredths is larger than 4 hundredths because 30 is larger than 4. But how do you decide whether .031 or .31 is larger?

EXAMPLE: Which decimal is larger, 0.031 or 0.31?

STEP 1. Give the decimals the same number of places. Put a zero to the right of 0.31 to give it three places. Remember, this zero does not change the value.

added zero

0.31 = 0.310

STEP 2. Compare the numbers to the right of the decimal point. Don't worry about the decimal point. 310 is greater than 31, so **0.31 is greater than 0.031**.

Rules for Comparing Decimals

1. Give the decimals the same number of places. You can put zeros to the right of a decimal without changing its value.
2. Now that the number of places is the same, decide which decimal is larger by comparing the numbers without worrying about place value.

Decimals Exercise 6

Directions: For items 1-4, circle the larger number in each pair. Remember that your first step is to add zeros so that each number has an equal place value.

1. a. 0.08 or 0.7 **b.** 0.62 or 0.062 **c.** 0.33 or 0.403

2. a. 0.0029 or 0.001 **b.** 0.01 or 0.101 **c.** 0.895 or 0.9

3. a. 0.8 or 0.098 **b.** 5.2 or 5.23 **c.** 0.4 or 0.0268

4. a. 0.31 or 0.295 **b.** 1.68 or 1.678 **c.** 3.004 or 3.04

5. Arrange in order from smallest to largest: 0.62, 0.062, 0.602, 0.26.

6. Arrange in order from smallest to largest: 0.43, 0.0034, 0.34, 0.403.

ANSWERS START ON PAGE 329.

Problem Solving for the GED

Comparing and Ordering

Ben has to stack cans on the shelves of the Corner Store where he works. He wants to put the heaviest cans on the lowest shelves and increasingly lighter cans on the upper shelves. Following are the weights of the different cans Ben has to shelve:

can A — 0.475 kilograms
can B — 0.5 kilograms
can C — 2 kilograms
can D — 1.75 kilograms
can E — 0.34 kilograms

Which of the following sequences lists the order in which the cans should be placed on the shelves, from lower to upper?

(1) D, A, E, B, C
(2) C, D, B, A, E
(3) D, B, E, C, A
(4) D, B, A, E, C
(5) C, E, B, D, A

This is an example of a GED-type problem where your skills in comparing decimals will come in handy. Here are some hints to help you solve problems like these.

1. Make sure that the units of measurement are all the same. Sometimes you'll see problems where you'll have to convert units of measurement so that you are comparing the same things.

   In this problem, all weights are given in kilograms, so no conversion is necessary. Don't be concerned if you are not familiar with kilograms. Treat these units of measurement as just labels, and concentrate on the numbers. You will learn more about units of measurement in Chapter 8.

2. Be sure of the *order* specified. Do they want least to greatest, or greatest to least?

   In this problem, Ben must stack *heaviest* to *lightest*, so order the numbers from largest to smallest.

3. Follow the steps you learned for comparing decimals.

   In this case, give all the numbers three decimal places.
   A = 0.475 = .475
   B = 0.5 = .500
   C = 2 = 2.000
   D = 1.75 = 1.750
   E = 0.34 = .340

4. Now put the numbers in order from heaviest to lightest.

   (C) 2.000
   (D) 1.750
   (B) .500
   (A) .475
   (E) .340

5. Look at the order you came up with and find the correct answer choice from the five you are given.

ANSWER: (2) C, D, B, A, E

★ GED PRACTICE ★

Comparing and Ordering

Following are the lengths of five pieces of galvanized pipe that Thelma needs to organize.

piece A — 0.8 meter
piece B — 0.95 meter
piece C — 0.85 meter
piece D — 0.09 meter
piece E — 0.085 meter

Which of the following gives these lengths in order of size from *smallest* to *largest*?

(1) A, E, D, C, E
(2) D, A, C, B, E
(3) B, C, A, D, E
(4) E, D, A, C, B
(5) A, D, C, E, B

SOLUTION IS ON PAGE 329.

Adding and Subtracting Decimals

Addition

Think about adding money. Suppose you bought oranges for $1 and milk for 65¢. Your total bill would be $1.65. Automatically, you lined up the decimal points.

$$\begin{array}{r} \$1.00 \\ +\ \ .65 \\ \hline \$1.65 \end{array}$$

Adding decimals is as easy as adding money or whole numbers.

Rules for Adding Decimals

1. Line up the decimals with point under point.
2. Add each column and bring the decimal point straight down into each answer. Carry if necessary. You can carry to the left of the decimal point.

EXAMPLE 1: Add .8 and .4.

$$\textbf{(1)}\ \begin{array}{r} .8 \\ +.4 \\ \hline \end{array} \qquad \textbf{(2)}\ \begin{array}{r} .8 \\ +.4 \\ \hline 1.2 \end{array}$$

STEP 1. Line up the numbers with point under point.
STEP 2. Add the numbers and bring the decimal point straight down into the answer.

TIP: Remember not to confuse the period at the end of a sentence with a decimal point, as in Example 1 above.

Notice in the last example that 8 tenths and 4 tenths give a sum of 12 tenths. Only one digit fits in the tenths place. *Carry* the digit 1 to the units place. This is just like the carrying you do when you add whole numbers.

EXAMPLE 2: Find the sum of 3.8, 47, .0092, and 1.83.

$$\textbf{(1)}\ \begin{array}{l} \ \ 3.8 \\ 47. \\ \ \ \ .0092 \\ +1.83 \\ \hline \end{array} \qquad \textbf{(2)}\ \begin{array}{r} 3.8000 \\ 47.0000 \\ .0092 \\ +1.8300 \\ \hline 52.6392 \end{array}$$

STEP 1. Line up the numbers with decimal point under decimal point. Notice that the whole number 47 is understood to have a decimal point at the right even though it is not written in the problem.

STEP 2. Write zeros in the empty columns to help keep digits aligned. Add each column and bring the decimal point straight down into the answer. Carry where necessary.

Subtraction

Suppose you buy items for $1.65 at the grocery and give the cashier a five-dollar bill. How much change should you receive? As in addition, you must line up the decimal points to subtract and find the answer.

$$\begin{array}{r} \$5.00 \\ -1.65 \\ \hline \$3.35 \end{array}$$

Rules for Subtracting Decimals

1. Use zeros to give each number the same number of decimal places. If necessary, compare the decimals to decide which is larger.
2. Line up the numbers with point under point. Be sure to put the larger number on top.
3. Subtract, borrowing if necessary. Bring the decimal point straight down into the answer.

EXAMPLE: What is the difference between .254 and .7?

(1) .254 = .254
.7 = .700

(2)
$$\begin{array}{r} .700 \\ -.254 \\ \hline .446 \end{array}$$

STEP 1. Add zeros to give both numbers the same number of decimal places.

STEP 2. .700 is the larger number. Put .700 on top, borrow, and subtract.

Decimals Exercise 7

Directions: Solve each problem.

1. .6 + .9 =

2. .57 + .8 =

3. 16 + 9.24 + 170.3 + .369 =

4. 15.23 + 4 + 1.816 + 9.4 =

5. .8 − .26 =

6. .5 − .345 =

7. 18 − .32 =

8. .09 − .075 =

9. Add 12.3, 0.016, 5, and 216.
10. What is the sum of 283.4, 87.49, and 107.3?
11. 4.036 + 23 + 2.19 + .084 =
12. Take .094 from .3.
13. Add 83 and 36.27.
14. From .008 take .0025.
15. Find the difference between .325 and .6.
16. How much more is $30 than $13.68?
17. $84 + $19.65 + $.23 + $1.56 =
18. Find the total of .0075, .00128, .004, and .03806.
19. Find the difference between 20 and 4.63.
20. Take .00925 from .03.

ANSWERS AND SOLUTIONS START ON PAGE 329.

Adding and Subtracting Decimals: Word Problems

The next problems will give you a chance to apply your skills in adding and subtracting decimals in word problems. These are the kinds of problems you are likely to see on the GED. Read each problem carefully to decide whether to add or subtract, and be sure you are answering exactly what is asked for. Remember the five-step approach to problem solving you saw in Chapter 2.

Decimals Exercise 8

Directions: Solve each problem.

1. Mr. Rigby is 1.7 meters tall. His son Jeff is 1.48 meters tall. How many meters taller is Mr. Rigby than his son?
2. Martha's empty suitcase weighs 4.5 pounds. Her large purse when empty, weighs 1.5 pounds. After she packed the suitcase, it weighed 16.25 pounds. Find the weight in pounds of the contents of the suitcase.
3. On Friday, Phil drove the distances shown on the map to the right. How many miles did Phil drive altogether?

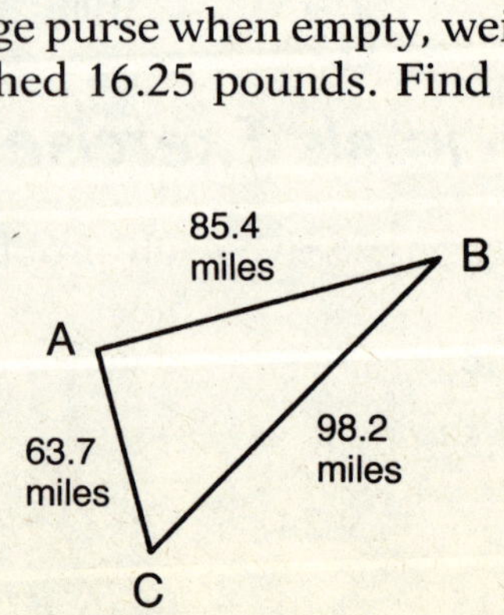

4. Last season Don Johnson had a batting average of .206, which was .35 below the second baseman's average. By the end of this season Don's hitting had improved, and his average rose .095. Find his batting average at the end of the season.

5. To pay for a new sports center, the town of Troy raised $1.35 million from the state government, $.85 million from the county government, and $1.05 million from local businesses and individuals. The total estimated cost of the project is $4 million. How much more money does the town need?

6. Matt is filling a carton for shipping. The sign on the carton says that the carton cannot support more than 50 pounds. So far Matt has put items weighing 26.9 pounds, 3.75 pounds, and 12.4 pounds into the carton. What additional weight can the carton hold?

7. A sign at a tunnel says, "height limit = 11.5 feet." When the air is low in the tires of Mark's truck, the truck is 10.65 feet tall. When the tires are full, the truck is an additional 0.2 feet high. By how much does Mark's truck clear the ceiling of the tunnel when the truck tires are full?

8. In 1980 the population of the U.S. was 226.8 million. In 1985 the population was 239 million. From 1980 to 1985 the population of the South and the West grew by 11.1 million. By how much did the population of the rest of the country grow from 1980 to 1985?

9. The year's goal for the Central City United Fund is $5 million. By June the fund had $2.45 million, and by September, it had an additional $0.95 million. Which expression tells how many millions the fund needs to reach the goal by the end of the year?
(1) 5 + 2.45 + 0.95
(2) (2.45 + 0.95) − 5
(3) 5 − (2.45 + 0.95)
(4) 2.45 − (5 − 0.95)

10. Caroline bought fourteen pints of milk for $.44 each. She also bought eight bags of chips for $1.09 each. What operations should you use to find how much she paid for all the milk and chips?
(1) Add .44 and 1.09.
(2) Add .44 and 1.09, then multiply by the sum of 14 and 8.
(3) Multiply 14 by the sum of 1.09 and .44.
(4) Multiply 14 by .44, multiply 8 by 1.09, and add the two products.

ANSWERS AND SOLUTIONS START ON PAGE 330.

Multiplying Decimals

Lou bought three pairs of socks at a cost of $1.75 a pair. To find the price of three pairs of socks, multiply the price of one pair by 3.

$$\begin{array}{r} \$1.75 \\ \times \quad 3 \\ \hline \$5.25 \end{array}$$

When you multiply a dollars and cents figure by a whole number you mark off two decimal places for the cents in the answer. Now look at this example:

EXAMPLE 1: What is the product of 3.26 and .4?

(1) $\begin{array}{r} 3.26 \\ \times \ .4 \\ \hline \end{array}$ **(2)** $\begin{array}{r} 326 \\ \times \ 4 \\ \hline 1304 \end{array}$ **(3)** $\begin{array}{r} 3.26 \\ .4 \\ \hline 1.304 \end{array}$ ← two places / ← one place / ← total of three places

STEP 1. Place the longer number on top. Put the other number beneath it, lining the numbers up at the right.

STEP 2. Multiply as you would with whole numbers.

STEP 3. Count the number of decimal places in *both* the numbers you have multiplied (3 total). Insert a decimal point so that there are 3 places to the right of it.

TIP: Notice the numbers in the problem are lined up at the right. This is not the same as in addition and subtraction.

Rules for Multiplying Decimals

1. Place the longer number on top. Put the other number underneath, lining the numbers up on the right.
2. Multiply as you would with whole numbers.
3. Count the number of decimal places in both the numbers you have just multiplied. (Decimal places are *to the right* of the decimal point.)
4. Counting from the right, insert the decimal point so that you have this total number of decimal places in your answer. Use zeros immediately after the decimal point if you need more places.

Carefully study the examples below. Look at the way each problem is written and notice where the decimal points are placed.

EXAMPLE 2: Multiply .76 by .05.

$$\begin{array}{rl} 76 & \leftarrow \text{no places} \\ \times\ .05 & \leftarrow \text{two places} \\ \hline 3.80 & \leftarrow \text{two places} \end{array}$$

Notice that in Example 2 the answer is 3.80 *or* 3.8 if you drop the unnecessary zero.

EXAMPLE 3: Find the product of 2.0413 and 0.006.

$$\begin{array}{rl} 2.0413 & \leftarrow \text{four places} \\ \times\ 0.006 & \leftarrow \text{three places} \\ \hline .0122478 & \leftarrow \text{add a zero after the decimal point to make seven places} \end{array}$$

Notice that in Example 3 the answer had only six digits, yet the problem requires a seven-place answer. You need to add a zero between the decimal point and the first number.

> **TIP:** There are shortcuts for multiplying a decimal or mixed decimal by 10, 100, or 1000:
>
> - To multiply by 10, move the decimal point one place to the right.
>   .34 × 10 = .34 = 3.4
> - To multiply by 100, move the decimal point two places to the right.
>   .34 × 100 = .34 = 34
> - To multiply by 1000, move the decimal point three places to the right.
>   .34 × 1000 = .340 = 340

Sometimes you will see a dot (·) instead of a times sign (×) to indicate multiplication. This is just another way to show the same operation.

EXAMPLE 4: Find the product of 34.2 and 100.

34.2 · 100 = 34.20 = 3420.

Put a zero to the right of the 2 in order to move the decimal point two places to the right.

Decimals Exercise 9

Directions: Solve each problem. Be careful to put the right number of decimal places in the answer.

1. .8 × .7 =
2. Multiply 20.6 by 0.3.
3. 2.09 × 0.4 =
4. 8.3 · 2.7 =
5. 0.378 × 0.6 =
6. Multiply .0004 by 30.
7. 0.8 × 0.01 =
8. Multiply 1.439 by 0.8.
9. .076 × 1.5 =
10. 4.2 · $24.80 =
11. $250 × 2.3 =
12. $1.16 × 1.25 =
13. 1.432 × 10 =
14. Multiply 0.056 by 100.
15. 0.083 × 10 =
16. Multiply 19.7 by 1000.
17. Find the product of 15 and 0.3.
18. What is the product of 143 and 2.5?
19. Find the product of 0.34 and 12.61.
20. Find the product of 1000 and 7.95.

ANSWERS AND SOLUTIONS START ON PAGE 330.

Dividing Decimals

Dividing Decimals by Whole Numbers

The bill for three people in a restaurant came to $14.55. If they shared the bill equally, how much did each person owe? To solve this, divide the total bill by the number of people who shared it.

When you divide a decimal by a whole number, put the decimal point in the answer *directly above its position in the problem.*

$$\text{divisor}\rightarrow 3\overline{)\,\$14.55}\ \leftarrow\text{dividend} \quad \text{with quotient } \$\ 4.85 \leftarrow \text{quotient}$$

Each person owes **$4.85.**

In the rules below, the term ***dividend*** means the number being divided. The ***divisor*** is the number being divided into the dividend. The term ***quotient*** is the answer to a division problem.

Rules for Dividing a Decimal by a Whole Number

1. Put the decimal point in the quotient directly above its position in the dividend.
2. Divide as you would for whole numbers.

Study these examples carefully.

EXAMPLE 1: .168 ÷ 2 = ?

$$\begin{array}{r} .084 \\ 2\overline{).168} \end{array}$$

Notice the 0 in the quotient. Since 2 does not divide into 1, you must put the 8 above the 6 in the hundredths place (since you are now dividing 2 into 16.) The 0 shows that there are no tenths and holds the 8 in the hundredths place.

EXAMPLE 2: Divide 29.72 by 5.

$$\begin{array}{r} 5.944 \\ 5\overline{)29.720} \end{array}$$

Notice the extra 0 that has been added to the dividend. This zero does not change the value of the dividend. Adding a zero may allow you to carry out the answer one more decimal place and eliminate a remainder.

> **TIP:** There are shortcuts for dividing a number by 10, 100, or 1000. These shortcuts will be useful in your work with percents and with metric measurements.
>
> - To divide a number by 10, move the decimal point one place to the left. 14.3 ÷ 10 = 1.4.3 = 1.43
> - To divide a number by 100, move the decimal point two places to the left. 14.3 ÷ 100 = .14.3 = .143
> - To divide a number by 1000, move the decimal point three places to the left. 14.3 ÷ 1000 = .014.3 = .0143

Decimals Exercise 10

Directions: Solve each problem.

1. Divide 52.608 by 8.
2. Divide 9 into 758.7.
3. $14\overline{)10.5}$
4. .039 split 3 ways =
5. Divide 2.375 by 25.
6. $7\overline{).0168}$
7. 9.6 ÷ 24 =
8. $24\overline{).96}$
9. $18\overline{)2.7}$
10. Divide 36.72 by 12.
11. 42.3 ÷ 100 =
12. Divide .04 by 10.
13. Divide 1000 into 19.5.
14. .65 ÷ 100 =

ANSWERS AND SOLUTIONS START ON PAGE 331.

Dividing by Decimals

To divide a number by a decimal, you must first change the divisor to a whole number.

EXAMPLE 1: Divide 1.842 by .06.

(1) $.06\overline{)1.842}$ **(2)** $.06.\overline{)1.84.2}$ **(3)** $\begin{array}{r} 30.7 \\ 6\overline{)184.2} \end{array}$

STEP 1. Set the problem up for long division.
STEP 2. Make the divisor a whole number. Move the decimal point two places to the right. Move the decimal point in the dividend two places to the right.
STEP 3. Bring the decimal point up in the quotient directly above its new position in the dividend. Divide.

Rules for Dividing by a Decimal

1. Make the divisor a whole number. Move the point to the right as far as it will go.
2. Move the point in the dividend to the right *the same number of places* you moved the point in the divisor. You may need to add zeros to the dividend in order to move the decimal point that far.
3. Bring the decimal point up in the quotient directly above its *new* position in the dividend and divide.

Study the next examples carefully.

EXAMPLE 2: Divide .8 by .002.

$$\begin{array}{r} \textbf{400.} \\ .002_{\smile}\overline{)\,.800_{\smile}} \end{array}$$

In this case we added two zeros.

EXAMPLE 3: Divide 40 by .008.

(1) $.008\overline{)\,40}$ **(2)** $.008_{\smile}\overline{)\,40.000_{\smile}}$ **(3)** $\begin{array}{r} \textbf{5,000} \\ 8\overline{)\,40,000} \end{array}$

STEP 1. Set the problem up for long division.
STEP 2. Make the divisor a whole number. Move the decimal point three places to the right.
STEP 3. Put a point to the right of 40. Put three zeros at the right of the point. Then move the point the same number of places to the right as you did in the divisor. Divide.

Decimals Exercise 11

Directions: Solve each problem.

1. 8.4 ÷ 0.2 =

2. Divide 5.25 by .5.

3. $.13\overline{)\,.338}$

4. .42 ÷ .028 =

5. $1.2\overline{)\,.324}$

6. $.04\overline{)\,83.6}$

7. Divide 101.2 by 25.3

8. 90.6 ÷ .03 =

9. 6.7 ÷ .134 =

10. Divide 64 by 1.28.

11. $.12\overline{)\,96}$

12. 261 ÷ 4.35 =

ANSWERS AND SOLUTIONS START ON PAGE 331.

Multiplying and Dividing Decimals: Word Problems

The next problems give you a chance to apply your skills in multiplying and dividing decimals. These are the kinds of problems you are likely to see on the GED.

> **TIP:** Remember that, in any division problem, the amount being divided up must go inside the division bracket.

Decimals Exercise 12

Directions: Solve each problem.

1. The electricity to operate a black-and-white television costs $.025 an hour. The Randolphs watch their television an average of 40 hours a week. How much do the Randolphs pay each week to watch television?

2. Colin cut a piece of electric cable 3.5 meters long into four equal pieces. Assuming no waste, what was the length in meters of each piece?

3. Fred's weekly salary is $240. He works 37.5 hours each week. How much does Fred make per hour?

4. The chart at right shows the distances that three athletes are able to jump. Which operations should you use to find the average distance that the *women* are able to jump?

Cindy	5.4 feet
Paul	7.4 feet
Carol	7.1 feet

(1) Multiply 5.4 by 3.
(2) Add 5.4, 7.4, and 7.1, then divide by 3.
(3) Add 5.4 and 7.1, then divide by 3.
(4) Add 5.4 and 7.1, then divide by 2.
(5) Add 5.4 and 7.1.

5. Alejo bought 18.5 yards of lumber which weighed 5.8 pounds per yard and cost $7.20 a yard. How much did he pay for the lumber?

6. Mrs. Goncalves talked to her sister in Portugal for twelve minutes on Tuesday night. The night rate for a call to Portugal is $3.15 for the first three minutes and $1.05 for each additional minute. How much did the call to Portugal cost Mrs. Goncalves?

7. On Monday Grace worked 7 hours at her regular rate of $5.20 an hour and 2.5 hours at her overtime rate of $7.80 an hour. How much did Grace make that day?

8. Senta drove for 2.5 hours at an average speed of 60 mph and for 1.5 hours at an average speed of 48 mph. What total distance did she drive?

9. Nick cut two lengths of copper tubing each 1.45 meters long from the piece shown at the right. How long was the remaining piece?

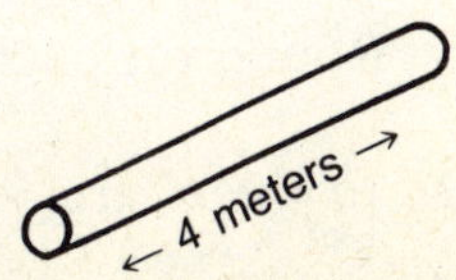

10. A seamstress wanted to sew together two pieces of cloth, then cut the new piece into four equal parts. The two pieces of cloth measured 1.3 yards and 3.5 yards. Which of the following expressions gives the size of each of the four pieces?
(1) $4 \div (1.3 + 3.5)$
(2) $4 \times 1.3 \div 3.5$
(3) $(1.3 + 3.5) \div 4$
(4) $(1.3 \times 4) + (3.5 \times 4)$

ANSWERS AND SOLUTIONS START ON PAGE 332.

Rounding Off Decimals

Suppose George works for \$5.75 an hour. How much will he make in 3.5 hours? To find this amount, you would multiply \$5.75 by 3.5.

$$\begin{array}{r} \$5.75 \\ \times \quad 3.5 \\ \hline 2\,875 \\ 17\,25 \\ \hline \$20.125 \end{array}$$

Our money system has only two decimal places. In this problem, you need to **round off** \$20.125 to the nearest hundredth. In some problems, you may be asked to find an answer to the nearest tenth, hundredth, thousandth, etc.

Rounding off decimals is almost the same as rounding off whole numbers.

Rules for Rounding Off Decimals

1. Underline the digit in the place you are rounding off to.
2. If the digit to the right is 5 or more, add 1 to the underlined digit.
3. If the digit to the right is less than 5, leave the underlined digit as it is.
4. Drop the digits to the right of the underlined digit.

EXAMPLE: Round off 8.1942 to the nearest thousandth.

STEP 1. Underline the digit in the thousandths place.
$8.19\underline{4}2$

STEP 2. Look at the digit to the right of the underlined digit. Since the digit is less than 5, keep the underlined digit and drop the digit at the right.
8.194

Decimals Exercise 13

Directions: Round off each of the following decimals to the place indicated.

1. \$0.328 to the nearest penny.

2. 6.527 to the nearest tenth.

3. 20.063 to the nearest tenth.

4. 8.0376 to the nearest thousandth.

5. .00825 to the nearest ten thousandth.

6. Adrian pays \$1.298 for a gallon of gasoline. To the nearest penny, what does he have to pay for 12 gallons of gas?

7. Lois drove for 3.5 hours at an average speed of 45 miles an hour. Find how far she drove to the nearest ten miles.

8. Frank works for 37.5 hours a week at a rate of \$7.85 an hour. To the nearest dollar, how much does he make in a week?

9. An inch is equal to 2.54 centimeters. How many centimeters are there in 12 inches? Round off your answer to the nearest centimeter.

10. Find the cost of 2.4 pounds of a cheese that sells for \$3.49 a pound. Round your answer to the nearest penny.

ANSWERS START ON PAGE 332.

Repeating Decimals

In the previous sections, decimal division problems worked out evenly. However, some problems don't work out evenly. For instance, if you divide .20 by 3, you get a repeating decimal, no matter how many zeros you add to the dividend.

$$3\overline{).2000}\quad \text{quotient } .0666$$

A common way to handle a repeating decimal is to round off your answer. Generally, a problem will tell you what place to round off to. The answer for the example above, rounded off to the nearest thousandth, is .067.

EXAMPLE: Find 1 ÷ .3 to the nearest hundredth.

(1) $.3\overline{)1}$ **(2)** $.3\overline{)1.0}$ quotient 3.333 **(3)** $3.3\underline{3}3 \rightarrow 3.33$

STEP 1. Set up the division problem.
STEP 2. Move the decimal point one place to the right in both divisor and dividend. Divide. Carry out the division to one more place than the answer requires so that you can round off.
STEP 3. Underline the digit in the hundredths place. Look at the digit to the right and round off. To the nearest hundredth, 3.333 is **3.33**.

Decimals Exercise 14

Directions: Divide and round off each problem to the decimal place indicated.

1. .4 ÷ 15 to the nearest hundredth.

2. .02 ÷ 3 to the nearest hundredth.

3. .5 ÷ 6 to the nearest tenth.

4. .3 ÷ .8 to the nearest tenth.

ANSWERS AND SOLUTIONS START ON PAGE 332.

Problem Solving for the GED

Introducing Item Sets

You will see some long passages on the GED Math Test. Each of these passages gives enough information to produce several questions. The questions are often not difficult, but sometimes it is a challenge to pick the information you need to answer each particular question. The skills you need to solve these "item sets" are an extension of the skills you developed on page 22 when you learned to sort out necessary and unnecessary information.

Below is a sample passage followed by several typical questions. The questions are answered as examples for you. Read the passage and the questions carefully.

Examples 1-4 refer to the following situation:

Wendy and Don have seen two apartments that they like. Apartment A has monthly rent of \$340.50 for the first year. The rent for the second year will be an additional \$35.50 a month. The rent for apartment A includes heat. The rent for apartment B is \$295 a month for two years. The rent for apartment B does not include heat. The landlord for apartment B says that the heating bills for the apartment average \$55.50 a month.

EXAMPLE 1: What is the rent for the first year of apartment A?

INFORMATION NEEDED. Monthly rent of \$340.50 and number of months in a year (12)

SOLUTION. 12 × \$340.50 = **\$4086**

EXAMPLE 2: Find the rent for the second year of apartment A.

INFORMATION NEEDED. First year's monthly rent of \$340.50, additional charge of \$35.50 a month, and number of months in a year (12)

SOLUTION. a. \$340.50 + \$35.50 = \$376
b. 12 × \$376 = **\$4512**

EXAMPLE 3: Assuming the landlord's estimate for the heating costs of apartment B are correct, what is the total cost of renting and heating apartment B for two years?

INFORMATION NEEDED. Monthly rent of \$295, monthly heating estimate of \$55.50, and number of months in two years (24)

SOLUTION. a. \$295 + \$55.50 = \$350.50
b. 24 × \$350.50 = **\$8412**

Sometimes information is introduced in a question which you must apply to the information in the passage you have read.

EXAMPLE 4: Don and Wendy take home \$1360 a month. How much would they have left over each month for the first year in apartment A?

INFORMATION NEEDED. Monthly take-home pay of $1360 and monthly rent of $340.50.

SOLUTION. $1360 − $340.50 = **$1019.50**

As you can see from these examples, it is very useful to think about and jot down the information you are going to need to solve a problem. You may want to try doing this as you practice with item sets below.

★ GED PRACTICE ★

Item Sets

Directions: Questions 1-5 refer to the following passage.

Store X sells a set of living room furniture for $1199.75. It offers a $100 rebate for paying cash. Store X will also sell the furniture for $150 down and installments of $45.50 a month for three years. Store Y sells the same furniture for $1279.99. Store Y offers no rebate for paying cash. Its terms for payment are $200 down and installments of $54.60 a month for two years.

1. If a buyer is considering paying cash, how much can he save by buying the furniture at store X rather than store Y?

   Information needed:

   Solution:

2. What is the total price of the furniture from store X if it is purchased in installments?

   Information needed:

   Solution:

3. What is the total price of the furniture purchased in installments from store Y?

   Information needed:

   Solution:

4. The Muellers decided to buy the furniture from the store with the lowest total cost under the installment plan. How much money did they save by buying from the store with the lowest total cost?

   Information needed:

   Solution:

5. How much more does store X get for the furniture from a buyer who pays in installments than from a buyer who pays cash?

   Information needed:

   Solution:

Directions: Questions 6-9 refer to the following passage.

> Ron is a builder. He tells the Smiths that he can build a garage for them for $15,000. He estimates that, working with an assistant, he can complete the job in 40 hours. Ron pays his assistant $8.50 an hour. Ron estimates that materials and expenses, not including his assistant, will cost $10,500.

6. If the work is completed in 40 hours as planned, how much will the assistant get paid?

Information needed:

Solution:

7. If Ron's estimates are correct, how much profit will he make on this job?

Information needed:

Solution:

8. What is Ron's profit per hour, assuming he also works 40 hours on the garage?

Information needed:

Solution:

9. In fact, by the end of the second day, Ron realized that he and his assistant could not complete the job in 40 hours. He hired an additional helper for 25 hours of work. Ron paid this helper $6.50 an hour. Ron kept the price of the job at $15,000. What was his total profit after he used the second helper?

Information needed:

Solution:

ANSWERS START ON PAGE 333.

Exercise 15: Review

Directions: For questions 1-10 simply answer each question.

1. Tell what place the 8 in 4026.385 is in.

2. Circle the number that is equal to seventy and sixteen thousandths: .7016 70.16 70.016.

3. Rewrite 060.04300 and omit unnecessary zeros.

4. Write the mixed decimal four hundred eight and fifteen ten-thousandths.

5. Arrange the following decimals in order from smallest to largest: .6, .06, .606, .066.

6. Find the sum of 36, 2.93, .065, and 4.

7. Take .0063 from .08.

8. Find the product of 3.2 and .005.

9. What is .405 divided by 15?

10. What is the quotient of 9 divided by .045?

★ GED PRACTICE ★

Directions: The following problems include whole numbers as well as decimals. Choose the correct answer to each of the following. Remember the problem-solving skills you have learned so far.

11. In 1970 the population of Florida was 6.79 million. By 1980 the population had increased 3.39 million. What was the population of Florida in 1980?
(1) 10.18 million
(2) 11.28 million
(3) 11.08 million
(4) 10.08 million
(5) 3.4 million

12. Allen took a board the length of Board A (at right) and cut off a piece the length of Board B. What was the length in yards of the remaining piece?
(1) 4.75
(2) 1.75
(3) 2.75
(4) 1.25
(5) 2.25

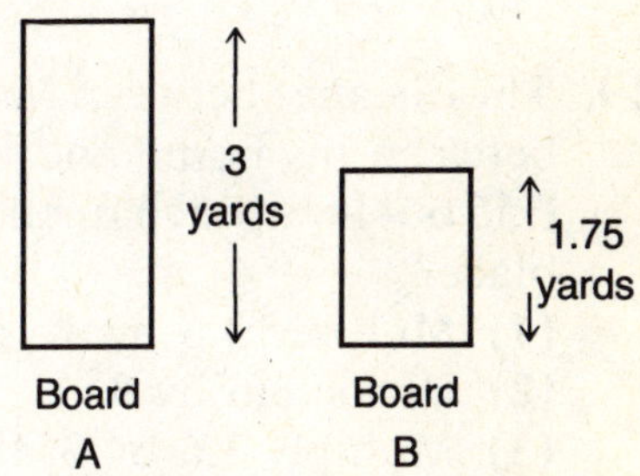

13. David drove 223 miles on 12 gallons of gasoline. To the nearest tenth, find the average number of miles he drove on one gallon of gasoline.
(1) 19.4
(2) 18.6
(3) 18.2
(4) 16.8
(5) 20.2

14. The rate schedule at the right shows the hourly electricity costs for various household appliances. The Walek family has their television turned on for an average of 52 hours a week. How much do the Waleks pay each week to operate their television?
(1) $1.77
(2) $1.98
(3) $2.60
(4) $5.40
(5) $8.60

Electricity *(cost per hour)*	
iron	$.005
television	.034
clock	.010
radio	.022

15. All but 56.3 pounds of topsoil were sold at a sale. If the store began with 100 pounds of the topsoil and charged $.59 a pound, how much money did it make?
(1) $25.00
(2) $25.78
(3) $44.29
(4) $59.00
(5) $257.60

16. The number of hits a baseball player gets is divided by the number of times he goes to bat to give his batting average. Last season Bill was at bat 75 times, and he made 22 hits. Find his batting average. Round off your answer to the nearest thousandth.
(1) .356
(2) .333
(3) .320
(4) .299
(5) .293

17. Arrange the following lengths of pipe in order from smallest to largest:

A — 1.8 yard
B — 0.75 yard
C — 1.75 yard
D — 0.125 yard
E — 1.125 yard

(1) A, B, D, C, E
(2) A, B, C, E, C
(3) D, B, C, E, A
(4) D, B, E, C, A
(5) B, D, E, C, A

18. Charlene bought 5 yards of 36-inch wide material for $7.80 a yard and 2.5 yards of 42-inch wide material for $8.40 a yard. How much did she pay altogether for the material?

(1) $81.00
(2) $63.36
(3) $60.00
(4) $55.80
(5) $50.00

19. The distance between Manny's workplace and his home is 3.6 miles. The distance between his home and his exercise class is three times that far. How can you find out how much further from home Manny's exercise class is than his workplace?

(1) Multiply 3.6 by 3.
(2) Divide 3.6 by 3.
(3) Multiply 3.6 by 3, then subtract 3.6.
(4) Divide 3.6 by 3, then subtract 3.6.
(5) Add 3.6 and 3.

20. Pete is a cabinetmaker. He charges $125 for a small bookcase and $190 for a large one. One month he sold 5 small bookcases and 3 large ones. Which expression represents the gross amount Pete got for the sale of the bookcases?

(1) (125 + 5)(190 + 3)
(2) 125(3 + 5) + 190
(3) 3(125) + 5(190)
(4) 5(125) + 3(190)
(5) 125(190) + 3(5)

Directions: Problems 21-24 refer to the following situation and mileage map.

Alfonso drove from El Paso to Dallas in 13 hours. Then he drove from Dallas to Houston in 4 hours. Finally he drove from Houston back to El Paso in 15.5 hours. Over the entire trip, Alfonso bought 70 gallons of gasoline at a cost of $1.38 per gallon.

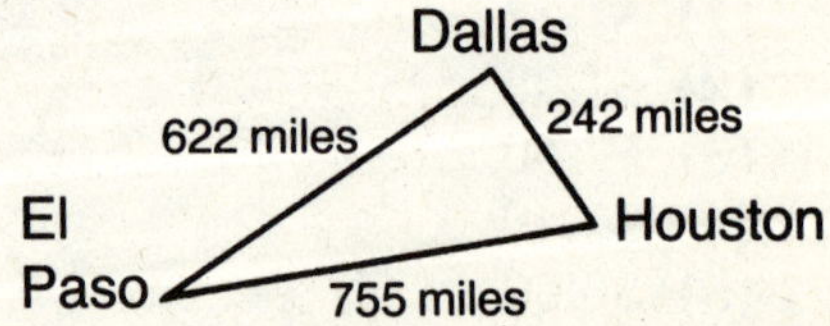

21. Find the total number of hours Alfonso spent driving on this trip.

(1) 17
(2) 19.5
(3) 28.5
(4) 30
(5) 32.5

22. How much did Alfonso spend for gas for the entire trip?
(1) \$66.60
(2) \$70
(3) \$85.60
(4) \$96.60
(5) \$138

23. The drive from Houston back to El Paso took how many hours less than the drive from El Paso to Houston by way of Dallas?
(1) 3
(2) 2.5
(3) 2
(4) 1.5
(5) 1

24. To the nearest unit, what was Alfonso's average speed in miles per hour for the trip from El Paso to Dallas?
(1) 42
(2) 48
(3) 50
(4) 55
(5) 58

ANSWERS AND SOLUTIONS START ON PAGE 333.

Decimals Review Evaluation

Problem	Section	Starting Page
1	Place Value	51
2	Reading Decimals	52
3	Zeros	53
4	Writing Decimals	54
5	Comparing Decimals	55
6–7	Adding and Subtracting Decimals	58
8	Multiplying Decimals	62
9	Dividing Decimals by Whole Numbers	64
10	Dividing by Decimals	65
11–12	Adding and Subtracting Decimals: Word Problems	60
13–15	Multiplying and Dividing Decimals: Word Problems	67
16	Rounding Off Decimals	68
17	Comparing and Ordering Decimals	56
18	Multiplying Decimals (*Finding Necessary Information*)	62/22*
19	Choosing the Operation	24*
20	Set-up	42*
21–24	Item Sets	70

**These skills are covered in earlier chapters.*

Passing score: 19 right out of 24 problems

Your score: _____ right out of 24 problems

If you had less than a passing score, review the sections for those problems you missed. Then redo these questions before going on to the fractions section.

If you had a passing score, correct any problem you got wrong, then go on to the fractions section.

5 FRACTIONS

Writing Forms of Fractions

A ***fraction***, like a decimal, is a way of showing a part of a whole. The picture below shows a circle divided into four equal parts. Three of these parts are shaded. We can say that three-fourths of the circle is shaded.

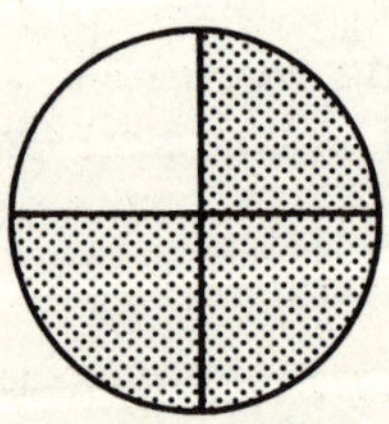

$\frac{3}{4}$

3 ← The ***numerator*** tells how many parts are shaded.
4 ← The ***denominator*** tells how many equal parts the whole is divided into.

With fractions, any number except 0 can be in the denominator. In other words, the whole can be divided into any number of equal parts. Notice how fractions are different from decimals. Decimals divide a whole into tenths, hundredths, thousandths, etc. Fractions are useful because they can divide a whole into thirds, eighths, seventeenths, or any other number.

EXAMPLE: Write a fraction that shows what part of the figure at the right is shaded.

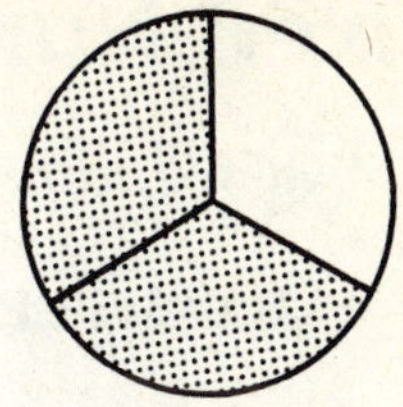

$\frac{2}{3}$ shaded parts (the numerator) / total equal parts (the denominator)

SOLUTION. Count the total number of equal parts that the figure is divided into. There are 3 equal parts. 3 is the denominator. Now count the number of shaded parts. 2 parts are shaded. 2 is the numerator. $\frac{2}{3}$ of the figure is shaded.

> **TIP:** You may see fractions written vertically, with the numerator over the denominator ($\frac{2}{3}$). You may also see fractions written horizontally, with a slash between the numerator and denominator (2/3). You should be familiar with both forms.

There are three uses of fractions you should know:

1. **as a part of a whole**
   $\frac{2}{3}$ of this figure is shaded

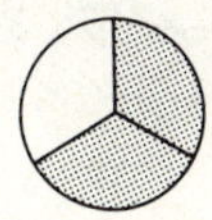

2. **as a part of a group**
   2 out of 3, or $\frac{2}{3}$ of this group, is shaded

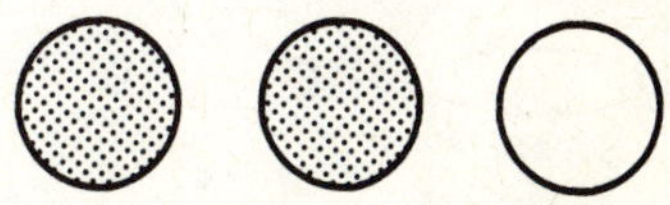

3. **as division**
   The fraction bar also means 2 divided by 3.

$$\frac{2}{3}$$

There are also three facts about fractions that you should know.

1. Any number over itself is equal to 1.
   $\frac{4}{4} = 1 \quad \frac{1342}{1342} = 1 \quad \frac{20}{20} = 1$
2. Any number with zero as a numerator equals zero.
   $\frac{0}{4} = 0 \quad \frac{0}{1002} = 0 \quad \frac{0}{90} = 0$
3. A fraction with a zero in the denominator has *no* meaning.
   $\frac{14}{0}$ (crossed out)

Fractions Exercise 1

Directions: Write a fraction that tells what part of each circle is shaded.

1. 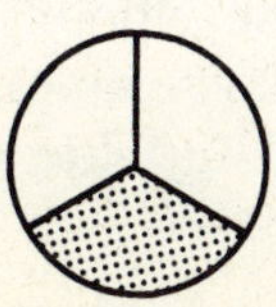
2. 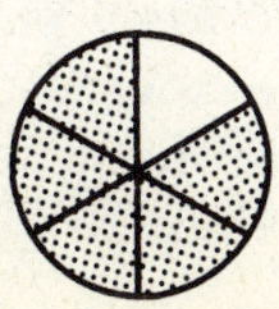
3. 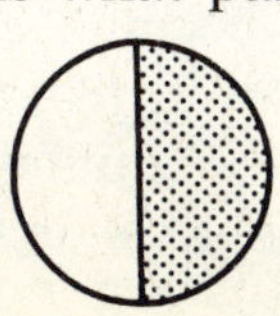
4. 
5. 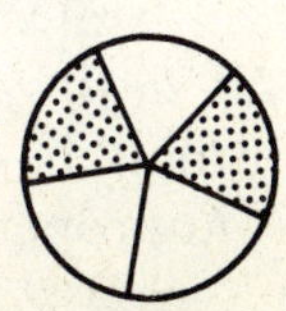

ANSWERS START ON PAGE 334.

Mixed Numbers

A ***mixed number*** is a whole number and a fraction.

EXAMPLE: Write a mixed number that shows how many circles at the right are shaded.

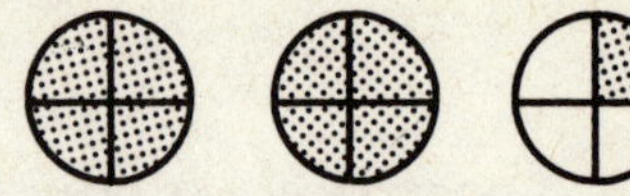

STEP 1. Count the number of completely shaded circles. 2 circles are completely shaded. 2 is the whole number.

STEP 2. Look at the partially shaded circle. The circle is divided into 4 equal parts. 4 is the denominator. 1 part of the circle is shaded. 1 is the numerator. Altogether, $\mathbf{2\frac{1}{4}}$ circles are shaded.

Fractions Exercise 2

Directions: Write a mixed number that tells how many figures in each set are shaded.

1. 

2. 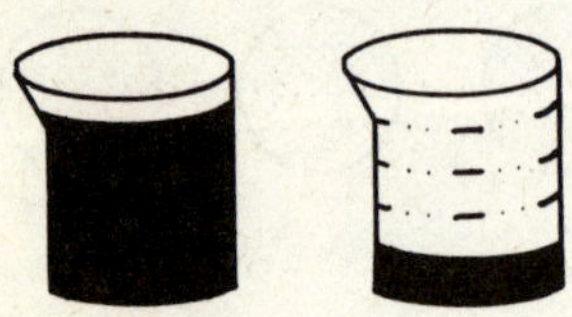

3. 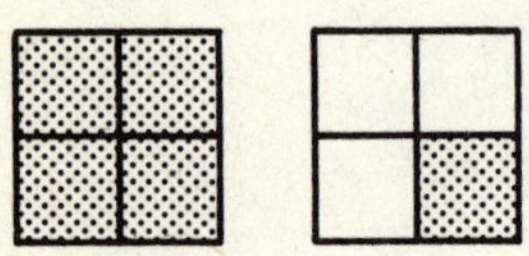

4. 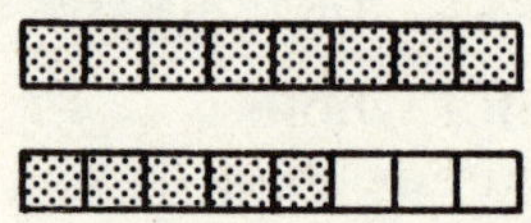

5. 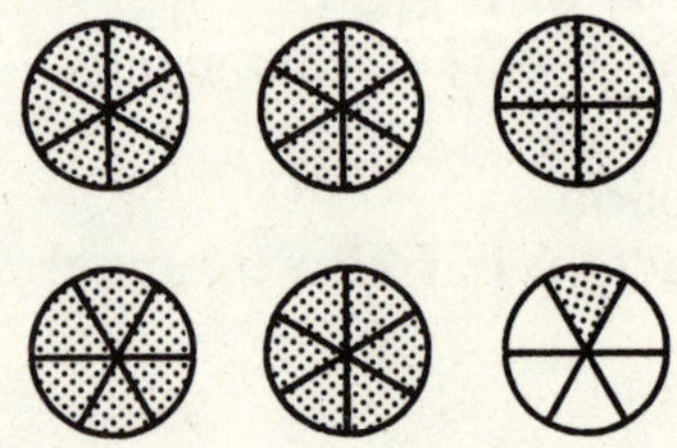

6. 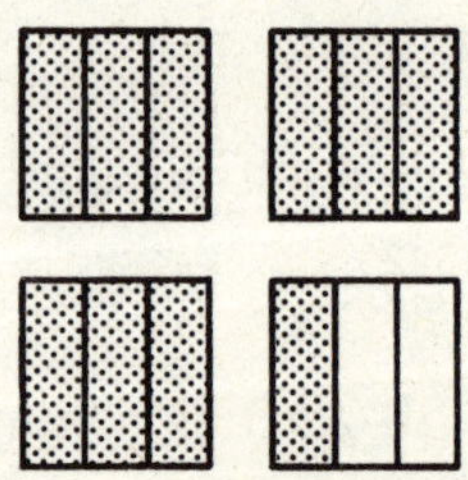

ANSWERS START ON PAGE 334.

Proper and Improper Fractions

A ***proper fraction*** is a fraction in which the numerator is smaller than the denominator. $\frac{2}{3}$, $\frac{99}{100}$, and $\frac{1}{5}$ are examples of proper fractions. The value of a proper fraction is always less than 1. This is because a fraction is a part of one whole.

An ***improper fraction*** is a fraction in which the numerator is as large as or larger than the denominator. $\frac{3}{3}$ and $\frac{6}{5}$ are examples of improper fractions. The value of an improper fraction is either equal to 1 or larger than 1.

EXAMPLE 1: Write an improper fraction that tells what part of the figure at the right is shaded.

$\frac{3}{3}$

SOLUTION. The figure is divided into 3 equal parts. 3 is the denominator. 3 parts of the figure are shaded. 3 is also the numerator.

EXAMPLE 2: Write an improper fraction that tells how many circles at the right are shaded.

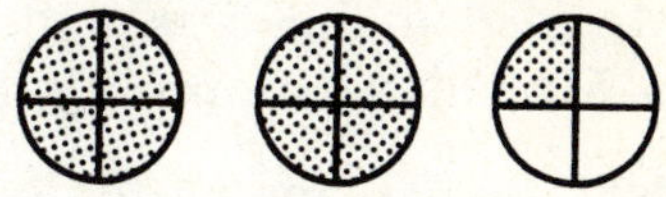

$\frac{9}{4}$

SOLUTION. Remember that the denominator tells how many parts each whole is divided into. Each circle here is divided into 4 equal parts. 4 is the denominator. Altogether, 9 pieces of the circles are shaded. 9 is the numerator. $\frac{9}{4}$ of the circles are shaded.

EXAMPLE 3: There are 25 students in Joe's math class. 13 of the students are women. What fraction of the class is women?

SOLUTION. In a problem like this, think of writing a fraction as putting a part over the whole. The denominator tells the total number of students in the class (the whole). 25 is the denominator. The numerator tells the number of women in the class (the part). 13 is the numerator. $\frac{13}{25}$ of the class is women.

Fractions Exercise 3

Directions: Write an improper fraction that tells how many parts in each set of figures are shaded.

1. 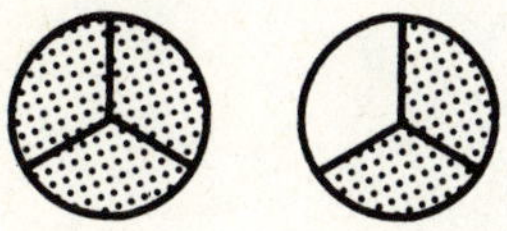

2. 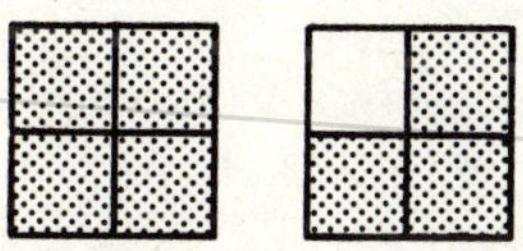

3. 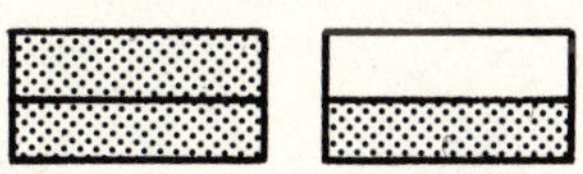

4. 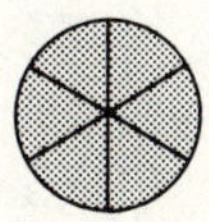

5. 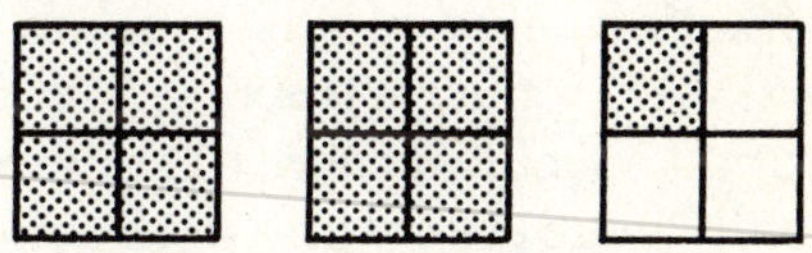

6. 

ANSWERS START ON PAGE 334.

Fractions Exercise 4

Directions: Solve each problem.

1. A yard has 3 feet. What fraction of a yard is one foot?
2. A foot has 12 inches. What fraction of a foot is 5 inches?
3. Mr. Allen makes \$800 a month. His rent is \$199 a month. Rent is what fraction of Mr. Allen's monthly salary?
4. Pete got 7 items wrong out of 50 on a test. What fraction of the total number of items did Pete get wrong?
5. Nina needs \$200 for a new cassette deck. So far she has saved \$137. What fraction of the price has she saved so far?
6. Frank ran in a 24-mile race. At the end of the first hour, he had completed 11 miles. What fraction of the race did Frank complete by the end of the first hour?
7. Together Saturday and Sunday make up what fraction of a week?

ANSWERS AND SOLUTIONS START ON PAGE 334.

Reducing Fractions

Reducing means expressing a fraction in a simpler way. Look at the two circles below. The circle at the left has six out of eight parts shaded. The circle at the right has three out of four parts shaded. But as you can see, both circles have the same amount of shaded space. This means that $\frac{6}{8}$ is equal to $\frac{3}{4}$.

$\frac{6}{8}$ shaded

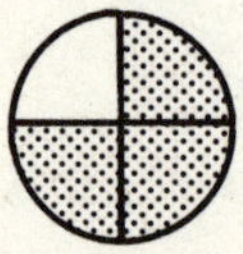

$\frac{3}{4}$ shaded

Reducing a fraction means writing it an easier way — with smaller numbers. Reducing a fraction does not change its value. A reduced fraction is equal to the original fraction, just as $\frac{6}{8}$ is equal to $\frac{3}{4}$ above.

How do you reduce a fraction like $\frac{6}{8}$? Look at both the numerator and the denominator and decide if there is a number that will divide evenly into each. As you know, 2 will divide evenly into both 6 and 8. Divide both numerator and denominator by 2:

$$\frac{6 \div 2}{8 \div 2} = \frac{3}{4}$$

While reducing changes the numbers in a fraction, it does not change its value. Two fractions with the same value are called ***equivalent fractions***. $\frac{6}{8}$ and $\frac{3}{4}$ are equivalent fractions. The fraction $\frac{3}{4}$ is the reduced form of $\frac{6}{8}$.

Rules for Reducing Fractions

1. Divide both the numerator and the denominator by the same number.
2. Check the result to see whether another number will divide evenly into both the new numerator and the new denominator.

When a fraction cannot be reduced further, it is reduced to **lowest terms**. Following are some hints for reducing fractions.

> **TIP:** If both the numerator and the denominator are even numbers, you can reduce by 2.

EXAMPLE 1: Reduce $\frac{6}{10}$ to lowest terms.

STEP 1. 6 and 10 are even numbers. Divide both 6 and 10 by 2.

$$\frac{6 \div 2}{10 \div 2} = \frac{3}{5}$$

STEP 2. Check $\frac{3}{5}$ to see whether another number will divide evenly into both. No other number divides evenly into both 3 and 5. $\frac{3}{5}$ is reduced to lowest terms.

EXAMPLE 2: Reduce $\frac{16}{24}$ to lowest terms.

STEP 1. 16 and 24 are even numbers. Divide both by 2.

$$\frac{16 \div 2}{24 \div 2} = \frac{8}{12}$$

STEP 2. $\frac{8}{12}$ can be reduced more. Divide both 8 and 12 by 4. $\frac{2}{3}$ is reduced to lowest terms.

$$\frac{8 \div 4}{12 \div 4} = \frac{2}{3}$$

Notice that in the last example we could have divided 16 and 24 by 8.

$$\frac{16 \div 8}{24 \div 8} = \frac{2}{3}$$

Finding the *largest* number that will divide evenly into both numerator and denominator can save time.

> **TIP:** If one number in a fraction ends with a 5 and the other ends with a 0, or if both numbers end with 5, you can reduce by 5.

EXAMPLE 3: Reduce $\frac{25}{70}$ to lowest terms.

$$\frac{25 \div 5}{70 \div 5} = \frac{5}{14}$$

Fractions Exercise 5

Directions: Reduce each fraction to lowest terms.

1. $\frac{6}{10} =$ $\frac{4}{6} =$ $\frac{14}{20} =$ $\frac{4}{32} =$ $\frac{18}{36} =$ $\frac{18}{24} =$

2. $\frac{25}{35} =$ $\frac{45}{50} =$ $\frac{15}{40} =$ $\frac{45}{60} =$ $\frac{30}{45} =$ $\frac{10}{25} =$

ANSWERS START ON PAGE 334.

> **TIP:** If both numbers in a fraction end with 0, you can reduce by 10 or a multiple of 10.

EXAMPLE 4: Reduce $\frac{30}{80}$ to lowest terms.

$$\frac{30 \div 10}{80 \div 10} = \frac{3}{8}$$

Fractions Exercise 6

Directions: Reduce each fraction to lowest terms.

$\frac{70}{100} =$ $\frac{50}{60} =$ $\frac{80}{160} =$ $\frac{40}{200} =$ $\frac{30}{200} =$ $\frac{50}{250} =$

ANSWERS START ON PAGE 335.

Sometimes it is hard to find a number that divides evenly into both the numerator and the denominator. In those cases, you may simply have to use the "trial and error" method. Just keep trying different numbers until you find one that divides evenly into both numerator and denominator. The better you know the multiplication tables, the easier you will be able to reduce fractions.

EXAMPLE 5: Reduce $\frac{49}{56}$ to lowest terms.

STEP 1. Look at the numerator and denominator and try to find a number that goes evenly into both. You may have to try a few different numbers before you come up with one that works. Since 7 works, divide both 49 and 56 by 7.

$$\frac{49 \div 7}{56 \div 7} = \frac{7}{8}$$

STEP 2. Check to see whether another number divides evenly into both 7 and 8. No other number divides evenly into both. $\frac{7}{8}$ is reduced to lowest terms.

Fractions Exercise 7

Directions: Reduce each fraction to lowest terms.

$\frac{7}{28} =$ $\frac{9}{30} =$ $\frac{27}{45} =$ $\frac{22}{33} =$ $\frac{17}{34} =$ $\frac{13}{39} =$

ANSWERS AND SOLUTIONS START ON PAGE 335.

Checking Your Work

A good way to check whether you have reduced correctly is to find the cross products. The ***cross products*** are the numerator of the first fraction multiplied by the denominator of the second and the denominator of the first fraction multiplied by the numerator of the second. When reducing is done correctly, the cross products are equal.

For example, look at $\frac{16}{20} = \frac{4}{5}$.

$\frac{16}{20} \times \frac{4}{5}$

The first cross product is $16 \times 5 = 80$.

The second cross product is $20 \times 4 = 80$.

Since the cross products are the same, $\frac{4}{5}$ is correct. Checking the cross products does not guarantee that a fraction is reduced to lowest terms, but it does guarantee that the reducing was correct.

Why Reduce Fractions to Lowest Terms?

When we use fractions in everyday life, they are most often reduced to lowest terms. You rarely see a recipe calling for $\frac{6}{12}$ of a cup of sugar; it asks for the same amount in lowest terms — $\frac{1}{2}$ cup. This is because smaller numbers are easier to work with. Likewise, on tests like the GED Math Test, the answer choices you will see are reduced to lowest terms. Sometimes, after doing your computation in a problem, you will need to reduce your answer in order to find the correct answer choice. This is why reducing is an important skill to practice.

Writing Fractions: Word Problems

Remember that a fraction expresses a part of some whole. The *whole* will be represented in the *denominator*; the *part* of the whole is the *numerator*.

Sometimes the whole is not given in a problem. Before you can write the fraction correctly, you need to determine what the whole is.

EXAMPLE: Rose got two problems wrong and eight problems right on a quiz. What fraction of the problems did she get right?

STEP 1. The whole in this problem is the total number of problems on the quiz. You can figure out that the total is the number of problems Rose got wrong plus the number she got right.
$2 + 8 = 10$

STEP 2. Make a fraction with the total number of problems (10) as the denominator. Now put the part that she got right (8) in the numerator.

$$\frac{\text{problems right}}{\text{total problems}} \quad \frac{8}{10}$$

STEP 3. Reduce to lowest terms. Rose got $\frac{4}{5}$ of the problems right.
$\frac{8}{10} = \frac{4}{5}$

> **TIP:** Your first step in writing fractions should always be to determine what *the whole* is and make that number your denominator.

Fractions Exercise 8

Directions: Solve each problem. Be sure each answer is reduced to lowest terms.

1. There are 6 women and 9 men in Martin's math class.
  a. What fraction of the class is women?
  b. What fraction of the class is men?
  c. One night three students were absent. What fraction of the class was absent?

2. Last season the Consolidated Electric softball team played 24 games and won 18.
   a. What fraction of the games did they win?
   b. What fraction of the games did they lose?

3. Ellen bought a coat on sale for \$60. This was \$30 off the original price.
   a. What was the original price?
   b. What fraction of the original price did Ellen save?
   c. What fraction of the original price did Ellen pay?

4. The Jacksons are driving from Center City to Westerville. They stop to eat after 175 miles. They still have another 125 miles to go.
   a. What is the total distance from Center City to Westerville?
   b. What fraction of the drive had they completed when they stopped to eat?
   c. What fraction of the drive did they have left to complete after they ate?

5. Carmen takes home \$1200 a month. She spends \$300 a month for rent, \$450 a month for food, and \$150 a month on car payments.
   a. What fraction of her income does Carmen spend for rent?
   b. What fraction of her income does she spend for food?
   c. What fraction of her income does she spend on car payments?
   d. Carmen saves \$120 every month. What fraction of her income does she save?

ANSWERS AND SOLUTIONS START ON PAGE 335.

Changing Improper Fractions to Whole or Mixed Numbers

The picture at the right shows a box with four parts shaded. We can say that $\frac{4}{4}$ of the box is shaded or that 1 whole box is shaded. The number $\frac{4}{4}$ is an improper fraction. 1 is a whole number.

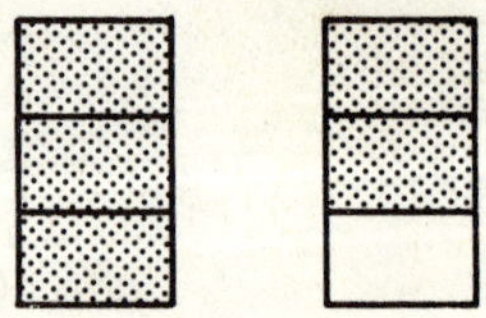

The picture at the right shows two boxes, each of which is divided into thirds. We can say that $\frac{5}{3}$ boxes are shaded or that $1\frac{2}{3}$ boxes are shaded.

To change an improper fraction to a mixed number or a whole number, divide. The fraction bar means to divide the numerator by the denominator.

EXAMPLE: Change $\frac{20}{6}$ to a mixed number.

(1)
```
   3
6)20
  18
   2
```

(2)
```
   3 2/6
6)20
  18
   2
```

(3) $3\frac{2}{6} = 3\frac{1}{3}$

STEP 1. Divide the denominator (6) into the numerator (20).
STEP 2. Write the remainder over the denominator.
STEP 3. Reduce $\frac{2}{6}$ to $\frac{1}{3}$.

Rules for Changing Improper Fractions to Whole or Mixed Numbers

1. Divide the denominator into the numerator.
2. Write the remainder as a fraction. Put the remainder over the original denominator.
3. Reduce the remaining fraction, if necessary.

Fractions Exercise 9

Directions: For problems 1-3, write an improper fraction and then a mixed number that expresses the parts of the figures that are shaded. Reduce each remaining fraction to lowest terms.

1. 

3. 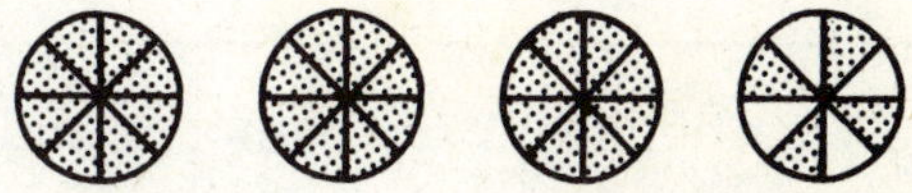

2. 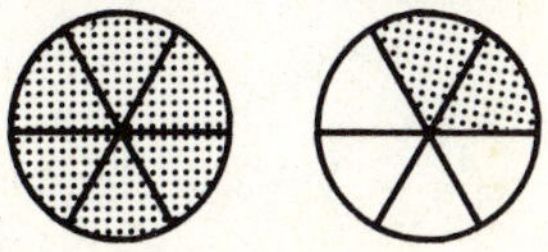

4. 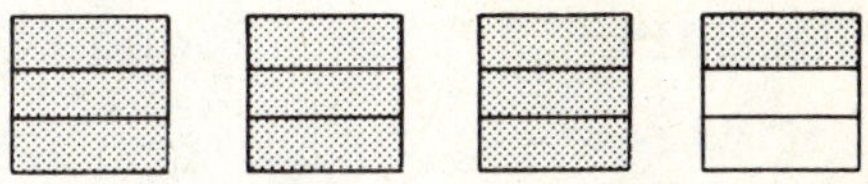

Directions: Change each improper fraction below to a whole or mixed number. Reduce any remaining fractions.

5. $\frac{15}{9} =$ $\qquad \frac{36}{8} =$ $\qquad \frac{13}{5} =$ $\qquad \frac{31}{6} =$

6. $\frac{6}{2} =$ $\qquad \frac{28}{16} =$ $\qquad \frac{15}{15} =$ $\qquad \frac{24}{18} =$

7. $\frac{7}{4} =$ $\qquad \frac{21}{3} =$ $\qquad \frac{50}{12} =$ $\qquad \frac{21}{6} =$

ANSWERS AND SOLUTIONS START ON PAGE 335.

Adding Like Fractions and Mixed Numbers

When you add decimals, you must add tenths to tenths, hundredths to hundredths, and so on. Fractions are similar. You must add like fractions. ***Like fractions*** are fractions with the same denominators. $\frac{1}{5}$ and $\frac{2}{5}$ are like fractions. $\frac{2}{3}$ and $\frac{3}{4}$ are unlike fractions. When you add like fractions, the answer has the denominator of the fractions you added.

EXAMPLE 1: Add $\frac{1}{5} + \frac{2}{5}$.

SOLUTION. Since the denominators are the same, add the numerators: $1 + 2 = 3$. Then put this total over the denominator. $\frac{3}{5}$ is reduced to lowest terms.

$\frac{1}{5} + \frac{2}{5} = \frac{3}{5}$

EXAMPLE 2: Find the sum of $\frac{7}{8}$ and $\frac{3}{8}$.

STEP 1. Since the denominators are the same, add the numerators and put the total over the denominator.

$\frac{7}{8} + \frac{3}{8} = \frac{10}{8}$

STEP 2. Change the improper fraction to a mixed number and reduce.

$\frac{10}{8} = 1\frac{2}{8} = 1\frac{1}{4}$

Rules for Adding Like Fractions

1. Add the numerators of each fraction and put the total over the denominator used in both fractions.
2. If the answer is an improper fraction, change it to a mixed number. Reduce if you can.

When adding mixed numbers, add the whole numbers and fractions separately, then combine them at the end.

EXAMPLE 3: Add $7\frac{5}{12}$ and $8\frac{11}{12}$.

(1)
$$\begin{array}{r} 7\frac{5}{12} \\ +\ 8\frac{11}{12} \\ \hline 15\frac{16}{12} \end{array}$$

(2) $\frac{16}{12} = 1\frac{4}{12}$

(3) $15 + 1\frac{4}{12} = 16\frac{4}{12} = 16\frac{1}{3}$

STEP 1. Since the denominators are the same, add the numerators and put the total over the denominator. Then add the whole numbers.

STEP 2. Change the improper fraction to a mixed number.

STEP 3. Add the whole number part of the answer to the mixed number and reduce.

TIP: Fraction addition problems can be written horizontally (with the numbers to the side of each other) or vertically (with one number under the other). In mixed number problems it is easier to solve the problems vertically. This makes it easy to separate the whole number parts of the problem from the fraction parts.

Fractions Exercise 10

Directions: Solve each problem.

1. $\frac{2}{5} + \frac{3}{5} =$ $\qquad$ $\frac{2}{9} + \frac{4}{9} =$ $\qquad$ $\frac{8}{15} + \frac{11}{15} =$

2. $\frac{5}{8} + \frac{5}{8} =$ $\qquad$ $9\frac{3}{10} + 4\frac{1}{10} =$ $\qquad$ $10\frac{2}{3} + 2\frac{1}{3} =$

3. $6\frac{5}{9} + 7\frac{8}{9}$ $\qquad$ $1\frac{7}{8} + 20\frac{3}{8}$ $\qquad$ $\frac{1}{16} + \frac{5}{16} + \frac{7}{16}$

4. $\frac{3}{4} + \frac{1}{4} + \frac{3}{4}$ $\qquad$ $2\frac{7}{12} + 4\frac{11}{12} + 9\frac{5}{12}$ $\qquad$ $5\frac{4}{9} + 4\frac{5}{9} + 3\frac{8}{9}$

ANSWERS AND SOLUTIONS START ON PAGE 335.

Subtracting Like Fractions

When you subtract decimals, you take tenths from tenths, hundredths from hundredths, and so on. The fractions you subtract must also have the same denominators. The difference between $\frac{4}{5}$ and $\frac{3}{5}$ is $\frac{1}{5}$.

EXAMPLE: Subtract $3\frac{9}{16}$ from $10\frac{15}{16}$.

$$\begin{array}{r} 10\frac{15}{16} \\ -\ 3\frac{9}{16} \\ \hline 7\frac{6}{16} = 7\frac{3}{8} \end{array}$$

STEP 1. Subtract the numerators and put the difference over the denominator.

STEP 2. Subtract the whole numbers and reduce.

Rules for Subtracting Like Fractions and Mixed Numbers, Without Borrowing

1. Subtract the numerators and put the difference over the denominator.
2. Subtract the whole numbers and reduce the fraction if you can.

Fractions Exercise 11

Directions: Solve each problem. Be sure every answer is reduced to lowest terms.

1. $\frac{5}{8} - \frac{1}{8} =$ $\qquad$ $8\frac{5}{6} - 2\frac{1}{6} =$ $\qquad$ $4\frac{7}{8} - 2\frac{3}{8} =$

2. $10\frac{4}{5} - 8\frac{1}{5} =$ $\qquad$ $\frac{19}{20} - \frac{13}{20} =$ $\qquad$ $6\frac{13}{15} - 2\frac{4}{15} =$

3. $\frac{25}{36} - \frac{17}{36} =$ $\qquad$ $8\frac{5}{6} - 1\frac{1}{6} =$ $\qquad$ $4\frac{17}{18} - 4\frac{5}{18} =$

ANSWERS AND SOLUTIONS START ON PAGE 336.

Borrowing with Like Denominators

Sometimes when you subtract fractions, you will need to borrow from a whole number to get an answer. For example, how do you subtract $8\frac{1}{4}$ from 10?

$$\begin{array}{r} 10 \\ -\ 8\frac{1}{4} \\ \hline \end{array}$$

As you know from your work with subtracting fractions, you need to subtract fractions before you subtract whole numbers. Since there is no fraction to subtract $\frac{1}{4}$ from, you must borrow 1 from the 10. Now you will be subtracting $\frac{1}{4}$ from 1.

In order to make this subtraction easier, write 1 as $\frac{4}{4}$. As you recall, any number over itself is equal to 1. We could rename this 1 as $\frac{8}{8}$, $\frac{2}{2}$, or $\frac{1000}{1000}$. However, we chose to write it as $\frac{4}{4}$ because we need to subtract $\frac{1}{4}$ from it, and you must always subtract *like fractions*. Now the problem looks like this:

$$\begin{array}{r} 9\frac{4}{4} \\ -8\frac{1}{4} \\ \hline \end{array}$$

Subtract as you would with any mixed numbers. Now look at the example below:

EXAMPLE 1: Take $4\frac{5}{6}$ from 8.

(1) $\begin{array}{rcl} 8 & = & 7\frac{6}{6} \\ -4\frac{5}{6} & = & 4\frac{5}{6} \\ \hline \end{array}$ **(2)** $\begin{array}{r} 7\frac{6}{6} \\ -4\frac{5}{6} \\ \hline 3\frac{1}{6} \end{array}$

STEP 1. Borrow 1 from 8 and rewrite the 1 as $\frac{6}{6}$ since 6 is the denominator of the bottom fraction.

STEP 2. Subtract the numerators and put the difference over the denominator. Subtract the whole numbers. $\mathbf{3\frac{1}{6}}$ is reduced to lowest terms.

Rules for Subtracting Mixed Numbers with Like Denominators When Borrowing Is Necessary

1. Borrow 1 from the top whole number. Rewrite the 1 with the same denominator as the fraction on the bottom. Add this rewritten 1 to the top fraction if there is one.
2. Subtract the numerators and put the difference over the denominator. Subtract the whole numbers and reduce the fraction.

Study the next example carefully. The top number has a fraction, but this fraction is not large enough to subtract from. Notice how we add this top fraction into the problem.

EXAMPLE 2: A plumber cut a piece of pipe $6\frac{7}{8}$ inches long from a piece that was $12\frac{5}{8}$ long. How long was the remaining piece?

(1) $\begin{array}{r} 12\frac{5}{8} = 11\frac{8}{8} + \frac{5}{8} = \\ -\ 6\frac{7}{8} = \\ \hline \end{array} \quad \begin{array}{r} 11\frac{13}{8} \\ 6\frac{7}{8} \\ \hline \end{array}$ **(2)** $\begin{array}{r} 11\frac{13}{8} \\ -\ 6\frac{7}{8} \\ \hline 5\frac{6}{8} = 5\frac{3}{4} \end{array}$

STEP 1. To find the length of the remaining piece, subtract. Notice that $\frac{5}{8}$ is too small to subtract $\frac{7}{8}$ from. Borrow 1 from 12 and rewrite the 12 as $11\frac{8}{8}$. Then add $11\frac{8}{8}$ to the $\frac{5}{8}$ we already had on top.

STEP 2. Subtract the fractions and the whole numbers and reduce. The remaining piece of pipe is $\mathbf{5\frac{3}{4}}$ inches long.

Fractions Exercise 12

Directions: Solve each problem.

1. $9 - 1\frac{2}{3} =$ $\qquad 11 - 3\frac{5}{8} =$ $\qquad 8 - 5\frac{7}{12} =$

2. $7 - 6\frac{1}{2} =$ $\qquad 9 - 1\frac{1}{4} =$ $\qquad 12 - 2\frac{5}{16} =$

3. $5\frac{1}{8} - 2\frac{5}{8} =$ $\qquad 8\frac{2}{5} - 1\frac{4}{5} =$ $\qquad 3\frac{1}{4} - 2\frac{3}{4} =$

4. $\begin{array}{r} 24\frac{11}{16} \\ -16\frac{15}{16} \\ \hline \end{array}$ $\qquad \begin{array}{r} 9\frac{1}{3} \\ -7\frac{2}{3} \\ \hline \end{array}$ $\qquad \begin{array}{r} 12\frac{3}{10} \\ -\ 4\frac{9}{10} \\ \hline \end{array}$

5. $\begin{array}{r} 11\frac{7}{12} \\ -\ 8\frac{11}{12} \\ \hline \end{array}$ $\qquad \begin{array}{r} 15\frac{7}{24} \\ -13\frac{19}{24} \\ \hline \end{array}$ $\qquad \begin{array}{r} 20\frac{5}{36} \\ -\ 7\frac{25}{36} \\ \hline \end{array}$

ANSWERS AND SOLUTIONS START ON PAGE 336.

Unlike Fractions

Finding Common Denominators

So far in this book you have added and subtracted like fractions — fractions that have the same denominator. Often the fractions in a problem will not have the same denominators. Suppose you want to add $\frac{1}{2}$ and $\frac{2}{3}$. The drawing below illustrates this problem.

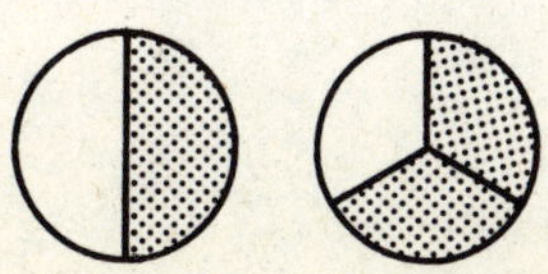

$\frac{1}{2} \quad + \quad \frac{2}{3} \quad = \quad ?$

The pieces are not the same size, and you cannot add them as they are written.

To add or subtract unlike fractions, you must first find a **common denominator**. In other words, you must make all fractions *like fractions* before you can add or subtract them. You must give all fractions the same, or common, denominator.

A ***common denominator*** is a number each denominator in a problem will divide into evenly. The lowest number that each denominator will divide into evenly is called the ***lowest common denominator***.

For example, for the fractions $\frac{1}{2}$ and $\frac{2}{3}$, the lowest common denominator is 6. 6 is the lowest number both 2 and 3 will divide into evenly.

For the fractions $\frac{5}{6}$ and $\frac{3}{4}$, the lowest common denominator is 12. 12 is the lowest number both 6 and 4 will divide into evenly.

Tips for Finding a Lowest Common Denominator

1. First test the largest denominator as the common denominator.
2. If that doesn't work, go through the multiplication table of the largest denominator.

EXAMPLE 1: Find the lowest common denominator for $\frac{3}{4}$ and $\frac{5}{8}$.

SOLUTION. Test the larger denominator, 8. 4 divides evenly into 8. **8** is the lowest common denominator.

EXAMPLE 2: Find the lowest common denominator for $\frac{7}{9}$, $\frac{1}{12}$, and $\frac{5}{6}$.

SOLUTION. Test the largest denominator, 12. 6 divides evenly into 12, but 9 does not. Go through the multiplication table of the largest denominator, 12, until you find the number that both 9 and 6 will divide into evenly.

$1 \times 12 = 12$, not evenly divisible by 9.
$2 \times 12 = 24$, not evenly divisible by 9.
$3 \times 12 = 36$, evenly divisible by both 9 and 6.

36 is the lowest common denominator for the three fractions because 6, 9, and 12 will all divide evenly into 36.

Fractions Exercise 13

Directions: Find the lowest common denominator for each set of fractions. Do not add or subtract the fractions.

1.	$\frac{2}{5}$ and $\frac{7}{10}$	$\frac{5}{6}$ and $\frac{3}{4}$	$\frac{3}{4}$ and $\frac{5}{9}$
2.	$\frac{1}{2}$ and $\frac{1}{3}$	$\frac{1}{2}$ and $\frac{5}{9}$	$\frac{5}{8}$ and $\frac{3}{20}$
3.	$\frac{2}{9}$ and $\frac{5}{6}$	$\frac{7}{12}$ and $\frac{5}{8}$	$\frac{1}{4}$ and $\frac{7}{10}$
4.	$\frac{1}{2}$, $\frac{3}{5}$, and $\frac{3}{4}$	$\frac{5}{6}$, $\frac{3}{4}$, and $\frac{2}{3}$	$\frac{3}{8}$, $\frac{5}{6}$, and $\frac{7}{12}$
5.	$\frac{4}{9}$, $\frac{5}{6}$, and $\frac{1}{2}$	$\frac{7}{8}$, $\frac{11}{20}$, and $\frac{1}{2}$	$\frac{8}{9}$, $\frac{7}{12}$, and $\frac{2}{3}$

ANSWERS START ON PAGE 336.

Raising Fractions to Higher Terms

After you have found a common denominator for the fractions in an addition or subtraction problem, you must change each fraction in the problem to a new fraction with the common denominator. For $\frac{1}{2}$ and $\frac{2}{3}$ the lowest common denominator is 6. The diagram below shows each fraction changed to sixths.

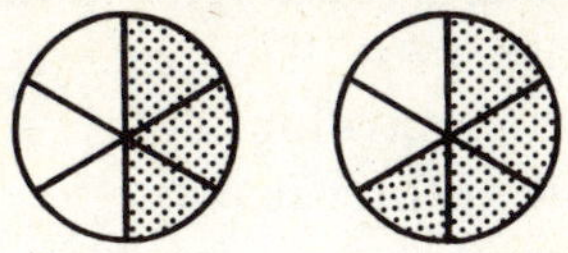

$\frac{1}{2} = \frac{3}{6}$ $\frac{2}{3} = \frac{4}{6}$

The process of changing $\frac{1}{2}$ to $\frac{3}{6}$ and $\frac{2}{3}$ to $\frac{4}{6}$ is called ***raising a fraction to higher terms***. It is the opposite of reducing a fraction. When you reduce a fraction, you *divide* both the numerator and the denominator by the same number. When you raise a fraction to higher terms, you *multiply* both the numerator and the denominator by the same number. You do this so that the fractions are easy to add and subtract.

To raise $\frac{1}{2}$ to sixths, ask yourself how many times the old denominator goes into the new one. 2 goes into 6 three times. Multiply both the numerator and the denominator by 3.

$$\frac{1 \times 3}{2 \times 3} = \frac{3}{6}$$

To raise $\frac{2}{3}$ to sixths, ask yourself how many times the old denominator goes into the new one. 3 goes into 6 two times. Multiply both the numerator and the denominator by 2.

$$\frac{2 \times 2}{3 \times 2} = \frac{4}{6}$$

Rules for Raising a Fraction to Higher Terms

1. Divide the old denominator into the new one to find the multiplier.
2. Multiply the old numerator by the new multiplier from step 1.

EXAMPLE 1: Raise $\frac{2}{3}$ to 15ths.

(1) $\begin{array}{r} 5 \\ 3\overline{)15} \end{array}$ **(2)** $\frac{2 \times 5}{3 \times 5} = \frac{10}{15}$

STEP 1. Divide the old denominator, 3, into the new denominator, 15.
STEP 2. Multiply the old numerator by 5.

TIP: Raising fractions to higher terms is an important skill. You will use this skill not only in adding and subtracting fractions but also later on when you study ratio and proportion and percents. As with reducing fractions, the better you know your multiplication tables, the easier you will find raising fractions to higher terms.

There are two ways to check your work when you raise a fraction to higher terms. One way is to reduce the new fraction. For the example above, reduce $\frac{10}{15}$ by 5. The result is $\frac{2}{3}$, which was the original fraction.

The other way to check these problems is to find the cross products. This means first to multiply the numerator of the original fraction by the denominator of the new fraction. Then multiply the denominator of the original fraction by the numerator of the new fraction. The products should be the same.

EXAMPLE 2: Find the cross products of $\frac{2}{3} = \frac{10}{15}$.

STEP 1. Multiply the numerator of the original fraction, 2, by the denominator of the new fraction, 15.
$2 \times 15 = 30$

STEP 2. Multiply the denominator of the original fraction, 3, by the numerator of the new fraction, 10. The products are both **30**.
$3 \times 10 = 30$

Fractions Exercise 14

Directions: Raise each fraction to higher terms. Use the new denominator shown in each problem. Check by finding the cross products.

1. $\frac{3}{5} = \frac{?}{20}$ $\quad \frac{5}{12} = \frac{?}{24}$ $\quad \frac{7}{8} = \frac{?}{40}$ $\quad \frac{3}{4} = \frac{?}{16}$

2. $\frac{5}{11} = \frac{?}{44}$ $\quad \frac{13}{20} = \frac{?}{40}$ $\quad \frac{7}{10} = \frac{?}{60}$ $\quad \frac{13}{20} = \frac{?}{100}$

3. $\frac{3}{4} = \frac{?}{100}$ $\quad \frac{5}{9} = \frac{?}{36}$ $\quad \frac{1}{8} = \frac{?}{24}$ $\quad \frac{1}{3} = \frac{?}{36}$

ANSWERS AND SOLUTIONS START ON PAGE 336.

Adding And Subtracting Unlike Fractions and Mixed Numbers

The diagram below illustrates the problem $\frac{1}{2} + \frac{2}{3}$. We cannot add these fractions until they are both changed to sixths, the lowest common denominator.

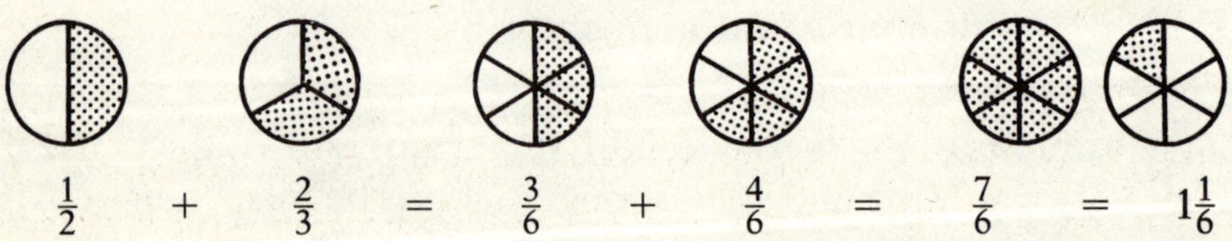

$$\frac{1}{2} + \frac{2}{3} = \frac{3}{6} + \frac{4}{6} = \frac{7}{6} = 1\frac{1}{6}$$

EXAMPLE 1: Add $5\frac{2}{3}$ and $3\frac{7}{9}$.

(1) $\begin{array}{r} 5\frac{2}{3} = 5\frac{6}{9} \\ +3\frac{7}{9} = 3\frac{7}{9} \end{array}$ (2) $\begin{array}{r} 5\frac{6}{9} \\ +3\frac{7}{9} \\ \hline 8\frac{13}{9} \end{array} = 8 + 1\frac{4}{9} = 9\frac{4}{9}$

STEP 1. Find the lowest common denominator for 3 and 9. Since 3 divides evenly into 9, 9 is the lowest common denominator. Raise $\frac{2}{3}$ to ninths.

STEP 2. Add the numerators and put the total over the common denominator. Then add the whole numbers. Simplify the answer. $\mathbf{9\frac{4}{9}}$ is already reduced to lowest terms.

Rules for Adding Unlike Fractions and Mixed Numbers

1. Find the lowest common denominator and raise each fraction to higher terms with the common denominator.
2. Add the numerators and put the total over the common denominator. Also add the whole numbers, if there are any.
3. Simplify the answer and reduce.

When you subtract unlike fractions, you must also find a common denominator.

EXAMPLE 2: Subtract $5\frac{3}{4}$ from $8\frac{1}{3}$.

(1) $\begin{array}{r} 8\frac{1}{3} = 8\frac{4}{12} \\ -5\frac{3}{4} = 5\frac{9}{12} \end{array}$ (2) $\begin{array}{r} 8\frac{4}{12} = 7\frac{4}{12} + \frac{12}{12} = 7\frac{16}{12} \\ -5\frac{9}{12} = \phantom{7\frac{4}{12} + \frac{12}{12} =} 5\frac{9}{12} \\ \hline 2\frac{7}{12} \end{array}$

STEP 1. Find the lowest common denominator, 12, and raise each fraction to twelfths.

STEP 2. Borrow 1 from 8 and rewrite the 1 as $\frac{12}{12}$. Then add $\frac{12}{12}$ to the $\frac{4}{12}$ you already had on top. Subtract the fractions and the whole numbers. $\mathbf{2\frac{7}{12}}$ is reduced to lowest terms.

Rules for Subtracting Unlike Fractions and Mixed Numbers

1. Find the lowest common denominator and raise each fraction to higher terms with the common denominator.
2. Borrow if necessary.
3. Subtract both the fractions and the whole numbers and reduce.

Fractions Exercise 15

Directions: Solve each problem. Watch out for subtraction problems where you have to borrow. Be sure each answer is reduced to lowest terms.

1. $\frac{5}{6} + \frac{7}{12} =$ $\quad \frac{2}{3} + \frac{3}{4} =$ $\quad \frac{2}{3} + \frac{7}{12} + \frac{5}{6} =$
2. $9\frac{2}{3} + 4\frac{1}{2} =$ $\quad 10\frac{1}{8} + 3\frac{1}{3} =$ $\quad 9\frac{1}{2} + 8\frac{3}{4} + 7\frac{3}{10} =$
3. $2\frac{5}{6} + 4\frac{1}{2} =$ $\quad \frac{1}{2} + \frac{1}{4} + \frac{3}{5} =$ $\quad 4\frac{1}{3} + 1\frac{5}{6} + 6\frac{5}{8} =$
4. $\frac{7}{10} - \frac{1}{2} =$ $\quad 2\frac{2}{3} - 1\frac{3}{4} =$ $\quad \frac{5}{8} - \frac{3}{5} =$
5. $8\frac{1}{5} - 3\frac{3}{4} =$ $\quad 11\frac{3}{5} - 9\frac{1}{2} =$ $\quad 9\frac{1}{2} - 5\frac{2}{3} =$
6. $16\frac{1}{4} - 7\frac{5}{8} =$ $\quad 7\frac{11}{12} - 3\frac{5}{8} =$ $\quad 8\frac{7}{20} - 2\frac{3}{4} =$

ANSWERS AND SOLUTIONS START ON PAGE 337.

Adding and Subtracting Fractions: Word Problems

The next problems will give you a chance to apply your adding and subtracting fractions skills to word problems. These are the kinds of problems you are likely to see on the GED. The situations that tell you to add or subtract are not hard to recognize, but read each problem carefully to be sure that you are answering exactly what is asked for. Remember to reduce every answer. An answer is not complete until it has been reduced to lowest terms.

Fractions Exercise 16

Directions: Solve each problem. Remember not only what you have learned about fractions but also what you know about problem solving.

1. From a piece of cable 5 yards long, Ray cut a piece measuring $3\frac{2}{3}$ yards. Find the length in yards of the remaining piece.
2. Mrs. Vega bought $2\frac{1}{2}$ pounds of chicken, $2\frac{5}{8}$ pounds of ground chuck, $1\frac{3}{4}$ pounds of sausage, 10 pounds of potatoes, and $3\frac{1}{4}$ pounds of oranges.
   - **a.** Find the weight in pounds of the meat (including chicken) products Mrs. Vega bought.
   - **b.** What was the combined weight of the potatoes and oranges?
   - **c.** Mrs. Vega refuses to carry more than 20 pounds of groceries. She asks the store personnel to deliver any order that is over 20 pounds. Will she have this order delivered?
3. Mr. and Mrs. Robinson are able to save $\frac{3}{16}$ of their income every month. What fraction of their income do they spend? (Hint: Think of their income as one whole.)

4. This problem refers to the following diagram:

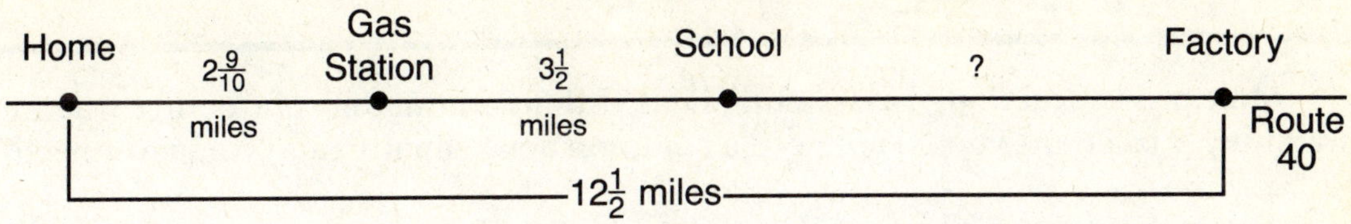

Mark's house, the gas station he usually goes to, his daughter's school, and the factory where he works are all on Route 40. One day on his way from home to work, Mark first stopped for gas. He then traveled the distance to his daughter's school. How far did he have to drive from the school to the factory?

5. At the beginning of the week, the cook at the Corner Kitchen had 100 pounds of flour and 45 pounds of sugar. Monday he used $16\frac{3}{4}$ pounds of flour. Wednesday he used $20\frac{1}{2}$ pounds, and Friday he used $24\frac{9}{16}$ pounds.
a. How many pounds of flour did the cook use on those three days?
b. How many pounds of flour were left when he finished on Friday?

6. By November, Paul's weight was down to $162\frac{1}{2}$ pounds. This was $43\frac{1}{2}$ pounds less than his weight had been in April. But, by the end of December of the same year, Paul gained 16 pounds over his November weight. What operations should you use to find how much Paul weighed in December?
(1) Add $162\frac{1}{2}$, $43\frac{1}{2}$, and 16.
(2) Add $43\frac{1}{2}$ and 16, then subtract the sum from $162\frac{1}{2}$.
(3) Subtract $43\frac{1}{2}$ from $162\frac{1}{2}$.
(4) Subtract 16 from $162\frac{1}{2}$.
(5) Add $162\frac{1}{2}$ and 16.

7. Following are the closing prices for a share of Central Utilities Stock:

Monday	Tuesday	Wednesday	Thursday	Friday
$19\frac{1}{8}$	$19\frac{5}{8}$	$20\frac{1}{8}$	$20\frac{5}{8}$	—

a. If the stock follows in the same pattern, what was the closing price on Friday?
b. If the stock has been following in the same pattern, what was the closing price the previous Friday? (Assume that the stock was not traded on Saturday or Sunday.)

8. Fred has the following lengths of $\frac{1}{2}$-inch copper pipe: $8\frac{5}{16}$ inches, $10\frac{1}{2}$ inches, $9\frac{3}{4}$ inches, and 26 inches.
a. If he welds together the three shortest lengths, what will be the length in inches of this new piece of pipe?
b. How many inches longer is the new welded piece than the longest piece Fred had?

ANSWERS AND SOLUTIONS START ON PAGE 338.

Multiplying Fractions

Multiplying fractions is less complicated than adding or subtracting fractions. To multiply fractions you do not need a common denominator, and you never need to borrow.

Think about this problem: A dozen eggs costs $.80. How much does a $\frac{1}{2}$ dozen eggs cost? To find the answer, you must multiply.

$$\frac{1}{2} \times \$.80 = \$.40.$$

> **TIP:** To find a fraction of some quantity, you must multiply.

EXAMPLE: Find the product of $\frac{3}{8}$ and $\frac{7}{10}$.

$$\frac{3}{8} \times \frac{7}{10} = \frac{21}{80}$$

STEP 1. Multiply the numerators together.
STEP 2. Multiply the denominators together.
STEP 3. Try to reduce. $\mathbf{\frac{21}{80}}$ is already reduced to lowest terms.

Rules for Multiplying Fractions by Fractions

1. Multiply the numerators together.
2. Multiply the denominators together.
3. Reduce the answer if possible.

Fractions Exercise 17

Directions: Solve each problem. Be sure each answer is reduced to lowest terms. Remember that a dot (·) indicates multiplication.

1. $\frac{3}{4} \times \frac{5}{7} =$	$\frac{2}{3} \times \frac{1}{2} =$	$\frac{1}{10} \cdot \frac{5}{8} =$	$\frac{7}{8} \times \frac{1}{5} =$
2. $\frac{1}{4} \times \frac{5}{16} =$	$\frac{3}{10} \cdot \frac{3}{5} =$	$\frac{2}{9} \times \frac{3}{5} =$	$\frac{3}{5} \times \frac{1}{6} =$

ANSWERS AND SOLUTIONS START ON PAGE 338.

Cancellation

There is a shortcut for multiplying fractions called cancellation. To cancel, find a number that divides evenly into the numerator of one fraction and the denominator of the other.

EXAMPLE 1: Find the product of $\frac{3}{4} \times \frac{8}{15}$.

STEP 1. Find a number that divides evenly into both the numerator of one fraction and the denominator of the other. Divide 3 and 15 by 3.

$$\frac{\overset{1}{\cancel{3}}}{4} \times \frac{8}{\underset{5}{\cancel{15}}}$$

STEP 2. Try the other numerator and denominator. Divide 4 and 8 by 4.

$$\frac{\overset{1}{\cancel{3}}}{\underset{1}{\cancel{4}}} \times \frac{\overset{2}{\cancel{8}}}{\underset{5}{\cancel{15}}}$$

STEP 3. Multiply the new numerators and the new denominators. Reduce if you can. $\frac{2}{5}$ is reduced to lowest terms.

$$\frac{\overset{1}{\cancel{3}}}{\underset{1}{\cancel{4}}} \times \frac{\overset{2}{\cancel{8}}}{\underset{5}{\cancel{15}}} = \frac{2}{5}$$

> **TIP:** Remember that cancellation is simply a shortcut. In some problems you will be able to cancel only once, and in other problems, not at all.

Cancellation makes it easy to multiply more than two fractions together. In these problems, you may be able to cancel several times. But remember to cancel only one numerator and one denominator at a time.

EXAMPLE 2: $\frac{5}{8} \times \frac{8}{9} \times \frac{7}{10} = ?$

STEP 1. Divide the 5 and the 10 by 5. Notice that you can "skip over" the middle fraction.

$$\frac{\overset{1}{\cancel{5}}}{8} \times \frac{8}{9} \times \frac{7}{\underset{2}{\cancel{10}}}$$

STEP 2. Divide 8 and 8 by 8.

STEP 3. Multiply numerators and denominators. Reduce if you can. $\frac{7}{18}$ is reduced to lowest terms.

$$\frac{\overset{1}{\cancel{5}}}{\underset{1}{\cancel{8}}} \times \frac{\overset{1}{\cancel{8}}}{9} \times \frac{7}{\underset{2}{\cancel{10}}} = \frac{7}{18}$$

Fractions Exercise 18

Directions: Solve each problem. Be sure each answer is reduced to lowest terms.

1. $\frac{3}{4} \times \frac{6}{7} =$ $\quad \frac{14}{15} \times \frac{3}{7} =$ $\quad \frac{4}{5} \cdot \frac{5}{6} =$ $\quad \frac{4}{9} \times \frac{3}{8} =$
2. $\frac{5}{8} \times \frac{2}{15} =$ $\quad \frac{9}{10} \times \frac{2}{3} =$ $\quad \frac{3}{20} \times \frac{1}{3} =$ $\quad \frac{4}{5} \times \frac{5}{24} =$
3. $\frac{2}{3} \cdot \frac{9}{20} =$ $\quad \frac{6}{7} \times \frac{7}{8} \times \frac{4}{5} =$ $\quad \frac{9}{10} \times \frac{1}{4} \times \frac{8}{9} =$ $\quad \frac{3}{4} \times \frac{2}{9} \times \frac{15}{16} =$

ANSWERS AND SOLUTIONS START ON PAGE 338.

Multiplying Fractions and Whole Numbers

To multiply a fraction by a whole number, rewrite the whole number as an improper fraction with a denominator of 1. The denominator 1 does not change the value of the whole number. Remember, any number divided by 1 equals itself. The 1 only pushes the whole number into the numerator position. For example, $\frac{24}{1}$ is equal to the whole number 24 because 24 divided by 1 equals 24.

EXAMPLE: There are 24 members in a club. At the last meeting $\frac{3}{4}$ of the members were present. How many people were at the meeting?

STEP 1. Rewrite the whole number 24 as an improper fraction with 1 as the denominator. $24 = \frac{24}{1}$.

$\frac{3}{4} \times \frac{24}{1}$

STEP 2. Divide the 4 and the 24 by 4.

$\frac{3}{\cancel{4}_1} \times \frac{\cancel{24}^6}{1}$

STEP 3. Multiply the new numerators and denominators.

$\frac{3}{\cancel{4}_1} \times \frac{\cancel{24}^6}{1} = \frac{18}{1}$

STEP 4. Change the improper fraction to a whole or mixed number. $\frac{18}{1} =$ **18.**

TIP: Always check your answer to make sure it makes sense. Here are a couple of clues to help you when multiplying fractions:

- When you multiply a whole number by a proper fraction, the result is smaller than the original number. For example, $12 \times \frac{1}{4} = 3$.
- When you multiply two proper fractions, the product is smaller than either fraction. For example, $\frac{1}{2} \times \frac{1}{2} = \frac{1}{4}$.

Fractions Exercise 19

Directions: Solve each problem. Be sure each answer is reduced to lowest terms.

1.	$\frac{1}{2} \times 16 =$	$10 \times \frac{2}{3} =$	$\frac{3}{8} \times 12 =$	$8 \cdot \frac{7}{10} =$
2.	$\frac{2}{5} \times 10 =$	$9 \times \frac{11}{20} =$	$\frac{5}{6} \times 9 =$	$15 \times \frac{7}{100} =$
3.	$18 \times \frac{2}{3} =$	$20 \cdot \frac{4}{5} =$	$\frac{4}{15} \times 36 =$	$24 \times \frac{3}{10} =$

ANSWERS AND SOLUTIONS START ON PAGE 339.

Changing Mixed Numbers to Improper Fractions

To multiply with mixed numbers, you must first change the mixed numbers into improper fractions. Look at the following example to see how this is done.

EXAMPLE: Change $4\frac{2}{3}$ to an improper fraction.

$4\frac{2}{3} = \frac{12}{3} + \frac{2}{3} = \frac{14}{3}$

STEP 1. Multiply the denominator (3) by the whole number (4). Put the product (12) over the denominator (3).

STEP 2. Add $\frac{12}{3}$ and $\frac{2}{3}$.

Rules for Changing a Mixed Number to an Improper Fraction

1. Change the whole number to an improper fraction with the same denominator as the fraction. To do this, multiply the denominator of the fraction and the whole number. Put the product over the denominator.
2. Add the two fractions.

Fractions Exercise 20

Directions: Change each mixed number to an improper fraction.

1. $2\frac{2}{3} =$	$1\frac{5}{8} =$	$8\frac{2}{5} =$	$3\frac{1}{4} =$	$3\frac{5}{6} =$
2. $5\frac{4}{7} =$	$3\frac{1}{2} =$	$7\frac{1}{3} =$	$6\frac{2}{9} =$	$10\frac{1}{3} =$
3. $12\frac{3}{5} =$	$9\frac{1}{4} =$	$13\frac{2}{3} =$	$15\frac{1}{3} =$	$4\frac{3}{8} =$

ANSWERS AND SOLUTIONS START ON PAGE 339.

Multiplying Mixed Numbers

Remember to change every mixed number to an improper fraction before you multiply. Look at how the following fraction and mixed number are multiplied:

EXAMPLE 1: $\frac{1}{3} \times 2\frac{1}{4} = ?$

STEP 1. Change $2\frac{1}{4}$ to an improper fraction.

$2\frac{1}{4} = \frac{9}{4}$

STEP 2. Cancel.

$\frac{1}{\cancel{3}_1} \times \frac{\cancel{9}^3}{4}$

STEP 3. Multiply the numerators and the denominators. $\frac{3}{4}$ is already reduced to lowest terms.

$\frac{1}{\cancel{3}_1} \times \frac{\cancel{9}^3}{4} = \frac{3}{4}$

Rules for Multiplying Fractions, Whole Numbers, and Mixed Numbers

1. Change every mixed number or whole number to an improper fraction.
2. Cancel.
3. Multiply the numerators together and multiply the denominators together.
4. If the answer is an improper fraction, change it to a mixed number and reduce if possible.

EXAMPLE 2: $3\frac{3}{10} \times 4\frac{1}{6} = ?$

STEP 1. Change the mixed numbers to improper fractions.

$\frac{33}{10} \times \frac{25}{6}$

STEP 2. Cancel. Then multiply the numerators and denominators.

$\frac{\overset{11}{\cancel{33}}}{\underset{2}{\cancel{10}}} \times \frac{\overset{5}{\cancel{25}}}{\underset{2}{\cancel{6}}} = \frac{55}{4}$

STEP 3. Change the improper fraction to a mixed number.

$\frac{55}{4} = 13\frac{3}{4}$

EXAMPLE 3: Find the product of $\frac{1}{3}$, 16, and $1\frac{1}{2}$.

STEP 1. Change the mixed and whole numbers to improper fractions.

$\frac{1}{3} \times \frac{16}{1} \times \frac{3}{2}$

STEP 2. Cancel. Then multiply the numerators and denominators.

$\frac{1}{\underset{1}{\cancel{3}}} \times \frac{\overset{8}{\cancel{16}}}{1} \times \frac{\overset{1}{\cancel{3}}}{\underset{1}{\cancel{2}}} = 8$

Estimating with Fractions

Estimating is a good way to check whether your answers to fraction multiplication problems are sensible. In Example 1 on page 99, you must find $\frac{1}{3}$ of $2\frac{1}{4}$. $2\frac{1}{4}$ is less than 3. $\frac{1}{3}$ of 3 is 1. Your answer should be less than 1. $\frac{3}{4}$ is a sensible answer.

In example 2 you are multiplying a number that is bigger than 3 by a number that is bigger than 4. Your answer should be bigger than 12. $13\frac{3}{4}$ is a sensible answer.

Fractions Exercise 21

Directions: Solve each problem. First try to estimate what your answer will be. Be sure each answer is reduced to lowest terms.

1. $1\frac{2}{3} \times 3\frac{1}{2} =$	$1\frac{1}{2} \times \frac{2}{3} =$	$6\frac{1}{2} \cdot 3\frac{1}{4} =$
2. $\frac{5}{8} \times 2\frac{2}{9} =$	$\frac{2}{3} \times 3\frac{3}{4} =$	$2\frac{5}{8} \times 1\frac{1}{3} =$
3. $1\frac{1}{2} \times \frac{4}{5} \times 2\frac{1}{4} =$	$\frac{5}{8} \cdot 4 \cdot 1\frac{2}{5} =$	$2\frac{1}{2} \times 1\frac{3}{5} \times 6 =$

4. $1\frac{7}{20} \cdot 5\frac{1}{3} \cdot 1\frac{1}{4} =$ $\quad$ $1\frac{7}{8} \times \frac{3}{4} \times 2\frac{2}{9} =$ $\quad$ $\frac{3}{8} \times 2\frac{5}{6} \times 1\frac{1}{3} =$

ANSWERS AND SOLUTIONS START ON PAGE 339.

Dividing Fractions

Suppose you want to divide five pies into quarters. Mathematically this problem is $5 \div \frac{1}{4}$. If you draw a picture, you can see what happens.

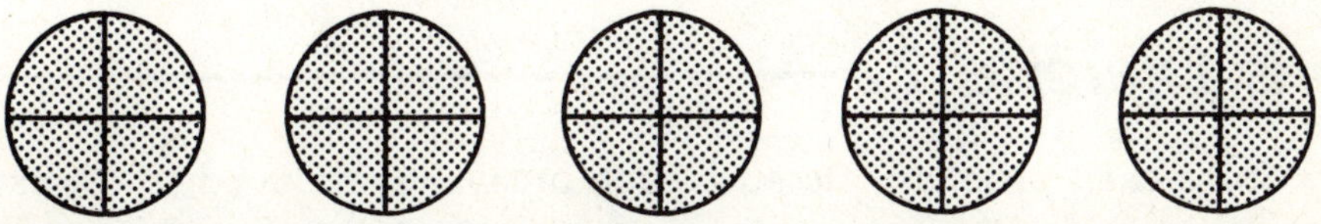

When you divide five pies into quarters, you get 20 pieces. Notice that the answer to $5 \div \frac{1}{4}$ gives the same answer as 5×4.

> **TIP:** It is very important to remember that in any fraction division problem the amount being divided up *must come first*.

Before you attempt to solve a fraction division problem, take a minute to make sure you have the problem set up properly. The first number (the number to the *left* of the ÷ sign) is the amount being divided up. The second number (the number to the *right* of the ÷ sign) is the divisor and tells what the first number is being divided *by*.

When dividing by a fraction, first invert (turn upside down) the fraction you are dividing by. The inverted number is called a ***reciprocal***. Then multiply.

EXAMPLE: $3\frac{1}{2} \div \frac{3}{8} = ?$

STEP 1. Change $3\frac{1}{2}$ to an improper fraction.

$\frac{7}{2} \div \frac{3}{8}$

STEP 2. Invert the divisor (the number you are dividing by) and change the ÷ sign to ×.

$\frac{7}{2} \times \frac{8}{3}$

STEP 3. Cancel and multiply the numerators and denominators.

$$\frac{7}{\cancel{2}_1} \times \frac{\cancel{8}^4}{3} = \frac{28}{3}$$

STEP 4. Change $\frac{28}{3}$ to a mixed number.

$\frac{28}{3} = 9\frac{1}{3}$

> **TIP:** When you divide by a proper fraction, the answer is always *larger* than the dividend.

Rules for Dividing by a Fraction

1. Change any mixed number or whole number into an improper fraction.
2. Invert the divisor (the number at the right) and change the ÷ sign to a × sign.
3. Cancel if possible.
4. Multiply the numerators and denominators.
5. Change an improper fraction answer to a mixed number and reduce.

Fractions Exercise 22

Directions: Solve each problem. Be sure each answer is reduced to lowest terms.

1. $\frac{1}{3} \div \frac{1}{6} =$	$5 \div \frac{5}{6} =$	$4\frac{1}{2} \div \frac{3}{4} =$	$2\frac{2}{3} \div \frac{2}{15} =$
2. $\frac{1}{3} \div \frac{2}{3} =$	$\frac{5}{7} \div \frac{5}{14} =$	$4 \div \frac{3}{8} =$	$\frac{5}{9} \div \frac{3}{4} =$
3. $5\frac{5}{6} \div \frac{7}{8} =$	$\frac{9}{10} \div \frac{3}{5} =$	$10 \div \frac{5}{6} =$	$3\frac{1}{3} \div \frac{1}{3} =$

ANSWERS AND SOLUTIONS START ON PAGE 339.

Dividing Fractions by Whole Numbers or Mixed Numbers

To divide a fraction by a mixed number or a whole number, first write all whole and mixed numbers as improper fractions. Then invert the divisor (the number following the ÷ sign) and multiply.

EXAMPLE 1: $4\frac{2}{3} \div 6 = ?$

STEP 1. Change $4\frac{2}{3}$ to an improper fraction and write 6 as an improper fraction with a denominator of 1.

$$\frac{14}{3} \div \frac{6}{1}$$

STEP 2. Invert the divisor and change the ÷ to ×. Cancel and multiply. $\frac{7}{9}$ is reduced to lowest terms.

$$\frac{\overset{7}{\cancel{14}}}{3} \times \frac{1}{\underset{3}{\cancel{6}}} = \mathbf{\frac{7}{9}}$$

EXAMPLE 2: How many bags each holding $1\frac{1}{2}$ pounds of nails can you fill with $4\frac{1}{2}$ pounds of nails?

STEP 1. You want to find how many times $1\frac{1}{2}$ will go into $4\frac{1}{2}$. In other words, divide $4\frac{1}{2}$ by $1\frac{1}{2}$. Change both mixed numbers to improper fractions. Remember that the amount being divided up comes first.

$$4\frac{1}{2} \div 1\frac{1}{2} = \frac{9}{2} \div \frac{3}{2}$$

STEP 2. Invert the divisor and change the ÷ to ×.

$$\frac{9}{2} \times \frac{2}{3}$$

STEP 3. Cancel, and multiply the numerators and denominators.

$$\frac{\overset{3}{\cancel{9}}}{\underset{1}{\cancel{2}}} \times \frac{\overset{1}{\cancel{2}}}{\underset{1}{\cancel{3}}} = \frac{3}{1}$$

STEP 4. Change the improper fraction to a whole number.

$\frac{3}{1} = \mathbf{3}$

Fractions Exercise 23

Directions: Solve each problem. Be sure each answer is reduced to lowest terms.

1. $10 \div 1\frac{1}{2} =$ $\quad 1\frac{1}{3} \div 3\frac{1}{5} =$ $\quad 6 \div 1\frac{1}{3} =$

2. $2\frac{1}{2} \div 3\frac{1}{4} =$ $\quad 21 \div 4\frac{1}{5} =$ $\quad 2\frac{2}{9} \div 2 =$

3. $\frac{9}{10} \div 3 =$ $\quad 1\frac{3}{4} \div 7 =$ $\quad 5\frac{5}{6} \div 7 =$

4. $\frac{3}{4} \div 3\frac{1}{5} =$ $\quad 3\frac{3}{4} \div 4 =$ $\quad 10 \div 1\frac{2}{3} =$

ANSWERS AND SOLUTIONS START ON PAGE 339.

Problem Solving for the GED

Multiplying and Dividing with Fractions

How many boards, each $\frac{1}{2}$ foot long, can be cut from a board that measures 15 feet?

(1) $7\frac{1}{2}$
(2) 15
(3) $15\frac{1}{2}$
(4) 30
(5) 45

With a problem like this, it is sometimes hard to know where to start. You may not know for sure whether to divide or multiply. Remember:

- A number *multiplied* by a proper fraction gets *smaller*.
  $12 \times \frac{1}{3} = 4$
- A number *divided* by a proper fraction gets *larger*.
  $12 \div \frac{1}{3} = 36$

This can be confusing because it is the *opposite* of what happens when multiplying and dividing by a whole number.

If a good way to solve the problem above does not immediately pop out at you, try making some basic decisions one at a time.

1. Should my answer be more or less than 15 (the number being divided up)?

   From your experiences, you can figure out that you can get *more* than 15 boards ($\frac{1}{2}$ foot long) out of a board that is 15 *whole* feet.

2. Should I multiply or divide?

   Step 1 above told you that you need an answer *larger* than 15. Remember from your work in this chapter that when you *divide* by a proper fraction (like $\frac{1}{2}$), you get an answer larger than the dividend (15). Therefore, in this problem, choose division as your operation.

3. If you plan to divide, which number should come first?

   As you learned in this chapter, setting up a fraction division problem is extremely important. The amount *being divided up* ALWAYS comes first. 15 boards are being divided up. Therefore, you would set this problem up this way: $15 \div \frac{1}{2} =$

4. Solve the problem.

   When you divide by a fraction, remember to invert the fraction and multiply: $15 \times \frac{2}{1} = 30$

5. Is your answer sensible?

   30 is larger than 15. It makes sense that you would have more than the original number of boards.

★ GED PRACTICE ★

Multiplying and Dividing with Fractions

Directions: This problem is followed by a series of questions that should help you set up the problem, just as you saw in the example above.

Marla has cut a 45-inch-wide piece of fabric into strips $\frac{3}{4}$ inch wide. How many strips does she have altogether?

(1) $11\frac{1}{4}$
(2) 15
(3) $33\frac{3}{4}$
(4) 60
(5) 75

Should answer be more or less than 45? ________________

Multiplication or division? ________________

If division, what number comes first? ________________

Answer: ________________

ANSWER AND SOLUTION START ON PAGE 340.

Multiplying and Dividing Fractions: Word Problems

The next problems give you a chance to apply your multiplying and dividing of fractions skills. These are the kinds of problems you are likely to see on the GED. It is often harder to recognize whether to multiply or divide than it is to recognize whether to add or subtract. Remember the five-step approach to solving word problems and use the method described above if it is helpful.

Fractions Exercise 24

Directions: Solve each problem. Be sure that every answer is reduced to lowest terms.

1. One cubic foot of water weighs $62\frac{1}{2}$ pounds. How many pounds do three cubic feet of water weigh?

   Should answer be more or less than $62\frac{1}{2}$ pounds?

   Multiplication or division?

   If division, what comes first?

   Answer:

2. Jeff paid $38 for $9\frac{1}{2}$ yards of lumber. Find the price of one yard.

   Should answer be more or less than $38?

   Multiplication or division?

   If division, what comes first?

   Answer:

3. How many jars each holding $1\frac{3}{4}$ pounds of tomatoes can Margaret fill with $17\frac{1}{2}$ pounds of tomatoes?

   Should answer be more or less than 17?

   Multiplication or division?

   If division, what comes first?

   Answer:

4. Don worked for $8\frac{1}{2}$ hours at $8.20 an hour. How much did he earn?

   Should answer be more or less than $8.20?

   Multiplication or division?

   If division, what comes first?

   Answer:

5. The MacDonalds want to split up 90 acres of their farm into $1\frac{1}{2}$-acre lots. How many lots can they get from the 90 acres?

   Should answer be more or less than 90?

   Multiplication or division?

   If division, what comes first?

   Answer:

6. In a recent election in Middletown only $\frac{2}{5}$ of the registered voters went to the polls. There are 45,000 registered voters in the town.

a. How many voters went to the polls?

Should answer be more or less than 45,000?

Multiplication or division?

If division, what comes first?

Answer:

b. How many registered voters did not go to the polls?

7. Last year Middletown received $\$1\frac{1}{2}$ million for community programs. The money was shared equally among six programs. How much did each program receive?

Should answer be more or less than $\$1\frac{1}{2}$ million?

Multiplication or division?

If division, what comes first?

Answer:

8. Nick cut a piece of metal tubing $6\frac{3}{4}$ inches long into 12 equal pieces. Assuming no waste, find the length in inches of each piece.

Should answer be more or less than $6\frac{3}{4}$?

Multiplication or division?

If division, what comes first?

Answer:

Directions: Problems 9-13 refer to the following situation. These problems include operations besides multiplication and division.

Hiro makes \$21,500 a year, and his wife Fumiko makes \$14,500. They spend $\frac{1}{5}$ of their combined income for rent, $\frac{2}{5}$ for food, and $\frac{1}{8}$ for car payments.

9. How much rent do they pay each month?

10. Rent, food, and car payments use up what fraction of Hiro and Fumiko's income?

11. How much do they spend in a year on car payments?

12. Assume that a month has 30 days. How much do Hiro and Fumiko spend each day for food?

13. Hiro and Fumiko save $\frac{1}{4}$ of Fumiko's salary. How much do they save in a year?

ANSWERS AND SOLUTIONS START ON PAGE 340.

Fractions and Decimals

We have mentioned decimals several times already in this chapter. Decimals and fractions are both ways of expressing parts of a whole. In this section you will learn how to change between decimals and fractions — a skill that will be very important on the GED Math Test.

Changing Decimals to Fractions

What is .321 expressed as a fraction?

To change a decimal to a fraction, first decide the number of places the decimal holds. For example, the place value of .321 is *thousandths*. Make this value the denominator of the fraction:

$$\frac{\quad}{1000}$$

The numerator of the fraction will be your original number without its decimal point:

$$\frac{321}{1000}$$

EXAMPLE: .25 is twenty-five hundredths. This can be written as a fraction: $\frac{25}{100} = \frac{1}{4}$

If you reduce this fraction as shown above, you get the fraction $\frac{1}{4}$.

Below are several examples of decimals changed to fractions.

Decimal		**Fraction**
.3	three tenths	$\frac{3}{10}$
.017	seventeen thousandths	$\frac{17}{1000}$
.35	thirty-five hundredths	$\frac{35}{100} = \frac{7}{20}$

Fractions Exercise 25

Directions: Change each decimal to a fraction and reduce to lowest terms.

1. .6 =	.5 =	.45 =	.80 =
2. .125 =	.065 =	.15 =	.96 =
3. .024 =	.0002 =	.010 =	.34 =

ANSWERS AND SOLUTIONS START ON PAGE 341.

Changing Fractions to Decimals

Remember that the fraction bar means to divide. $\frac{1}{4}$ means 1 divided by 4. To change a fraction to a decimal, divide the denominator into the numerator.

$$4\overline{)1} \qquad \begin{array}{r} .25 \\ 4\overline{)1.00} \end{array}$$

The numerator and the denominator of a fraction are both whole numbers. When you divide the denominator into the numerator, remember to put a decimal point to the right of the numerator and to bring the point up into the answer.

Study the next examples carefully. In each example there is at least one zero to the right of the decimal point in the dividend. Sometimes one zero is enough for the division to come out evenly. Other times you'll need more than one.

Fraction	**Decimal**
$\frac{3}{10}$	$\begin{array}{r} .3 \\ 10\overline{)3.0} \end{array}$
$\frac{9}{20}$	$\begin{array}{r} .45 \\ 20\overline{)9.00} \\ \underline{8\ 0} \\ 1\ 00 \\ \underline{1\ 00} \end{array}$
$\frac{3}{8}$	$\begin{array}{r} .375 \\ 8\overline{)3.000} \\ \underline{2\ 4} \\ 60 \\ \underline{56} \\ 40 \\ \underline{40} \end{array}$

Fractions Exercise 26

Directions: Change each of these fractions to decimals.

1. $\frac{3}{4} =$ $\qquad$ $\frac{1}{3} =$ $\qquad$ $\frac{7}{10} =$ $\qquad$ $\frac{5}{8} =$

2. $\frac{1}{2} =$ $\qquad$ $\frac{1}{20} =$ $\qquad$ $\frac{5}{6} =$ $\qquad$ $\frac{4}{9} =$

3. $\frac{1}{12} =$ $\qquad$ $\frac{2}{3} =$ $\qquad$ $\frac{3}{25} =$ $\qquad$ $\frac{1}{6} =$

ANSWERS AND SOLUTIONS START ON PAGE 341.

Common Fractions and Decimals

On the next page is a chart that you should fill in. Check your answers and memorize the chart. These are frequently used fractions and decimals. It will be convenient if you can change from one form to the other without having to make a calculation. For example, in some problems you may find it easier to use $\frac{3}{4}$. In other problems .75 may be easier.

Fractions Exercise 27

Directions: Fill in the blanks on this chart. Notice that the first line is complete.

Fraction	Decimal
$\frac{1}{4}$	.25
$\frac{1}{2}$	
$\frac{1}{8}$	
	.75
	.2
	.6
	.8
	.1
	.9
$\frac{1}{3}$	
$\frac{2}{3}$	

ANSWERS START ON PAGE 341.

Problem Solving for the GED

When Both Decimals and Fractions Appear in Word Problems

If $4\frac{1}{2}$ pounds of grapefruit cost $3.69, how much does grapefruit cost per pound?

(1) $.78
(2) $.82
(3) $.92
(4) $8.30
(5) $16.61

In problems where both a fraction and a decimal appear, you need to change one so that both are in the same form. In other words, change the decimal to a fraction or the fraction to a decimal. Then perform whatever math operation is needed. Here are some hints to help you decide how to go about this.

- If you find fractions *a lot* easier than decimals, consider putting the decimal in fraction form. Likewise, if you are more comfortable with decimals than fractions, try changing all fractions to decimals.

- Look at the answer choices given. Are they in fraction or decimal form? It is a good idea to convert the numbers in your problem to the same form the answer choices are in. This will save you from having to convert again to find the correct answer.

  In this problem, answers are in decimals. Change $4\frac{1}{2}$ to 4.5 and divide 4.5 into \$3.69 to find your answer.

  (2) \$.82

★ **GED PRACTICE** ★

Decimals and Fractions in Word Problems

Find the average weight in pounds of three crates that weigh $4\frac{1}{3}$ pounds, 9.5 pounds, and $5\frac{1}{6}$ pounds respectively.

(1) $6\frac{1}{3}$
(2) $8\frac{1}{3}$
(3) $9\frac{1}{2}$
(4) 18
(5) 19

ANSWER AND SOLUTION IS ON PAGE 341.

The next exercise gives you a chance to practice using both fractions and decimals in word problems.

Fractions Exercise 28

Directions: Solve each problem.

1. From a board $8\frac{1}{2}$ feet long, Al cut a piece that was 3.75 feet long. How long was the remaining piece?
a. Express the answer as a mixed decimal.
b. Express the answer as a mixed number.

2. Find the average weight, to the nearest tenth of a pound, of three packages that weigh $2\frac{1}{2}$ pounds, $5\frac{3}{4}$ pounds, and 6.3 pounds respectively.

3. Fran bought $\frac{3}{4}$ pound of cheese for \$2.50 a pound and $1\frac{1}{2}$ pounds of coffee for \$3.90 a pound. Which expression represents the amount she paid for these items?

(1) $(\frac{3}{4} + 1\frac{1}{2}) + (2.50 + 3.90)$
(2) $.75(2.50) + 1.5(3.90)$
(3) $.75(2.50) + 3.90$
(4) $2.50 + 1.5(3.90)$
(5) $\frac{3}{4}(3.90) + \frac{3}{2}(2.50)$

4. From a ten-pound bag of flour Janina first took $3\frac{1}{2}$ pounds and then another .75 pound. Which expression represents the amount of flour (in pounds) that was left in the bag?

(1) (.75 + 10) − 3.5
(2) 3.5 − 10 − .75
(3) 10 + 3.5 + .75
(4) 10 − 3.5 − .75
(5) 10 × 3.5 × .75

ANSWERS AND SOLUTIONS START ON PAGE 341.

Comparing Fractions

Which is larger, $\frac{3}{8}$ or $\frac{2}{5}$? If you know the decimal values of these fractions, you can compare them easily:

$$\frac{3}{8} = .375 \qquad \frac{2}{5} = .4 = .400$$

If you do not know the decimal values for the fractions, find a common denominator. Then raise each fraction to higher terms with the common denominator.

$$\frac{3}{8} = \frac{15}{40} \qquad \frac{2}{5} = \frac{16}{40}$$

$\frac{16}{40}$ is more than $\frac{15}{40}$. Therefore, $\frac{2}{5}$ is larger.

Fractions Exercise 29

Directions: For items 1-3 circle the larger fraction in each pair.

1. a. $\frac{1}{2}$ or $\frac{4}{9}$ **b.** $\frac{2}{3}$ or $\frac{3}{5}$ **c.** $\frac{3}{4}$ or $\frac{5}{6}$

2. a. $\frac{13}{20}$ or $\frac{7}{10}$ **b.** $\frac{5}{8}$ or $\frac{3}{5}$ **c.** $\frac{1}{6}$ or $\frac{2}{9}$

3. a. $\frac{5}{6}$ or $\frac{7}{10}$ **b.** $\frac{5}{12}$ or $\frac{5}{9}$ **c.** $\frac{5}{8}$ or $\frac{11}{20}$

4. Arrange in order from smallest to largest: $\frac{5}{8}$, $\frac{5}{6}$, $\frac{7}{12}$.

5. Arrange in order from smallest to largest: $\frac{7}{10}$, $\frac{4}{5}$, $\frac{3}{4}$, $\frac{13}{20}$.

6. Kevin wants to pack four boxes in order from lightest to heaviest. Following are the weights of the boxes.

Box A — $\frac{9}{16}$ pound
Box B — 0.75 pound
Box C — $\frac{5}{8}$ pound
Box D — 0.5 pound

Which of the following represents the boxes in order from *lightest* to *heaviest*?

(1) C, B, A, D
(2) C, D, B, A
(3) D, C, A, B
(4) D, C, B, A
(5) D, A, C, B

7. To put together her videotape, Veronica first needs to put each piece of film in order from longest to shortest. The pieces are the following lengths:

A — $5\frac{1}{2}$ seconds

B — 5.30 seconds

C — 5.2 seconds

D — $5\frac{2}{3}$ seconds

E — $5\frac{7}{8}$ seconds

In order from *longest* to *shortest*, how should Veronica line up the pieces of film?

ANSWERS AND SOLUTIONS START ON PAGE 342.

Exercise 30: Review

Directions: For questions 1-16, simply solve each problem.

1. There are 20 students in Andrew's Spanish class. 11 of the students are men. Men make up what fraction of the class?

2. Reduce $\frac{48}{60}$ to lowest terms.

3. Change $\frac{52}{8}$ to a mixed number and reduce.

4. $7\frac{3}{8} + 3\frac{5}{8} + 4\frac{7}{8} =$

5. $12\frac{1}{5} - 8\frac{4}{5} =$

6. $5\frac{3}{4} + 6\frac{1}{2} + 8\frac{7}{10} =$

7. $8\frac{5}{12} - 3\frac{2}{3} =$

8. $\frac{3}{8} \times \frac{1}{6} \times \frac{4}{5} =$

9. $5 \times \frac{3}{10} =$

10. Change $7\frac{2}{3}$ to an improper fraction.

11. $3\frac{3}{4} \times 3\frac{1}{3} =$

12. $\frac{3}{4} \div \frac{1}{8} =$

13. $24\frac{1}{2} \div 3\frac{1}{2} =$

14. Change .65 to a fraction and reduce.

15. Change $\frac{6}{25}$ to a decimal.

16. Which fraction is smaller: $\frac{5}{9}$ or $\frac{2}{3}$?

★ GED PRACTICE ★

Directions: The next problems include whole numbers and decimals as well as fractions. Choose the correct answer to each of the following. Remember what you have learned about problem solving.

17. On Friday Paul drove 524.7 miles; on Saturday, 208.6 miles; and on Sunday, 380.9 miles. Find, to the nearest mile, the average distance Paul drove in a day.
(1) 557
(2) 525
(3) 381
(4) 371
(5) 279

18. One day Mark swept floors for $2\frac{3}{4}$ hours, earning $7.20 per hour. He also earned a flat fee of $35.00 for yardwork. What operations should you use to find how much more Mark earned doing yardwork than he earned sweeping floors?
(1) Add $7.20 and $35.00, then divide by $2\frac{3}{4}$.
(2) Add $7.20 and $35.00, then multiply by $2\frac{3}{4}$.
(3) Multiply $2\frac{3}{4}$ by $7.20, then subtract the product from $35.00.
(4) Divide $7.20 by $2\frac{3}{4}$, then subtract the quotient from $35.00.
(5) Subtract $7.20 from $35.00.

Directions: Questions 19-21 refer to the following situation.

Alfredo's employer withholds $\frac{1}{5}$ of his gross wages for federal tax, $\frac{1}{8}$ for state tax, and $\frac{1}{10}$ for social security. Alfredo's weekly gross pay is $360.

19. What fraction of his gross wages does Alfredo take home?
(1) $\frac{23}{40}$
(2) $\frac{17}{40}$
(3) $\frac{9}{10}$
(4) $\frac{7}{8}$
(5) $\frac{4}{5}$

20. How much of Alfredo's salary is withheld each week for federal tax?
(1) $23
(2) $36
(3) $40
(4) $50
(5) $72

21. Alfredo spends $\frac{1}{3}$ of his take-home pay for food. How much does he spend on food each week?
(1) $27
(2) $69
(3) $80
(4) $90
(5) $120

22. If $4\frac{1}{2}$ yards of material cost $41.40, what is the cost of one yard of the material?
(1) $2.30
(2) $4.60
(3) $8.60
(4) $9.20
(5) $10.20

23. A stock that sold for $28\frac{3}{8}$ on Monday had dropped $3\frac{1}{4}$ by Wednesday. By Friday afternoon, the stock had dropped a total of $6\frac{1}{2}$ points. Find the price of the stock on Friday afternoon.
(1) $21\frac{7}{8}$
(2) $22\frac{1}{8}$
(3) $22\frac{7}{8}$
(4) $20\frac{1}{8}$
(5) $18\frac{5}{8}$

24. Assuming no waste, how many strips of metal each .45 inch wide can be cut from a sheet that is 90 inches wide?
(1) 2000
(2) 200
(3) 20
(4) 5
(5) 2

25. Roger bought 3 quarts of oil for $12.50 each and got a $5 discount for paying cash. Which expression represents the amount he paid for the oil?
(1) $(3 \times 12.50) - 5$
(2) $5 - (3 \times 12.50)$
(3) $5 - 3 - 12.50$
(4) $12.50 - (3 \times 5)$
(5) $(3 + 12.50) - 5$

ANSWERS AND SOLUTIONS START ON PAGE 342.

Fractions Review Evaluation

Problem	Section	Starting Page
1	Writing Fractions	76
2	Reducing Fractions	80
3	Changing Improper Fractions to Whole or Mixed Numbers	84
4	Adding Like Fractions and Mixed Numbers	85
5	Subtracting Like Fractions	87
6, 7	Adding and Subtracting Unlike Fractions and Mixed Numbers	92
8	Multiplying Fractions	96
9	Multiplying Fractions and Whole Numbers	98
10	Changing Mixed Numbers to Improper Fractions	98
11	Multiplying Mixed Numbers	99
12	Dividing Fractions	101
13	Dividing Fractions by Whole Numbers or Mixed Numbers	102
14, 15	Fractions and Decimals	107
16	Comparing Fractions	111
17	Mean and Median	34*
18	Fractions and Decimals (*Choosing the Operation*)	107/*24**
19	Writing Fractions: Word Problems	83
20, 21	Multiplying and Dividing Fractions (*Item Sets*)	104/*70**
22	Decimals and Fractions in Word Problems	109
23	Adding and Subtracting Fractions (*Finding Necessary Information*)	94/*22**
24	Dividing by Decimals	65*
25	Properties of Numbers (*Set-up Questions*)	38*/*42**

**These skills are covered in earlier chapters.*

Passing score: 21 right out of 26 problems.

Your score: ______ right out of 26 problems.

If you had less than a passing score, review the sections for those problems you missed. Then redo these questions before going on to the next section.

If you had a passing score, correct any problem you got wrong. Then go on to the next section.

6 PROBABILITY, RATIO, AND PROPORTION

Probability

Probability is the chance of something happening. Every day, weathermen tell us what the "chances of rain" are. Gamblers use probability when they try to figure out what their chances of winning are. Probability, believe it or not, is an application of your fractions skills. We define probability as the following fraction:

$$\text{probability of an event} = \frac{\text{number of favorable outcomes}}{\text{total number of possible outcomes}}$$

What is a "favorable outcome?" This depends on what you are trying to figure out. If you are playing a slot machine in Las Vegas and want to know your chances of winning, your numerator, or ***favorable outcome***, would be the number of wins. The denominator would be your total number of tries. If you want to know the probability of picking a yellow marble out of a group of marbles, the *favorable outcome* is the number of yellow marbles in the group. The denominator would be the total number of marbles in the group.

The next three examples refer to the picture on the right. Imagine that the arrow is free to spin. Each time it spins, it lands on one of the sections of the circle. Assume that it never lands on a line.

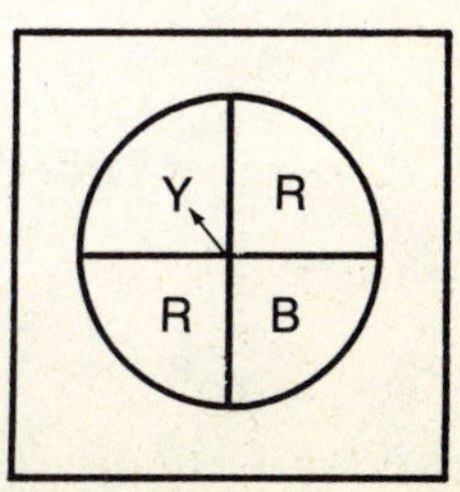

EXAMPLE 1: What is the probability that the arrow will land on yellow?

(1) $\frac{\text{favorable outcomes}}{\text{total outcomes}} \qquad \frac{\_}{4}$ **(2)** $\frac{1}{4}$

STEP 1. Count all the possible outcomes: red, yellow, red, blue. There are four possible outcomes. 4 is the denominator.

STEP 2. Find the number of favorable outcomes, in this case, the number of times yellow appears on the circle. Yellow appears only once. 1 is the numerator.

The probability that the arrow will land on yellow is $\mathbf{\frac{1}{4}}$.

EXAMPLE 2: What is the probability that the arrow will land on red?

(1) $\frac{\text{favorable outcomes}}{\text{total outcomes}} \qquad \frac{\_}{4}$ **(2)** $\frac{2}{4} = \frac{1}{2}$

STEP 1. There are four possible outcomes. 4 is the denominator.

STEP 2. Find the number of favorable outcomes, the number of times red appears in the circle. Red appears twice. 2 is the numerator. Reduce $\frac{2}{4}$ to $\frac{1}{2}$.

The probability that the arrow will land on red is $\mathbf{\frac{1}{2}}$.

EXAMPLE 3: What is the probability that the arrow will land on green?

(1) $\frac{\text{favorable outcomes}}{\text{total outcomes}} \qquad \frac{\_}{4}$ **(2)** $\frac{0}{4}$

STEP 1. There are four possible outcomes. 4 is the denominator.

STEP 2. Find the number of times green appears in the circle. Green never appears. 0 is the numerator.

The probability that the arrow will land on green is **0**.

TIP: Remember to reduce a probability to lowest terms.

In some probability problems, you may have to find the total number of possible outcomes in order to complete the problem.

EXAMPLE 4: A box contains 6 nickels, 4 dimes, and 2 quarters. Find the probability of picking a nickel from the box.

(1)

number of nickels	6
number of dimes	4
number of quarters	+2
total	12

(2) $\frac{\text{favorable outcomes}}{\text{total outcomes}} \qquad \frac{6}{12} = \frac{1}{2}$

STEP 1. Find the total number of outcomes, in this case, the total number of coins in the box. Add 6, 4, and 2. 12 is the denominator.

STEP 2. The number of favorable outcomes is the number of nickels in the box, 6. Reduce the fraction. The probability of picking a nickel is $\mathbf{\frac{1}{2}}$.

Probability Exercise 1

Directions: The following cards are lying face down on a table.

1. What is the probability of picking a jack?
2. Find the probability of picking a king.
3. What is the probability of picking a face card (king, queen, or jack)?

Directions: A die has 6 sides. Each side has a different number of dots, ranging from 1 to 6.

4. What is the probability that you would roll a 6?
5. Find the probability of rolling an odd number.
6. Ted's sock drawer contains 6 blue socks, 3 black socks, 1 white sock, and 2 plaid socks. If he chooses one at random in the dark, what is the probability that he will pick a black sock?

ANSWERS AND SOLUTIONS START ON PAGE 343.

Dependent Probability

Suppose the first coin you pick from the box in Example 4 is a dime. You take that dime and set it aside from the group of coins. Now you would like to find the probability that the *next* coin you pick will be a nickel. Since you have removed one coin from the group, the total number of outcomes (denominator) has changed.

number of nickels	6
number of dimes	3
number of quarters	+2
total	11

$$\frac{\text{favorable outcomes}}{\text{total outcomes}} \quad \frac{\quad}{11} \quad \frac{6}{11}$$

STEP 1. Find the total number of outcomes, this time, the total number of coins left in the box. Now there are only 3 dimes. Add 6, 3, and 2. 11 is the denominator.

STEP 2. The number of favorable outcomes is the number of nickels left in the box, 6.

The probability that the next coin will be a nickel is $\frac{6}{11}$.

Probability Exercise 2

Directions: Solve each problem.

1. A deck of cards has 52 cards. There are four aces in a deck. What is the probability of picking an ace from a deck?

2. There are 20 marbles in a bag. 16 of them are black. What is the probability of picking a black marble from the bag?

3. Bob received a shipment of shirts to sell in his store. He ordered 12 blue shirts and 6 white shirts. What is the probability that the first shirt he unwraps will be blue?

4. Mr. Robinson bought two raffle tickets. His wife bought three tickets, and his daughter bought one. Altogether, 1000 tickets were sold. What was the probability that Mr. Robinson won?

5. What is the probability that someone in the Robinson family in problem 4 won?

6. Out of every 200 tires that a factory produces, 15 are defective. What is the probability that a tire produced at the factory will be defective?

Directions: Questions 7-10 refer to the picture at the right.

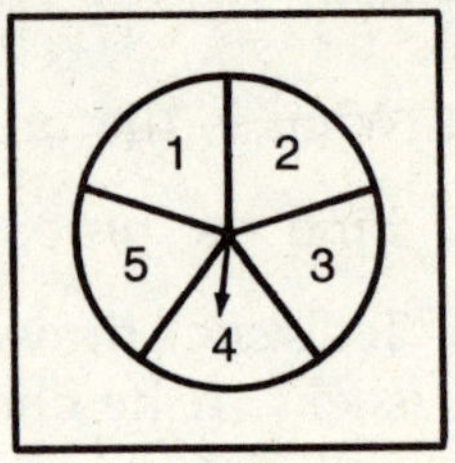

7. What is the probability that the arrow will land on 1?

8. What is the probability that the arrow will land on 2?

9. What is the probability that the arrow will land on 1 or 5?

10. What is the probability that the arrow will land on an odd number?

Directions: Questions 11-13 refer to the following situation.

Luba bought 1 can of white paint, 3 cans of gray paint, and 4 cans of green paint. When she got home, she realized that the cans had not been labeled.

11. What is the probability that the first can she opens will be white?

12. In fact, the first can she opened was gray. What is the probability that the second can will be white?

13. The second can turned out to be gray as well. What is the probability that the third can will be green?

Directions: Questions 14-16 refer to the following situation.

Sam has a bag that contains 12 red apples, 8 yellow apples, and 6 green apples.

14. What is the probability that the first apple he picks from the bag will be yellow?

15. In fact, the first apple Sam picked was green. What is the probability that the next apple he picks will be green?

16. The second apple turned out to be yellow. What is the probability that the third apple he picks will be red?

ANSWERS AND SOLUTIONS START ON PAGE 343.

Ratio and Proportion

Ratio

In mathematics we often compare numbers. For example, suppose there are 150 men working in a factory and 75 women. One way to compare these facts is to subtract. There are 75 more men working in the factory than there are women. $(150 - 75 = 75.)$

Another way to compare these numbers is to use a **ratio**. The ratio of men to women working in the factory is 150 to 75 or, in reduced form, 2 to 1.

There are three ways to write ratios: with the word *to*, with a colon (:), or as a fraction. Like fractions, ratios should always be reduced. Following are the three ways to write the ratio of the number of men to the number of women working in the factory. Notice that each way is reduced.

- 150 to 75 = 2 to 1
- $150 : 75 = 2 : 1$
- $\frac{150}{75} = \frac{2}{1}$

> **TIP:** When a ratio is written as a fraction, it keeps the denominator 1. There are always two numbers in a ratio.

The numbers in a ratio *must* be written in the order the problem asks for. Mixing up the order will result in an incorrect answer. Look at the following example, where we want to compare a length of wood to a length of rope.

To build a child's swing, Clay uses 5 feet of rope for every 3 feet of wood. What is the ratio of wood to rope?

The answer to this problem is $\frac{3}{5}$. Because the problem asked for wood first, the number 3 goes in the numerator. If you answered $\frac{5}{3}$, your answer would be wrong. $\frac{5}{3}$ is the ratio of *rope to wood*.

Now follow this example:

EXAMPLE: Evelyn earns $1200 a month. She pays $480 in rent. What is the ratio of her income to her rent?

$$\frac{\text{income}}{\text{rent}} \quad \frac{\$1200}{480} = \frac{5}{2} \text{ or } 5{:}2 \text{ or } 5 \text{ to } 2$$

SOLUTION. Make a ratio with the income on top (numerator) and the rent on the bottom (denominator). Then reduce.

> **TIP:** Remember: a ratio, like a fraction, should *always* be reduced to lowest terms. However, a ratio that is written as an improper fraction should *not* be changed to a mixed number.

Two-Step Ratio Problems

Ratio problems often involve more than one step. You may not be given both numbers to include in the ratio as you were in the example above. Instead, you may have to determine one of the numbers. Look at the example below, where you are comparing questions right to questions wrong.

EXAMPLE: On a test of 20 questions. Maceo got 2 questions wrong. What is the ratio of the number he got right to the number he got wrong?

(1)

total questions	20
number wrong	−2
number right	18

(2) $\frac{\text{right}}{\text{wrong}} \quad \frac{18}{2} = \frac{9}{1}$ or 9:1 or 9 to 1

STEP 1. One number you are asked for is the number of problems Maceo got right. You are not given this number, so you will have to figure it out. Subtract the number he got wrong from the total. This gives you the number of *correct* answers.

STEP 2. Make a ratio with the number of questions right on top and the number of questions wrong on the bottom. Then reduce.

Ratio and Proportion Exercise 1

Directions: Solve each problem.

1. Simplify the ratio 12:15.

2. In a GED class with 20 students, there are 12 women.
 a. What is the ratio of the number of women to the total number of students?
 b. What is the ratio of the number of men to the total number of students?
 c. What is the ratio of the number of men to the number of women?
 d. What is the ratio of the number of women to the number of men?

3. In a shop there are 105 union workers and 45 nonunion workers.
 a. What is the ratio of the number of union workers to the total number of workers?
 b. What is the ratio of the number of nonunion workers to the total number of workers?
 c. What is the ratio of the number of union workers to the number of nonunion workers?
 d. What is the ratio of the total number of workers to the number of union workers?

4. From a total yearly budget of $18,000,000, Central City spends $3,000,000 on education. What is the ratio of the amount spent on education to the amount not spent on education?

5. On a test with 50 questions, there were 15 fraction problems and 5 decimal problems. What is the ratio of the total number of fraction and decimal problems to the number of problems on the test?

6. In a day care center with 80 children, there are 48 girls.
   a. What is the ratio of the number of boys to the number of girls?
   b. What is the ratio of the number of girls to the total number of children?
   c. What is the ratio of the number of girls to the number of boys?
   d. What is the ratio of the number of boys to the total number of children?
7. One season a basketball team won 62 games and lost 20. What was the ratio of the number of games the team won to the total number of games they played?
8. Another team played the same number of games as the team in problem 7 and lost 41. What was the ratio of the number of games that team won to the number of games they played?

ANSWERS AND SOLUTIONS START ON PAGE 343.

Proportion

A ***proportion*** is made up of two equal ratios. For example, 2:4 = 1:2 (or $\frac{2}{4} = \frac{1}{2}$) is a proportion. Note that in the second form a proportion looks very much like the equivalent fractions you worked with in Chapter 5.

When we talked about equivalent fractions, you learned that the cross products of equal fractions are equal to each other. The ***cross product*** of equivalent fractions is the product of a number in one upper corner multiplied by the number in the opposite lower corner. In the example below, notice that both cross products are equal to 4.

$$\frac{2}{4} \times \frac{1}{2} \qquad \text{cross products: } 4 \times 1 = 4, \quad 2 \times 2 = 4$$

Each of the four numbers in a proportion is called an ***element*** or a ***term***. In proportion problems, one term is often missing. Your job is to find this missing term. Look at the example below. Notice that the missing term is shown by using the letter n.

EXAMPLE 1: Find the missing term in $\frac{n}{8} = \frac{9}{12}$.

(1) $8 \times 9 = 72$ **(2)** $72 \div 12 = 6$

STEP 1. Find the cross product of 8 and 9.
STEP 2. Divide the cross product, 72, by the other term in the proportion, 12. The value of the missing term is **6**.

Rules for Solving a Proportion

1. Find the cross product for the numbers that are given in the proportion.
2. Divide the cross product by the remaining number in the proportion.

EXAMPLE 2: Solve for c in $\frac{3}{7} = \frac{8}{c}$.

(1) $\begin{array}{r} 7 \\ \times 8 \\ \hline 56 \end{array}$ (2) $3\overline{)56}$ = $18\frac{2}{3}$

STEP 1. Find the cross product of 7 and 8.
STEP 2. Divide the cross product, 56, by the other number in the proportion, 3. The value of c is $\mathbf{18\frac{2}{3}}$.

EXAMPLE 3: Solve for y in the proportion $5{:}y = 2{:}8$.

(1) $\frac{5}{y} = \frac{2}{8}$ (2) $\begin{array}{r} 5 \\ \times 8 \\ \hline 40 \end{array}$ (3) $2\overline{)40}$ = 20

STEP 1. Rewrite the proportion in fractional form. Notice that the first term on each side of the equal sign becomes a numerator.
STEP 2. Find the cross product of 5 and 8.
STEP 3. Divide the cross product, 40, by the other number in the proportion, 2. The value of y is **20**.

Ratio and Proportion Exercise 2

Directions: Find the missing term in each of the following proportions.

1. $\frac{m}{6} = \frac{10}{15}$ $\frac{3}{a} = \frac{5}{6}$ $\frac{4}{9} = \frac{y}{3}$ $\frac{8}{7} = \frac{4}{x}$
2. $\frac{1}{3} = \frac{s}{5}$ $\frac{3}{6} = \frac{w}{5}$ $\frac{2}{11} = \frac{4}{p}$ $\frac{2}{8} = \frac{9}{x}$
3. $4{:}e = 6{:}8$ $3{:}7 = 4{:}y$ $15{:}40 = x{:}60$ $30{:}a = 12{:}16$

ANSWERS AND SOLUTIONS START ON PAGE 344.

Proportion Word Problems

You can use proportions to solve many word problems. The key to using proportion is *setting up a problem carefully*. Like ratios, proportions are used to *compare* two numbers. It is extremely important to make sure that the numbers being compared are in the right order.

Study the next examples closely. Notice how the parts of the proportion are set up.

EXAMPLE 1: If 12 yards of lumber cost \$40, how much do 30 yards cost?

(1) $\frac{\text{yards}}{\text{cost}}$ $\frac{12}{40} = \frac{30}{c}$ (2) $\begin{array}{r} 40 \\ \times 30 \\ \hline 1200 \end{array}$ (3) $12\overline{)1200}$ = \$100

STEP 1. In this problem, you are *comparing* yards to cost. Set up two ratios — yards are on the top in each ratio. Cost is on the bottom. Here c stands for the cost you are looking for.
STEP 2. Find the cross product of 40 and 30.
STEP 3. Divide 1200 by the other number in the proportion, 12. The cost of 30 yards of lumber is **\$100**.

> **TIP:** Although x is often used to represent an unknown, you may want to use the first letter of the quantity you are looking for. In the example above, c is used to stand for *cost*. You will learn more about using letters to stand for numbers in the algebra chapter.

EXAMPLE 2: The ratio of the number of men to the number of women working in a certain hospital is 2:3. If 480 women work in the hospital, how many men work there?

(1) $\frac{\text{men}}{\text{women}} \quad \frac{2}{3} = \frac{m}{480}$ **(2)** $\begin{array}{r} 480 \\ \times\ 2 \\ \hline 960 \end{array}$ **(3)** $\begin{array}{r} 320 \text{ men} \\ 3\overline{)960} \end{array}$

STEP 1. Make a proportion comparing men to women. The top number in each ratio refers to men. The bottom number refers to women. Here m stands for the unknown number of men.

STEP 2. Find the cross product of 2 and 480.

STEP 3. Divide 960 by the other number in the proportion, 3. There are **320 men** working in the hospital.

In some cases, you may have to recognize how a proportion problem should be set up rather than find the solution itself. Example 3 shows this type of problem.

EXAMPLE 3: Manny drives 110 miles in 2 hours. Which expression below shows the distance he can go in 5 hours if he drives at the same speed?

(1) $\frac{5 \times 2}{110}$

(2) $\frac{110 \times 2}{5}$

(3) $\frac{110 \times 5}{2}$

(4) $\frac{110 + 5}{2}$

STEP 1. Make a proportion. The top number is miles, and the bottom number is hours. Here m stands for the unknown number of miles.

$$\frac{\text{miles}}{\text{hours}} \quad \frac{110}{2} = \frac{m}{5}$$

STEP 2. Show how you would multiply to find the first cross product. 110×5

STEP 3. Show how you would divide to find the missing number.

$$\frac{110 \times 5}{2}$$

STEP 4. Choose the solution that shows the cross product of 110 and 5 divided by the other number in the proportion, 2. This is an example of a "set up" question that you learned about in Chapter 3. Choice **(3)** $\frac{110 \times 5}{2}$ is correct. Notice that choice (4) is wrong because it shows the sum instead of the product of 110 and 5.

Problem Solving for the GED

Using Proportion to Solve Word Problems

A tree 12 feet tall casts a shadow 15 feet long. The tree next to it casts a shadow that is 50 feet long. How many feet tall is the second tree?
(1) 27
(2) 40
(3) 47
(4) 50
(5) $62\frac{1}{2}$

At first glance, you may not see that proportion will help you solve this problem. You don't see the word ratio or proportion mentioned anywhere. However, if you look closely, you will see that numbers are *being compared*. The height of a tree is being compared to the length of its shadow. This is a good clue that you may be able to use proportion to solve the problem.

The following steps should help you solve a problem like the one above.

1. Set up a ratio comparing the two quantities you are given. Remember to decide which category goes on the top of the ratio and which category goes on the bottom.

   $$\frac{\text{tree}}{\text{shadow}} \qquad \frac{12}{15}$$

2. Now set up a proportion using the third number you are given. Be sure to put this number in the right place, using the same order as the ratio you have already set up.

   In this problem, you are given the length of another tree's shadow. Put this length in the denominator of your second ratio since this is the position of the other shadow length.

   $$\frac{\text{tree}}{\text{shadow}} \qquad \frac{12}{15} = \frac{s}{50}$$

3. Complete the proportion using the cross multiplication method you learned in this chapter.

   $12 \times 50 = 600$ $\qquad$ $15\overline{)600}$ = 40

4. The tree is **40** feet long. **(2)** is the correct answer.

★ **GED PRACTICE** ★

Using Proportions to Solve Word Problems

The Towsons planted a 35-acre field that yielded 3150 bushels of wheat. At the same rate, how many acres would they need to produce 1890 bushels?

(1) 17
(2) 18
(3) 21
(4) 30
(5) 36

ANSWER AND SOLUTION IS ON PAGE 344.

Ratio and Proportion Exercise 3

Directions: Solve each problem with a proportion.

1. If 6 feet of wire cost \$3.40, how much do 9 feet of wire cost?
2. The scale on a map says that 2 inches = 150 miles. If two cities are actually 325 miles apart, how far apart will they be on the map?
3. If 30 chickens lay 75 eggs a month, how many eggs will 50 chickens lay in a month?
4. How long will it take to go 1200 miles if a plane travels 450 miles in two hours?
5. To make a certain color of paint, the ratio of blue paint to white paint is 5:2. How many gallons of blue paint are required to mix with 14 gallons of white paint?
6. On a certain bar graph, a line $2\frac{1}{2}$ inches long represents 75 degrees. How many inches of bar are required to represent 110 degrees?
7. A snapshot that was 3 inches wide and 5 inches long was enlarged to be 12 inches long. Which expression below represents the width of the enlargement?
   (1) $(3 + 12)/5$
   (2) $(5 \cdot 12)/3$
   (3) $(3 \cdot 12)/5$
   (4) $(3 \cdot 5)/12$
8. A worker can make 16 parts in 2 hours. Which expression below represents the time the worker needs to make 100 parts?
   (1) $(2 \times 100)/16$
   (2) $(16 \times 100)/2$
   (3) $(16 \times 2)/100$
   (4) $(16 + 2)/100$
9. A recipe calls for two cups of sugar for every three cups of flour. Which expression below shows the number of cups of sugar a cook needs with 12 cups of flour?
   (1) $(2 \times 3)/12$
   (2) $(3 + 12)/2$
   (3) $(3 \times 12)/2$
   (4) $(2 \times 12)/3$
10. Apples cost 90 cents a dozen. Which expression below represents the cost of 8 apples?
    (1) $(12 \times 8) \times 90$
    (2) $(90 \times 8)/12$
    (3) $90/(12 \times 8)$
    (4) $(90 \times 12)/8$

ANSWERS AND SOLUTIONS START ON PAGE 344.

Two-Part Proportion Word Problems

Your first step in solving any problem involving proportion is to decide which quantity goes on the top of each ratio and which quantity goes on the bottom. This order *must remain consistent* throughout your work in the problem.

For some proportion word problems, one of the quantities to include in a ratio may not be given. You will need to find this quantity before you can continue to solve the problem. Read the next example carefully.

EXAMPLE: Carlos got 2 problems wrong for every 5 problems right on a test. How many problems did Carlos get wrong if there were 35 problems altogether on the test?

STEP 1. The question in this problem asks you to compare the number of wrong answers to the total number of problems. First set up a ratio using a letter to stand for the number you are being asked to find.

$$\frac{\text{wrong}}{\text{total}} \quad \frac{w}{35}$$

STEP 2. You need to set up a proportion to solve the problem. Therefore, you will need another ratio that compares wrong answers to total answers. If you look at the numbers given, you find a number corresponding to *wrong* answers (2) but no number corresponding to *total* answers. You need to find a total to complete the proportion. Carlos got 2 problems wrong for every 5 problems right. This means that he got 2 wrong out of a *total* of 7 problems (wrong + right = total).

$$\frac{\text{wrong}}{\text{total}} = \frac{2}{7}$$

STEP 3. Now set up your proportion and solve for the missing element.

$$\frac{2}{7} = \frac{w}{35} \qquad 35 \times 2 = 70 \qquad 7\overline{)70}\ \text{quotient } 10$$

Carlos got **10 problems** wrong on the test.

> **TIP:** Before you set up a proportion, read the problem over carefully to find out whether you need to add or subtract to set up a ratio.

Ratio and Proportion Exercise 4

Directions: Solve each problem.

1. For every $10 Helen takes home, her employer withholds $3 for taxes and social security. Helen's gross pay each month is $1950. How much does she take home each month?

2. Pat's softball team won 5 games for every 3 games they lost. Altogether the team played 32 games. How many games did they win?

3. The ratio of the number of men to the number of women working at Apex, Inc., is 7:2. Altogether there are 360 workers at the company. How many of the workers are women?

4. The ratio of good parts to defective parts coming off the assembly line at Apex, Inc., is 20:1. Every day the factory produces 10,500 parts. How many of these parts are defective?

5. Recently 300 people in Central County took a civil service examination. For every five people who passed the examination, one person failed. How many people passed?

6. The ratio of the workers who voted to strike to the workers who voted not to strike at Apex was 3:2. 360 workers voted. How many voted to strike?

7. A ladder is placed against a building 40 feet from the ground. The bottom of the ladder is 13 feet from the side of the building. If another ladder is placed against the building at the same angle 20 feet from the ground, how far from the building will the bottom of the ladder be? *Hint:* Set up a proportion comparing the height of the ladder to the distance on the ground.

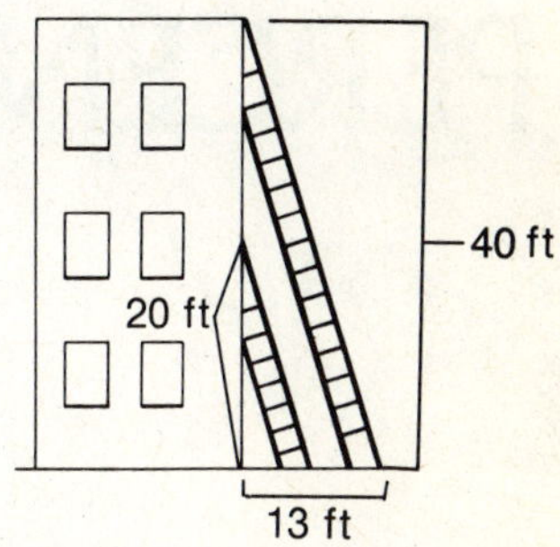

ANSWERS AND SOLUTIONS START ON PAGE 345.

7
PERCENTS

Understanding Percents

Percents are commonly used in the business world. Percents are used to measure commissions, taxes, interest, markups, and discounts. Like the decimals and fractions you have already studied, ***percents*** are another way of *describing the parts of a whole*.

With decimals, one whole is divided into tenths, hundredths, thousandths, etc. With fractions, one whole can be divided into *any* number of parts (2, 3, 4, 5, etc.). With percents, one whole is always divided into 100 parts. The percent sign, %, means "per 100" or "out of 100."

Each square below is divided into 100 equal parts.

A

B
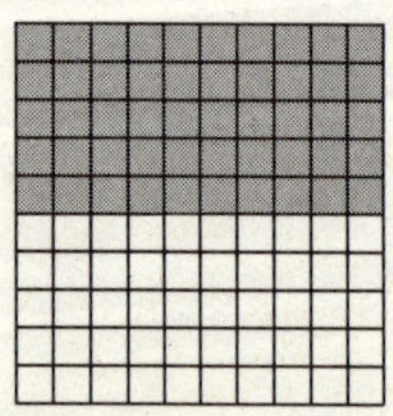

C
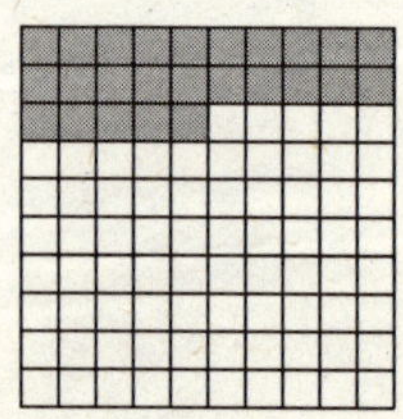

D
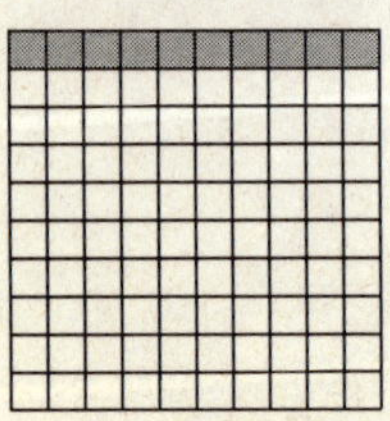

E
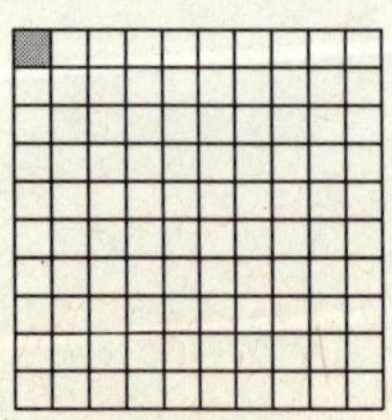

F
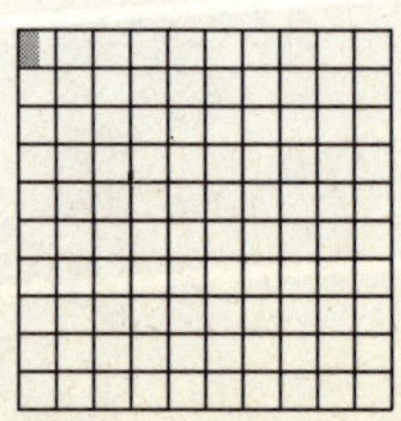

- In square A, all **100** parts are shaded. We can say that **100%** of the square is shaded. 100% is the same as one whole.
- In square B, **50** of the 100 parts are shaded. **50%** of the square is shaded. 50% is the same as one-half.
- In square C, **25** of the 100 parts are shaded. **25%** of the square is shaded. 25% is the same as one-quarter.
- In square D, **10** of the 100 parts are shaded. **10%** of the square is shaded. 10% is the same as one-tenth.
- In square E, only **1** of the 100 parts is shaded. **1%** of the square is shaded. 1% is the same as $\frac{1}{100}$.
- In square F, $\mathbf{\frac{1}{2}}$ of 1 part of the square is shaded. $\mathbf{\frac{1}{2}}$**%** of the square is shaded. We can also say that $\frac{1}{2}$ of 1 percent of this square is shaded. You know that $\frac{1}{2}$ can be written as the decimal .5. Therefore, we can also say that .5% ($\frac{5}{10}$ of one percent) of the square is shaded. Any percent less than 1% has a value of less than $\frac{1}{100}$.

Mixed decimals like 2.6 and 1.05 are used to represent amounts greater than one whole. Mixed numbers like $1\frac{1}{2}$ or $10\frac{3}{4}$ also represent amounts greater than one whole. With percents, one whole is represented as 100%. Percents greater than 100% represent amounts greater than one whole. For example, 150% and 1000% are greater than one whole.

> **TIP:** Remember that any percent less than 1% has a value less than $\frac{1}{100}$, and any percent greater than 100% has a value greater than 1.

Percent Exercise 1

Directions: Use the following list of percents to answer the next questions.

60%	110%	99%	$\frac{1}{4}$%	6.5%	0.9%
2%	$83\frac{1}{3}$%	.65%	240%	0.1%	650%

1. From the list above choose the percents that have a value between $\frac{1}{100}$ (1%) and one whole (100%).
2. From the list above choose the percents that have a value greater than one whole (100%).
3. From the list above choose the percents that have a value less than $\frac{1}{100}$ (1%).

ANSWERS START ON PAGE 345.

Decimals and Percents

Changing a Percent to a Decimal

In some problems you may need to change a percent to a decimal in order to make the amounts easier to work with.

Rule for Changing a Percent to a Decimal

Drop the percent sign (%) and move the decimal point two places to the *left*.

TIP: Remember that a whole number written without a decimal point is understood to have a decimal point to the right of the units digit.

Notice that changing a percent to a decimal is like dividing by 100.

EXAMPLES: 18% = .18 250% = 2.50. = 2.5

You may have to write extra zeros to the left of the digits in order to move the point two places.

EXAMPLES: 3.2% = .03.2 = .032 .25% = .00.25 = .0025

When a percent contains a fraction, remember that a decimal point is understood to be to the right of the last *whole number digit*.

EXAMPLES: $33\frac{1}{3}\%$ = $.33.\frac{1}{3}$ = $.33\frac{1}{3}$ $6\frac{2}{3}\%$ = $.06.\frac{2}{3}$ = $.06\frac{2}{3}$

In these examples, $\frac{1}{3}$ and $\frac{2}{3}$ *are not place values*. The decimal point is understood to be after the 3 and after the 6 in the units place, the *last whole number digit*.

Percent Exercise 2

Directions: Change each percent to a decimal or a whole number.

1. 9% =

2. 24% =

3. 100% =

4. .3% =

5. $87\frac{1}{2}\%$ =

6. $8\frac{1}{3}\%$ =

7. .15% =

8. 275% =

9. 2.7% =

10. 3.95% =

11. 57% =

12. 1000% =

ANSWERS AND SOLUTIONS START ON PAGE 345.

Changing a Decimal to a Percent

In other problems you may need to change a decimal to a percent, again in order to make amounts easier to work with.

Rule for Changing a Decimal to a Percent

Move the decimal point two places to the *right* and write the percent sign after the last digit.

EXAMPLES: .25 = .25. = 25%

.19 = .19. = 19%

3.65 = 3.65. = 365%

.625 = .62.5 = 62.5%

Notice that you do not have to write the decimal point when it moves to the end of the number.

You may have to write extra zeros in order to move the decimal point two places.

EXAMPLES: .6 = .60. = 60%

2.7 = 2.70. = 270%

36 = 36.00. = 3600%

You do not write the decimal point when it moves between a digit and a fraction.

EXAMPLES: $.37\frac{1}{2} = .37.\frac{1}{2} = 37\frac{1}{2}\%$

$.14\frac{1}{4} = .14.\frac{1}{4} = 14\frac{1}{4}\%$

$.05\frac{1}{3} = .05.\frac{1}{3} = 5\frac{1}{3}\%$

You can drop unnecessary zeros after you move the decimal point.

EXAMPLES: .045 = .04.5 = 4.5%

.003 = .00.3 = .3%

.0008 = .00.08 = .08%

TIP: If you are unsure about which way to move the decimal point in a problem, think about the diagram at the right. It is in alphabetical order. The D is to the left of P. To change a percent to a decimal, move the point to the *left*. To change a decimal to a percent move the point to the *right*.

D P
.25 ↔ 25%

Percent Exercise 3

Directions: Change each decimal to a percent.

1. .81 =	**5.** .0009 =	**9.** 2.1 =
2. $.37\frac{1}{2}$ =	**6.** .217 =	**10.** 4.85 =
3. .5 =	**7.** .03 =	**11.** 3.924 =
4. .004 =	**8.** $.33\frac{1}{3}$ =	**12.** .015 =

ANSWERS AND SOLUTIONS START ON PAGE 345.

Fractions and Percents

Changing a Fraction to a Percent

Suppose you work five days a week. Your *whole* work week, expressed as a fraction, is $\frac{5}{5}$, or 5 out of 5 days. If you are out sick for one day, you miss one out of five days of work or $\frac{1}{5}$ of the work week.

In percent, your *whole* work week is 100% of the week. You can say that you were sick for $\frac{1}{5}$ of 100%. You know from studying fractions that to find a "fraction of" something means to multiply.

$$\frac{1}{5} \times \frac{100\%}{1} = \frac{20}{1} = 20\%$$

In other words, you were sick **20%** of your work week.

Rule for Changing a Fraction to a Percent

Method 1: Multiply the fraction by 100%.

Method 2: Divide the denominator of the fraction into the numerator. Then change the decimal point.

METHOD 1: Change $\frac{3}{4}$ to a percent.

$$\frac{3}{4} \times \frac{100\%}{1} = 75\%$$

METHOD 2: Change $\frac{3}{4}$ to a percent.

$$4\overline{)3.00} = .75 \qquad .75 = 75\%$$

When the denominator of the fraction does not divide evenly into 100, the percent will contain a fraction.

EXAMPLE 3: Change $\frac{5}{7}$ to a percent.

METHOD 1

$$\frac{5}{7} \times \frac{100\%}{1} = \frac{500}{7} = 71\frac{3}{7}\%$$

METHOD 2

$$7\overline{)5.0} = .71\frac{3}{7} = 71\frac{3}{7}\%$$

```
   .71 3/7
7) 5.0
   4 9
     10
      7
      3
```

To change any proper or improper fraction to a percent, be sure the fraction is reduced. This makes your work easier. Look at Example 4 on the next page.

EXAMPLE 4: Change $\frac{30}{25}$ to a percent.

METHOD 1

$$\frac{30}{25} = \frac{6}{5} \quad \frac{6}{5} \times \frac{100\%}{1} = 120\%$$

METHOD 2

$$5\overline{)6.00} = 1.20 = 120\%$$

STEP 1. Reduce the improper fraction.

STEP 2. Multiply the reduced fraction by 100%. (Or, if you are using Method 2, divide the numerator by the denominator and change the decimal to a percent.)

Percent Exercise 4

Directions: Change each fraction to a percent. Use whichever method you prefer.

1. $\frac{1}{5} =$

2. $\frac{5}{6} =$

3. $\frac{3}{8} =$

4. $\frac{2}{3} =$

5. $\frac{14}{8} =$

6. $\frac{9}{10} =$

7. $\frac{5}{12} =$

8. $\frac{6}{7} =$

9. $\frac{1}{6} =$

10. $\frac{10}{5} =$

11. $\frac{1}{12} =$

12. $\frac{2}{11} =$

ANSWERS AND SOLUTIONS START ON PAGE 346.

Changing a Percent to a Fraction

A percent can be expressed as a fraction with a denominator of 100. For example, 20% is the same as $\frac{20}{100}$.

Rule for Changing a Percent to a Fraction or a Mixed Number

Replace the % sign with a denominator of 100 and reduce.

EXAMPLE 1: Change 75% to a fraction.

$$75\% = \frac{75}{100} = \frac{3}{4}$$

EXAMPLE 2: Change 125% to a mixed number.

$$125\% = \frac{125}{100} = 1\frac{25}{100} = 1\frac{1}{4}$$

Some percents, like $16\frac{2}{3}\%$, are hard to reduce when you replace the percent sign with 100. To simplify these percents, divide by 100 instead of trying to reduce the fraction. It is possible to divide by 100 because a fraction bar means "divided by." $\frac{16\frac{2}{3}}{100}$ means $16\frac{2}{3}$ divided by 100.

EXAMPLE 3: Change $16\frac{2}{3}\%$ to a fraction.

STEP 1. Write $16\frac{2}{3}$ over a denominator of 100.

$$\frac{16\frac{2}{3}}{100}$$

STEP 2. Rewrite the problem as a division problem.

$$16\frac{2}{3} \div 100$$

STEP 3. Change $16\frac{2}{3}$ to an improper fraction, invert 100 to $\frac{1}{100}$, and multiply.

$$\frac{50}{3} \div \frac{100}{1} \qquad \frac{\overset{1}{\cancel{50}}}{3} \times \frac{1}{\underset{2}{\cancel{100}}} = \frac{1}{6}$$

Percents containing decimals can also be changed. First change these percents to decimals. Then change the decimals to fractions and reduce.

EXAMPLE 4: Change 2.5% to a fraction.

STEP 1. Change 2.5% to a decimal by dropping the percent sign and moving the decimal point two places to the left.

$$2.5\% = .02.5 = .025$$

STEP 2. Change .025 to a fraction. Three places are thousandths. 25 is the numerator, and 1000 is the denominator. Reduce the fraction to lowest terms.

$$.025 = \frac{25}{1000} = \frac{1}{40}$$

Percent Exercise 5

Directions: Change each percent to a fraction or a mixed number and reduce.

1. 45% =

2. $37\frac{1}{2}\%$ =

3. $6\frac{2}{3}\%$ =

4. 8% =

5. 2% =

6. $83\frac{1}{3}\%$ =

7. 24% =

8. $33\frac{1}{3}\%$ =

9. $28\frac{4}{7}\%$ =

10. 80% =

11. 150% =

12. $12\frac{1}{2}\%$ =

13. 1.5% =

14. .09% =

15. .6% =

16. 1% =

ANSWERS AND SOLUTIONS START ON PAGE 346.

Common Fractions, Decimals, and Percents

The next exercise is a chart that you should fill in. When you finish, check your answers. The values on this chart are some of the most commonly used fractions, decimals, and percents. You will save time on many problems if you know these equivalencies. For example, if a problem includes several fractions, you may find it easier to use the fraction $\frac{1}{4}$ than to use .25 or 25%. The first line of this exercise has been filled in as an example.

Percent Exercise 6

Directions: Fill in the missing numbers in the chart.

Fraction	Decimal	Percent
$\frac{1}{4}$	.25	25%
$\frac{1}{2}$		
		75%
	.125 or $.12\frac{1}{2}$	
		20%
	.4	
$\frac{3}{5}$		
		80%
	.1	
$\frac{3}{10}$		
		70%
	.9	
	$.33\frac{1}{3}$	
$\frac{2}{3}$		

ANSWERS START ON PAGE 346.

Proportion and Percents

In this section you will learn how your work with proportions in Chapter 6 can help you in your work with percents. As you recall, the most important part of working with proportions is setting them up carefully. Keep this in mind as you look at the following example.

Three cans are 50% of a six-pack. The statement "3 is 50% of 6" contains three facts:

6 is the whole
3 is a part.
50 is the percent.

Each number in a percent statement fits into a proportion:

$$\frac{\text{part}}{\text{whole}} = \frac{\%}{100}$$

Whenever you use a proportion to solve a percent problem, you will use this setup. This makes sense if you think of the righthand ratio as comparing a *part* of a percent to the *whole* percent. The righthand side of the proportion is showing 50% as a part of a whole, 100%. The proportion for the six-pack problem is:

$$\frac{3 \text{ cans}}{6 \text{ cans}} = \frac{50}{100}$$

Notice that the cross products of the proportion are equal.

$$3 \times 100 = 300, \text{ and } 6 \times 50 = 300.$$

You can use proportion to solve many percent problems.

Rules for Using Proportion with Percents

$$\frac{\text{part}}{\text{whole}} = \frac{\%}{100}$$

1. Fill in the information given in a percent problem as shown in the diagram above. Use a letter to stand for the missing term.
2. Find the cross product in the proportion.
3. Divide by the remaining number in the proportion.

The first step in setting up the proportion will be determining which number is the *part*, which number is the *whole*, and which number is the *percent*. Finding the percent is usually easy because either the word *percent* or the symbol % will appear after the number. The part and the whole are a little more difficult to identify. One clue is to look for the word *of*. The *whole* usually follows this word. Once you have found the percent and the whole, the *part* will be the only number left in the problem.

Once you have identified the percent, the part, and the whole, set up a proportion. Remember that the number 100 (which stands for 100%) *always* goes in the bottom of the second ratio.

Percent Exercise 7

Directions: In this exercise, practice identifying the percent, the part, and the whole. Then practice setting up proportions. The first problem is done for you.

1. 18 is fifty percent of 36.
2. 80% of 25 is 20.
3. 45 is 10% of 450.
4. 16% of 25 equals 4.
5. 30% of 200 is 60.
6. 13 is equal to twenty percent of 65.
7. 107 is 100% of 107.
8. 50 is 50 percent of one hundred.

ANSWERS START ON PAGE 346.

Percent Problems

In the next examples you will learn how to use proportion to solve percent problems. In percent problems, one of the three elements needed to set up a proportion will be missing. In other words, you will be trying to find a percent, a whole, or a part. As with any proportion, be sure you put each number in the correct position when solving the problem.

Finding the Percent

EXAMPLE: 27 is what percent of 45?

(1) $\frac{\text{part}}{\text{whole}} \quad \frac{27}{45} = \frac{x}{100}$

(2) $\begin{array}{r} 27 \\ \times 100 \\ \hline 2700 \end{array}$

(3) $\begin{array}{r} 60\% \\ 45\overline{)2700} \\ 270 \\ \hline 00 \end{array}$

STEP 1. Fill in the terms of the proportion. Always put 100 in the lower right. 45 follows the word *of*. 45 is the whole. Put 45 in the lower left. 27 is the part. Put 27 in the upper left. You are looking for the percent. Let x stand for the percent.

STEP 2. Find the cross product of 27 and 100.

STEP 3. Divide 2700 by the remaining number in the proportion, 45. 27 is **60%** of 45.

Percent Exercise 8

Directions: Solve each problem. Be sure to set up a proportion first.

1. 9 is what percent of 36?
2. 14 is what percent of 35?
3. 50 is what percent of 75?
4. 40 is what percent of 16?
5. 120 is what percent of 160?

ANSWERS START ON PAGE 346.

Finding the Part

EXAMPLE 1: What is 15% of 60?

(1) $\frac{\text{part}}{\text{whole}} \quad \frac{p}{60} = \frac{15}{100}$ (2) $\begin{array}{r} 15 \\ \times 60 \\ \hline 900 \end{array}$ (3) $100\overline{)900}$ = 9

STEP 1. Fill in the terms of the proportion. Always put 100 in the lower right. Put 15 in the upper right. 60 follows the word *of*. 60 is the whole. Put 60 in the lower left. You are looking for the part. Let *p* stand for the part.
STEP 2. Find the cross product of 60 and 15.
STEP 3. Divide 900 by the remaining number in the proportion, 100. 15% of 60 is **9**.

EXAMPLE 2: Find $62\frac{1}{2}$% of 56.

(1) $\frac{\text{part}}{\text{whole}} \quad \frac{p}{56} = \frac{62\frac{1}{2}}{100}$ (2) $56 \times 62\frac{1}{2} =$ $\frac{\overset{28}{\cancel{56}}}{1} \times \frac{125}{\underset{1}{\cancel{2}}} = 3500$ (3) $100\overline{)3500}$ = 35

STEP 1. Fill in the terms of the proportion. Put 100 in the lower right. Put $62\frac{1}{2}$ in the upper right. 56 follows the word *of*. 56 is the whole. Put 56 in the lower left. You are looking for the part. Let *p* stand for the part.
STEP 2. Find the cross product of 56 and $62\frac{1}{2}$.
STEP 3. Divide 3500 by the remaining number in the proportion, 100. $62\frac{1}{2}$% of 56 is **35**.

In some problems, the number we call the *part* is larger than the whole. This happens when you are finding more than 100% of a number.

EXAMPLE 3: What is 180% of 250?

(1) $\frac{\text{part}}{\text{whole}} \quad \frac{p}{250} = \frac{180}{100}$ (2) $\begin{array}{r} 250 \\ \times 180 \\ \hline 20\,000 \\ 25\,0 \\ \hline 45{,}000 \end{array}$ (3) $100\overline{)45{,}000}$ = 450

STEP 1. Fill in the terms of the proportion. 180 goes in the upper right, and 100 goes in the lower right. 250 follows the word *of*. 250 is the whole. Put 250 in the lower left. You are looking for the part. Let *p* stand for the part.
STEP 2. Find the cross product of 250 and 180.
STEP 3. Divide 45,000 by the remaining number in the proportion, 100. 180% of 250 = **450**.

Percent Exercise 9

Directions: Solve each problem using proportions.

1. Find 30% of 90.

2. Find 65% of 140.

3. What is $16\frac{2}{3}\%$ of 72?

4. Find 8.5% of 500.

5. What is 300% of 60?

ANSWERS AND SOLUTIONS START ON PAGE 347.

Finding the Whole

You will often know that you need to find the whole because there is no number following the word *of*. Following the word *of*, there will be a phrase such as *what number*.

EXAMPLE: 8 is 50% of what number?

(1) $\frac{\text{part}}{\text{whole}} \quad \frac{8}{w} = \frac{50}{100}$ **(2)** $\begin{array}{r} 100 \\ \times\ 8 \\ \hline 800 \end{array}$ **(3)** $50\overline{)800}$ = 16

STEP 1. Fill in the terms of the proportion. 50 goes in the upper right, and 100 goes in the lower right. The words *what number* follow the word *of*. The *whole* is missing. Let *w* stand for the whole. 8 is the part. Put 8 in the upper left.
STEP 2. Find the cross product of 8 and 100.
STEP 3. Divide 800 by the remaining number in the proportion, 50. 8 is 50% of **16**.

Percent Exercise 10

Directions: Solve each problem using proportions.

1. 18 is 50% of what number?

2. 24 is 15% of what number?

3. 90 is 4.5% of what number?

4. 2 is $12\frac{1}{2}\%$ of what number?

5. 86 is 200% of what number?

ANSWERS AND SOLUTIONS START ON PAGE 347.

Types of Percent Problems

You now know how to use proportion to solve three kinds of percent problems: (1) finding the part, (2) finding the percent, and (3) finding the whole. Remember that the whole usually follows the word *of*, and when the percent is given, it has the percent sign (%).

Your first step in any percent problem is *to find what element is missing*.

EXAMPLE 1: What are you looking for (the *part*, the *percent*, or the *whole*) in the problem: What is 60% of 45?
SOLUTION. 60 is the percent because it has the percent sign. 45 is the whole because it follows the word *of*. You are looking for the **part**.

EXAMPLE 2: What are you looking for (the *part*, the *percent*, or the *whole*) in the problem: 12 is 50% of what number?
SOLUTION. The words *what number* follow the word *of*. You are looking for the **whole**. 50 is the percent, and 12 is the part.

EXAMPLE 3: What are you looking for (the *part*, the *percent*, or the *whole*) in the problem: 16 is what percent of 64?

SOLUTION. No number has a percent sign. You are looking for the **percent**. 16 is the part, and 64 is the whole.

Percent Exercise 11

Directions: First tell what is being asked for (the part, the percent, or the whole) in each problem. Then go back and solve each problem.

1. 16 is what percent of 32?

2. Find 80% of 90.

3. 30 is 60% of what number?

4. What is $4\frac{1}{2}$% of 800?

5. What percent of 50 is 14?

6. $12\frac{1}{2}$% of what number is 15?

7. 15 is what percent of 45?

8. What is 3.6% of 900?

9. 120 is what percent of 80?

10. $33\frac{1}{3}$% of what number is 45?

11. What is 8.6% of 200?

12. 100% of 38.7 equals what?

ANSWERS AND SOLUTIONS START ON PAGE 347.

Percent Word Problems

To solve percent word problems, first decide whether you are looking for the part, the percent, or the whole. Study the next examples carefully.

EXAMPLE 1: Mr. Gomez pays $80 for a suit. He puts a 30% markup on each suit that he sells in his store. Find the amount of the markup.

(1) $\frac{\text{part}}{\text{whole}} \quad \frac{m}{8} = \frac{30}{100}$ **(2)** $\begin{array}{r} 80 \\ \times 30 \\ \hline 2400 \end{array}$ **(3)** $\begin{array}{r} \$\ 24 \\ 100\overline{)2400} \end{array}$

STEP 1. Decide whether you are looking for the part, the percent, or the whole. 30 is the percent, and $80, the cost of the suit, is the whole. You are looking for the part, the amount of the markup. Fill in the terms of the proportion. Let m stand for the markup.

STEP 2. Find the cross product of 80 and 30.

STEP 3. Divide 2400 by the remaining number in the proportion, 100. The amount of the markup is **$24**.

EXAMPLE 2: Mrs. Jackson makes $200 a week. She spends $60 a week for food. Food represents what percent of Mrs. Jackson's income?

(1) $\frac{\text{part}}{\text{whole}} \quad \frac{60}{200} = \frac{x}{100}$ **(2)** $\begin{array}{r} 60 \\ \times\ 100 \\ \hline 6000 \end{array}$ **(3)** $\begin{array}{r} 30\% \\ 200\overline{)6000} \end{array}$

STEP 1. Decide whether you are looking for the part, the percent, or the whole. $60 is the *part* Mrs. Jackson spends on food. $200 is her *whole* income. You are looking for the percent of her income she spends for food. Fill in the terms of the proportion. Let x stand for the percent.

STEP 2. Find the cross product of 60 and 100.
STEP 3. Divide 6000 by the remaining number in the proportion, 200. Mrs. Jackson pays **30%** of her weekly income for food.

EXAMPLE 3: Lois got a 6% commission for selling a house. Her commission was $3000. Find the selling price of the house.

(1) $\frac{\text{part}}{\text{whole}} \quad \frac{3000}{w} = \frac{6}{100}$ (2) $3000 \times 100 = 300{,}000$ (3) $300{,}000 \div 6 = \$50{,}000$

STEP 1. Decide whether you are looking for the part, the percent, or the whole. 6 is the percent. $3000 is the part Lois received as a commission. You are looking for the whole, the selling price of the house. Fill in the terms of the proportion. Let w stand for the whole price of the house.
STEP 2. Find the cross product of 3000 and 100.
STEP 3. Divide 300,000 by the remaining number in the proportion, 6. The selling price of the house was **$50,000**.

Percent Exercise 12

Directions: First write down what is being asked for in each problem — the *part*, the *percent*, or the *whole*. Then set up a proportion and solve each problem.

1. A jacket originally selling for $40 was on sale for 15% off the original price. How much is saved by buying the jacket on sale?
2. Mr. and Mrs. Shin need $8000 for a down payment on a house. So far they have saved $6000. What percent of the total amount have they saved?
3. Alfredo earns $250 a week. His employer deducts 12% of his earnings for taxes and social security. How much is deducted from Alfredo's weekly pay?
4. John now weighs 172 pounds. This is 80% of what John weighed a year ago. How much did John weigh a year ago?
5. Fiona makes $600 a month and pays $150 a month for rent. Rent is what percent of her income?
6. The sales tax in Muhammed's state is 6%. How much tax does Muhammed have to pay for a television that costs $240?
7. Eighteen people showed up for David's evening math class. This represents 75% of the number registered for the class. How many people are registered for the class?
8. Mr. Kee pays $20 for a pair of shoes. He puts a $6 markup on every pair of shoes in his store. The markup is what percent of the price Mr. Kee pays?
9. Eva sells cosmetics for a 9% commission. In November her sales were $3840. How much did she make in commissions in November?
10. A tent sold originally for $150. It was on sale for $90 off the original price. The discount was what percent of the original price?
11. Alejo worked for $250 a week. He received a $30 a week raise. His raise was what percent of his old salary?

12. A day care center has 60 boys. Boys make up 40% of the total. Find the total number of children at the center.

ANSWERS AND SOLUTIONS START ON PAGE 348.

Multi-Step Percent Problems

In many percent problems, you must do more than simply find the missing element. Once you have set up a proportion and found the part, the whole, and the percent, you may have to use this information to find yet another number. With this kind of problem, remember that you will need to perform a series of steps to get your answer. The first step *always* is setting up a proportion.

Finding the Part

In this example, first you need to find the part, then you must add this amount to the whole in order to get a total amount.

EXAMPLE 1: The owner of the Victoria Clothing Store pays $75 each for men's suits. He puts a 40% markup on each suit. How much does he charge for each suit?

(1) $\frac{\text{part}}{\text{whole}} \quad \frac{p}{75} = \frac{40}{100}$ (2) $\begin{array}{r} 75 \\ \times 40 \\ \hline 3000 \end{array}$ (3) $100\overline{)3000}$ = 30 (4) $\begin{array}{r} \$75 \\ +30 \\ \hline \$105 \end{array}$

STEP 1. You want to find the selling price of the suit, but first you need to find the part, the amount of the markup. Fill in the terms of the proportion. 40 is the percent, and 75 is the whole. Let p stand for the part.

STEP 2. Find the cross product of 75 and 40.

STEP 3. Divide 3000 by the remaining number in the proportion, 100. The part, or markup, is $30.

STEP 4. Although you have found the missing element in the proportion, you are not finished. The question asks how much the owner charges for a suit. This is his cost plus the markup. Add $30 to the original price. The selling price of the suit is **$105**.

In other problems, you may have to subtract the part from the original amount.

EXAMPLE 2: A chair originally sold for $130. It was on sale at a 20% discount. Find the sale price of the chair.

(1) $\frac{\text{part}}{\text{whole}} \quad \frac{p}{130} = \frac{20}{100}$ (2) $\begin{array}{r} 130 \\ \times 20 \\ \hline 2600 \end{array}$ (3) $100\overline{)2600}$ = 26 (4) $\begin{array}{r} \$130 \\ -26 \\ \hline \$104 \end{array}$

STEP 1. You want to find the sale price of the chair, but first you need to find the part, the amount of the discount. Fill in the terms of the proportion. 20 is the percent, and 130 is the whole.

STEP 2. Find the cross product of 130 and 20.

STEP 3. Divide 2600 by the remaining number in the proportion, 100. The part, or discount, is $26.

STEP 4. The question asks you for the sale price. To get this, subtract the discount from the original price. Subtract $26 from $130. The sale price of the chair is **$104**.

> **TIP:** One key to solving a percent problem is to read the question carefully and *understand what you are being asked to find.*

Percent Exercise 13

Directions: Solve each problem.

1. The owner of Gordon's Shoe Store pays $12 for a pair of children's shoes. He puts a 45% markup on each pair. Find the selling price of a pair of children's shoes at Gordon's.
2. A farm with a market value of $120,000 was assessed for 60% of its market value. The farm is taxed at 2% of the assessed value. Find the yearly tax on the farm.
3. Floria makes $15,600 a year. She spends 25% of her salary on rent. How much rent does Floria spend each month?
4. Fran bought new furniture for $1800. She paid 15% down and $50 a month for 36 months. Find the total amount she paid for the furniture.
5. The 100th Street Block Association wants to raise $500 to plant trees on their street. By the end of two weeks the association raised 65% of the amount they wanted. How much more did they need to raise?
6. Arthur's gross salary is $1040 a month. His employer withholds 9% for federal tax, 5% for social security, and 4% for state tax. Find Arthur's net salary for the month.
7. The population of Pleasant Hill was 2200 in 1970. By 1980 the population was 125% more than the 1970 population. How many people lived in Pleasant Hill in 1980?

Directions: Questions 8–10 refer to the following situation.

The George Street School bought 100 new desks originally listed at $65 each. The desks were on sale at 10% off the list price. The school received an additional 5% discount off the list price for buying in large quantity and another 3% discount off the list price for placing the order early in the spring.

8. Find the total percent of the discount.
9. How much does the school save on each desk?
10. Find the total amount the school must pay for 100 desks.

ANSWERS AND SOLUTIONS START ON PAGE 348.

Finding the Percent: Rate of Increase and Rate of Decrease

In some percent problems you will have to find the rate, in percent, that an amount changes over a period of time. You must compare the change (or difference) to the original amount. In the proportions for these problems, *the part is always the change, and the whole is always the original amount*. When an amount goes up, the percent of change is called the ***rate of increase***. When an amount goes down, the percent of change is called the ***rate of decrease***.

EXAMPLE 1: Nina bought some living room furniture for \$2500: Only a month later, she found that she had to sell it. Because it was not brand new, she received only \$2100 from the sale. What was the rate of decrease on the furniture?

STEP 1. Subtract to find the difference between the two amounts.

$$\begin{array}{r} \$2500 \\ -\ 2100 \\ \hline \$\ 400 \end{array}$$

STEP 2. Fill in the terms of the proportion. \$2500, the *original* amount, is the whole. \$400, the difference, is the part. Let x stand for the percent.

$$\frac{\text{part}}{\text{whole}} \quad \frac{400}{2500} = \frac{x}{100}$$

STEP 3. Find the cross product of 400 and 100.

$$\begin{array}{r} 400 \\ \times\ 100 \\ \hline 40{,}000 \end{array}$$

STEP 4. Divide 40,000 by the remaining number in the proportion, 2500. The rate of decrease was **16%**.

$$\begin{array}{r} 16 \\ 2500\overline{)40{,}000} \end{array}$$

> **TIP:** Whenever you are finding percent of increase or decrease, you are comparing an amount of change to the *original* number. Therefore, the bottom number in the proportion is always the *original* number.

EXAMPLE 2: Over the past ten years, the population of Little Lake, Minnesota, has increased from 1200 to 1500 people. What was the rate of increase in the population?

STEP 1. Subtract to find the amount of the increase (or difference).

$$\begin{array}{r} 1500 \\ -\ 1200 \\ \hline 300 \end{array}$$

STEP 2. Fill in the terms of the proportion. The part, or change (or difference), is 300. The whole, or original, is 1200. Let x stand for the percent.

$$\frac{\text{part}}{\text{whole}} \quad \frac{300}{1200} = \frac{x}{100}$$

STEP 3. Find the cross product of 300 and 100.

$$\begin{array}{r} 300 \\ \times\ 100 \\ \hline 30{,}000 \end{array}$$

STEP 4. Divide 100 by the remaining number in the proportion, 4. The rate of increase is **25%**.

$$\begin{array}{r} 25 \\ 1200\overline{)30{,}000} \\ 24\ 00 \\ \hline 6\ 000 \\ 6\ 000 \\ \hline \end{array}$$

Percent Exercise 14

Directions: Solve each problem.

1. The price of a dozen eggs increased from $.84 to $.90. What was the percent of increase in the price of the eggs?
2. Sal wants to know the rate of decrease in the value of his motorcycle. He bought it for $1200 and can get only $900 if he sells it now. What is the rate of decrease?
3. The unemployment rate rose from 10% to 12%. By what percent did the unemployment rate rise?
4. A company's workers have been asked to take a pay cut. An assembly line worker's weekly wages would decrease from $300 to $275. What is the percent of decrease?
5. The Porter family bought their house in 1960 for $25,000. They sold it in 1987 for $65,000. Find the rate of increase in the market value of the house.
6. The Vernon Tool and Die Company bought a new lathe for $3600. One year later the lathe was worth $3312. By what percent did the value of the lathe depreciate?
7. Alberto borrowed $1250 and paid back $1400 including interest one year later. What rate of interest did he pay on the loan?
8. The selling price of a portable typewriter is $90. The store owner originally paid the dealer $75 for the typewriter. Find the markup rate on the typewriter.

ANSWERS AND SOLUTIONS START ON PAGE 349.

Finding the Whole

In some percent problems, finding the whole is only the first step to solving the problem. Then you must use that whole to find an amount of change. Study the next example carefully.

EXAMPLE: Max wants to buy a used car. He has saved $1200 which is 40% of the price of the amount he needs. How much more does Max have to save?

STEP 1. Set up the proportion. 40 is the percent, and 1200 is the part he has saved. Let *w* stand for the whole.

$$\frac{\text{part}}{\text{whole}} \quad \frac{1200}{w} = \frac{40}{100}$$

STEP 2. Find the cross product of 1200 and 100.

$$\begin{array}{r} 1200 \\ \times\ 100 \\ \hline 120{,}000 \end{array}$$

STEP 3. Divide 120,000 by the remaining number in the proportion, 40. The total price of the car is $3000.

$$120{,}000 \div 40 = 3{,}000$$

STEP 4. Subtract the amount Max has saved from the total price of the car. He still needs **$1800**.

total	$3000
saved	− 1200
needs	$1800

Percent Exercise 15

Directions: Solve each problem.

1. Fred has a car worth \$6000. This is 80% of the amount he paid for the car last year. Find how much the value of the car depreciated.
2. During the last election in Central County, 4800 voters went to the polls. This represents 60% of the registered voters. How many registered voters in Central County did not go to the polls?
3. Joe bought a stereo on sale for \$342. This was 90% of the original price. How much did Joe save by buying the stereo on sale?
4. The Acevedos have saved \$6900 which is 75% of the amount they need for a down payment on a house. How much more do they need?
5. Membership in the Uptown Tenants Association is 120% of what it was one year ago. There are currently 1800 members in the association. How many people joined since this time last year?

ANSWERS AND SOLUTIONS START ON PAGE 349.

Successive Percents

In some problems, you will need to find a percent of a number, then find a percent of this *new* number. This is called finding ***successive percents***.

EXAMPLE 1: A total of 25% of the 1500 employees of Grant Machine Tool Company belong to the union. Of these union members, 20% voted in favor of a strike. How many union members voted to strike?

STEP 1. Figure out how many employees are in the union by setting up a proportion as you learned earlier. Find the cross product of 1500 and 25, then divide by the remaining term. 375 is the number of union members at Grant.

$$\frac{x}{1500} = \frac{25}{100} \qquad \begin{array}{r} 1500 \\ \times\ 25 \\ \hline 37500 \end{array} \qquad 100\overline{)37500}\ \ 375$$

STEP 2. Now find 20% of 375 to figure out how many union members voted to strike. Set up a proportion and solve for the missing number. **75** union members voted to strike.

$$\frac{x}{375} = \frac{20}{100} \qquad \begin{array}{r} 375 \\ \times\ 20 \\ \hline 7500 \end{array} \qquad 100\overline{)7500}\ \ 75$$

Notice that you *cannot* solve this problem by adding 25% to 15% and taking 40% of 1500. This type of problem requires you to figure out each percent separately. How is this different from Example 2 on the next page?

EXAMPLE 2: The manager at Hank's Appliances gives a 15% discount off the list price for refrigerators. He gives an additional 5% discount off the list price if the customer pays in cash. If a refrigerator's list price is $800, and it is paid for in cash, what is the total discount on the refrigerator?

(1)
$$\begin{array}{r} 15\% \\ +\ 5\% \\ \hline 20\% \end{array}$$

(2)
$$\frac{x}{800} = \frac{20}{100} \qquad \begin{array}{r} 800 \\ 20 \\ \hline 1600 \end{array} \qquad 100\overline{)1600}\ \ 160$$

STEP 1. A total discount of 20% is given if the customer pays in cash. Notice that the additional 5% is taken off the *original list price*, not the discounted price.

STEP 2. Find 20% of $800 by using the proportion method. The total discount is **$160**.

In Example 2, both discounts (15% and 5%) are taken off the list price. Therefore, you add them together, then find the total percent discount. However, in Example 1, the first percent you had to find was of *all employees*. The second percent you had to find was of *union members*. Therefore, it would be incorrect to add the two percents together.

> **TIP:** When two or more percents are given in a problem, read it carefully to determine whether you should add the percents together or use successive percents.

Percent Exercise 16

Directions: Solve the problems below. First decide whether you must add percents together or find successive percents.

1. A cut-rate toy store gives a discount of 18 percent off all marked prices. During a clearance sale, it gave an additional 10% off the discounted price. What would the clearance price be of a rocking horse marked $50?
2. Ms. Vitale, a piano teacher, reduces her regular lesson rate by 12% for senior citizens. She also takes 5% off her regular rate for students who pay before the 15th of the month. If her regular rate is $15 per hour, how much would an hour of piano lessons cost a senior citizen who pays before the 15th?
3. Out of a class of 40 students, 30 percent are black, 25 percent are white, and the rest are Hispanic. How many students in this class are not Hispanic?
4. Out of the 1400 people scheduled to tour an automobile factory one day, 14% did not show up. 75% of these "no-shows" were from the Toronto Tour Group. How many people who did not make it to the auto factory tour were from this tour group?
5. Twenty-five percent of all household accidents are caused by carelessness. Of these, 35% involve children. Out of an estimated 240,000 household accidents per year, how many are caused by carelessness and also involve children?

ANSWERS AND SOLUTIONS START ON PAGE 350.

Problem Solving for the GED

Working with Item Sets

Miriam lives near the border between two states. She is planning to purchase a lawn mower that is priced at $149 in both states. State A charges a sales tax of 7%; state B charges no sales tax. If she buys the lawn mower in the state with no sales tax, she will need to have the mower delivered. The store's delivery charge is either $21 or 6% of the purchase price, whichever amount is higher. Miriam wants to spend as little money as possible to get the lawn mower.

1. How much will the mower cost in state A, including tax?
(1) $10.43
(2) $149.00
(3) $159.43
(4) $180.83
(5) $1043.00

2. How much will the mower cost in state B, including delivery?
(1) $21.00
(2) $157.94
(3) $170.00
(4) $180.43
(5) $204.08

As you know, there will be some "item sets" on the GED Math Test. An item set consists of a passage, graph, or chart that is followed by several questions. In order to answer the questions, you must be able to sort through all of the information given and choose the specific information you need. One good way to organize information is to jot down a rough chart.

As you read through the passage above, you can see that Miriam is comparing two purchases — one made in one state and another made in a different state. Notice how this simple chart can help you keep track of the facts. Jot down the information where it belongs; the work has been started for you.

	State A	**State B**
Cost of mower	$149	
Tax		
Delivery Charge		

Now that you have sorted the information, go back and answer the questions above. From which store can Miriam get the mower for less money?

Since the mower costs $159.43 in State A and $170 in State B, it will cost her less money in State A.

★ GED PRACTICE ★

Item Sets

Vicki and Brad Bachman are deciding which of two homes to buy. They can purchase a home in Lincoln for $47,700 with a 12% down payment. They can also purchase a home in Milwaukee for $58,900 with an 8% down payment. The Bachmans figured out that they would need to spend an additional $6200 for home improvements in the Lincoln home and $2945 in the Milwaukee home.

1. How much more of a down payment would the Bachmans have to pay for the Lincoln home than the Milwaukee home?
(1) $912
(2) $1012
(3) $4712
(4) $5724
(5) $11,200

2. What percent of the price of the Milwaukee home would the Bachmans have to pay for home improvements?
(1) two
(2) three
(3) five
(4) forty
(5) ninety

3. How much would the Bachmans pay for the Lincoln home, including the home improvements?
(1) $8669
(2) $11,924
(3) $50,645
(4) $53,900
(5) $54,912

ANSWERS AND SOLUTIONS START ON PAGE 350.

Simple Interest

Interest for Whole Years

Interest is money that money makes. You *are paid* interest when your money is in a savings account because the bank is using your money to make other investments. You *pay* interest when you borrow money. You not only pay your loan back, but you give an additional amount of money (interest) in payment for the bank's help. Interest is one of the most common applications of percents.

Rule for Finding Interest

To find interest you need three things:

1. the principal — the amount of money borrowed or saved,
2. the rate — the percent used to find the interest, and
3. the time — the number of years (or part of a year) that the money is borrowed or saved. Multiply the principal by the rate by the time.

This rule is often written as the formula **I** = **prt** where **I** is the interest, **p** is the principal, **r** is the rate, and **t** is the time. When letters are written side by side with no arithmetic sign between them, they should be multiplied together. To use this formula, set up a fraction multiplication problem. $I = p \times r \times t$

EXAMPLE 1: Find the interest on $800 at 6% annual interest for one year.

(1) $800 \times \frac{6}{100} \times 1 = ?$ **(2)** $\frac{\overset{8}{\cancel{800}}}{1} \times \frac{6}{\underset{1}{\cancel{100}}} \times \frac{1}{1} = \$48$

STEP 1. Set the problem up like a fraction problem. $800 is the principal. 6% is the rate. Write 6% as the fraction $\frac{6}{100}$. One year is the time.

STEP 2. Cancel and multiply across. The interest is **$48**.

TIP: If the time in an interest problem is exactly one year, you do not have to put 1 in the problem. The answer will be exactly the same.

EXAMPLE 2: Find the interest on $500 at $8\frac{1}{2}$% annual interest for one year.

STEP 1. Set the problem up like a fraction problem. $500 is the principal. $8\frac{1}{2}$% is the rate. Write $8\frac{1}{2}$% as a fraction with a denominator of 100.

$$500 \times \frac{8\frac{1}{2}}{100} = ?$$

STEP 2. Cancel 500 and 100 by 100. You now have the simplified problem $5 \times 8\frac{1}{2}$.

$$\frac{\overset{5}{\cancel{500}}}{1} \times \frac{8\frac{1}{2}}{\underset{1}{\cancel{100}}}$$

STEP 3. Change $8\frac{1}{2}$ to an improper fraction and multiply by 5. The interest is **$42.50**.

$$\frac{5}{1} \times \frac{17}{2} = \frac{85}{2} = \$42.50$$

Multi-Step Interest Problems

In some problems you will have to find a new amount. The new amount is the interest plus the original principal.

EXAMPLE 3: Roger borrowed $6000 at a rate of 12% for one year to make home improvements. How much did he have to repay at the end of the year?

STEP 1. $6000 is the principal, and 12% is the rate. Write 12% as a fraction $\frac{12}{100}$.

$$6000 \times \frac{12}{100} =$$

STEP 2. Cancel 6000 and 100 by 100 and multiply across. The interest for one year is $720.

$$\frac{\overset{60}{\cancel{6000}}}{1} \times \frac{12}{\underset{1}{\cancel{100}}} = \$720$$

STEP 3. Add the interest to the principal. Roger must repay **$6720**.

$$\begin{array}{r} \$6000 \\ +\ 720 \\ \hline \$6720 \end{array}$$

TIP: Read interest problems carefully to see whether you are being asked for just the interest or the interest plus the principal.

Interest for Parts of a Year

The time in a percent problem is measured in years. If the time in a problem is not one whole year, make a fraction that expresses the time as a part of a year. For example, six months is $\frac{6}{12}$ or $\frac{1}{2}$ year. Two years and four months is $2\frac{4}{12} = 2\frac{1}{3}$ years.

EXAMPLE: Find the interest on \$500 at 9% annual interest for eight months.

(1) 8 months = $\frac{8}{12} = \frac{2}{3}$ year **(2)** $\frac{\overset{5}{\cancel{500}}}{1} \times \frac{\overset{3}{\cancel{9}}}{\underset{1}{\cancel{100}}} \times \frac{2}{\underset{1}{\cancel{3}}} = \$30$

STEP 1. Change 8 months to a fraction of a year.
STEP 2. \$500 is the principal. 9%, or $\frac{9}{100}$, is the rate, and $\frac{2}{3}$ year is the time. Cancel and multiply across. The interest is **\$30**.

Percent Exercise 17

Directions: Solve each problem.

1. Find the interest on \$3000 at 12.5% annual interest for one year.
2. How much money will Sara have at the end of one year on \$800 deposited in a savings account earning $5\frac{1}{4}$% interest?
3. What is the interest on \$5000 at 9% annual interest for two years?
4. Find the interest on \$800 at 6% annual interest for nine months.
5. How much interest did Emilita pay on \$900 at 11.5% annual interest for six months?
6. The Millers paid interest on \$500 borrowed at 14% annual interest for one year and six months. How much interest did they pay?
7. Sally had \$2000 deposited in her savings account for two years and six months. If she earned 6% interest, how much was in the account at the end of that time?
8. The Lewis family borrowed \$900 at 13% annual interest for one year and eight months. How much did they repay at the end of that time?
9. To the nearest penny, find the interest on \$4000 at 10% annual interest for two years and four months.
10. If you borrow \$1200 at 14.75% annual interest for nine months, how much do you have to repay at the end of that period?

ANSWERS AND SOLUTIONS START ON PAGE 350.

Problem Solving for the GED

When Not Enough Information Is Provided

Greg wants to buy a coat that costs $89.50 plus clothing tax. He has $106.00 in his wallet, but he wants some of this cash to pay for lunch, which will cost $5.80 plus a 4% meal tax. How much money will Greg have left over after he buys the coat but before he buys lunch?

(1) $6.09
(2) $12.92
(3) $16.50
(4) $21.50
(5) Not enough information is provided.

The last answer choice in this item is one that will appear in some problems on the GED Math Test. It means that you are unable to answer the question because the passage has left out some information that you need to solve the problem.

In the example above, you cannot tell how much money Greg would have left over because you do not know the *total* cost of the coat. The situation states that the coat costs $89.50 *plus clothing tax*. Is the clothing tax 5%? 6%? You are not given this information. You cannot assume that the tax is the same rate that you have in your state. Also, do not be fooled by the *meal tax* of 4% given in the problem. This will help you figure out tax on food, but not on clothing.

Here are some hints to help you with problems in which *not enough information is provided* is listed as a possible answer choice:

1. Read the situation through once carefully, then read the question. As you learned in Chapter 2, this is always the first step in problem solving. Your second step is to decide what information you need to answer the question.

   In the problem above, you need to know the cost of the coat plus the tax.

2. If you find that you may be missing some necessary information, skim your answer choices to see if *not enough information* is listed. If it is, **do not** immediately choose this as your answer! Instead, jot down the fact that you are missing.

   In this problem, you are not given the amount of the clothing tax. Though this is a clue that (5) may be the right answer, you shouldn't choose it too quickly. Jot down *amount of tax on coat* as the information you are missing.

3. Look at the problem again to see if there is another way to get the missing information. If this is a multi-step problem, you may have another step in

order to find the information you need. Also see if some numbers appear as words or appear in a chart with the problem. Sometimes information expressed this way is harder to find.

In the problem above, there is no other way to find the clothing tax.

4. Once you feel that you understand the problem and are definitely missing some necessary information, choose answer (5).

WARNING: Don't choose *not enough information* if you are confused by a problem. Be sure you can identify *what information is missing* before you choose this option.

★ GED PRACTICE ★

Not Enough Information Is Provided

Directions: Questions 1–4 refer to the following situation.

Fred is a mason. He has been offered a new job. For the first three months he will make \$1250 a month. Then he will get a 10% raise for the remainder of the year. At the end of a year, he will receive an additional 8% raise if he does well on his performance review. Fred's employer withholds 15% of his gross salary for federal tax, state tax, and social security.

1. How much will Fred make the first year?
(1) \$15,000
(2) \$15,275
(3) \$15,750
(4) \$16,125
(5) Not enough information is provided.

2. How much is withheld each month, for the first three months, from Fred's salary?
(1) \$87.50
(2) \$127.50
(3) \$187.50
(4) \$206.25
(5) Not enough information is provided.

3. How much is withheld for the first year from Fred's gross salary for social security?
(1) \$1875
(2) \$1250
(3) \$ 875
(4) \$ 750
(5) Not enough information is provided.

4. What will be Fred's monthly gross salary for the second year if he does well on his performance review?
(1) $1385
(2) $1485
(3) $1350
(4) $1450
(5) Not enough information is provided.

ANSWERS AND SOLUTIONS START ON PAGE 351.

Percent Exercise 18

Directions: Questions 1–3 refer to the following situation.

Paul and Ruth plan to renew the lease for their apartment. Their current rent is $250 a month. They have a choice of a two-year renewal at an 8% increase or a three-year renewal at an increase of 10%. If they take the two year renewal and stay on for a third year, the rent will go up an additional 10%.

1. If they choose the two-year renewal, what will the monthly rent be for the second year?
(1) $258
(2) $266
(3) $270
(4) $278
(5) Not enough information is provided.

2. If they choose the three-year renewal, what will the monthly rent be for the second year?
(1) $275
(2) $270
(3) $265
(4) $260
(5) Not enough information is provided.

3. If they choose the two-year renewal, but stay on for a third year at an additional 10% rent, what will the monthly rent be for the third year?
(1) $307
(2) $297
(3) $277
(4) $270
(5) Not enough information is provided.

ANSWERS AND SOLUTIONS START ON PAGE 351.

Exercise 19: Review

Directions: Solve each problem.

1. Write the ratio 18:24 in simplest form.
2. Simplify the ratio $\frac{20}{36}$.
3. Find s in $\frac{s}{15} = \frac{6}{20}$.
4. What is the value of x in 5:12 = x:84?
5. Change the decimal .95 to a percent.
6. Write 3.2% as a decimal.
7. Write $\frac{5}{8}$ as a percent.
8. Change $37\frac{1}{2}\%$ to a fraction.
9. Find 2.7% of 360.
10. What percent of 35 is 21?
11. 75 is 20% of what number?
12. 150% of what number is 48?
13. Find the interest on $450 at 8% annual interest for nine months.

★ GED PRACTICE ★

Directions: Choose the correct answer for each of the following.

14. Pete's socks are not stored together in pairs. In his drawer there are 2 blue socks, 5 brown socks, 4 black socks, and 1 white sock. What is the probability that the first sock Pete grabs from the drawer will be black?
   (1) $\frac{1}{12}$
   (2) $\frac{1}{4}$
   (3) $\frac{1}{8}$
   (4) $\frac{1}{3}$
   (5) $\frac{2}{3}$

15. Consolidated Utilities' softball team won 12 games and lost 8 last season. What was the ratio of the number of games the team won to the number of games they played?
   (1) 3:2
   (2) 2:5
   (3) 3:5
   (4) 5:3
   (5) 5:2

16. Kate worked 40 hours at $4.50 per hour. She also worked 6 extra hours and earned $38.40 in overtime. At the same rate, how much can she make for 10 hours of overtime work?
   (1) $32
   (2) $51.20
   (3) $64
   (4) $76.80
   (5) $96

17. For every $5 in her budget, Mrs. Murphy spends $2 on food. Her weekly budget is $215. Which expression below represents the amount she spends each week on food?

(1) (5 × 215)/2
(2) (5 + 2)/215
(3) (5 × 2)/215
(4) (5 × 2 × 215)
(5) (2 × 215)/5

18. The ratio of the number of people who voted to the number of voters who stayed home was 3:5 in the last election in Midvale. Out of 38,000 registered voters, how many actually voted in the last election?

(1) 10,133
(2) 14,250
(3) 23,750
(4) 28,500
(5) 47,500

19. Mr. Santiago sells shoes for a 5% commission. He earned $115 in commissions one week. Find the total value of the shoes he sold that week.

(1) $5.75
(2) $57.50
(3) $575.00
(4) $580.75
(5) $2300.00

20. Adrian wants to buy a used car that costs $1200. He has already saved $1000. What percent of the total price has he already saved?

(1) $16\frac{2}{3}\%$
(2) 20%
(3) 80%
(4) $83\frac{1}{3}\%$
(5) $16\frac{2}{3}\%$

21. The price of a gallon of gasoline went from $1.20 down to $1.05. By what percent did the price of a gallon drop?

(1) $12\frac{1}{2}\%$
(2) 15%
(3) 20%
(4) 85%
(5) $87\frac{1}{2}\%$

22. A stereo listed for $350 is on sale for 20% off the list price. The sales tax is 6%. What total price does a customer pay for the stereo if he buys it on sale?

(1) $324.00
(2) $259.00
(3) $262.20
(4) $296.80
(5) $301.00

Directions: Questions 23 through 25 refer to the following situation.

Mr. Allen wants to buy a new band saw. The saw is on sale in his town for $695. In his town he must pay 6% sales tax and a $50 shipping charge. The saw is for sale in another town across the state line. In that town the price of the saw is $749, but there is no sales tax in that state. The out-of-state dealer charges a 10% shipping charge. Mr. Allen estimates that if he borrows a friend's truck and picks up the saw himself, gas will cost $12 each way.

23. Find the total cost of the band saw including tax and shipping in Mr. Allen's hometown.
(1) $816.70
(2) $786.70
(3) $745
(4) $736.70
(5) Not enough information is provided.

24. What is the total cost of the band saw in the other state including gas and tolls if Mr. Allen picks up the saw himself?
(1) $761
(2) $773
(3) $827.90
(4) $847.90
(5) Not enough information is provided.

25. To the nearest dollar, what is the shipping cost on the band saw if it is purchased from the out-of-state dealer?
(1) $80
(2) $75
(3) $70
(4) $42
(5) Not enough information is provided.

ANSWERS AND SOLUTIONS START ON PAGE 351.

Review Evaluation

Problem	**Section**	**Starting Page**
1, 2, 15	Ratio	119
3,4	Proportion	121
5, 6	Decimals and Percents	129
7, 8	Fractions and Percents	132
9–12	Proportion and Percents	135
13	Simple Interest	149
14	Probability	115
16	Proportion (*Finding Necessary Information*)	121/*22**
17	Proportion (*Set-up Questions*)	121/*42**
18	Two-Part Proportion Problems	126
19, 20, 21	Percent Word Problems	140
22	Multi-step Percent Problems	142
23–25	Item Sets	148

**These skills are covered in earlier chapters.*

Passing score: 20 right out of 25 problems.

Your score: ______ right out of 25 problems.

If you had less than a passing score, review the sections for those problems you missed. Then redo these questions before going on to the measurements section.

If you had a passing score, correct any problem you got wrong. Then go on to the measurements section.

8 MEASUREMENT

Standard or Imperial Measurements

Below are some standard units of measurement for length, time, liquid measure, and weight. These are the units commonly used in the U.S. This system is also called the ***imperial system*** because it was the standard in the British empire.

In the chart below, the larger unit of measurement is on the left. Next to it on the right is an equivalent in smaller units. Abbreviations are in parentheses.

Take the time to memorize any of the units and equivalencies that you do not know. You will need to know these units for problems later in this section and on the GED test. They are also helpful to know for everyday life!

Measures of Length

1 foot (ft)	=	12 inches (in)
1 yard (yd)	=	36 inches
1 yard	=	3 feet
1 mile (mi)	=	5280 feet
1 mile	=	1760 yards

Measures of Time

1 minute (min)	=	60 seconds (sec)
1 hour (hr)	=	60 minutes
1 day	=	24 hours
1 week (wk)	=	7 days
1 year (yr)	=	365 days

Liquid Measures

1 pint (pt)	=	16 ounces (oz)
1 cup	=	8 ounces
1 pint	=	2 cups
1 quart (qt)	=	2 pints
1 gallon (gal)	=	4 quarts

Measures of Weight

1 pound (lb)	=	16 ounces (oz)
1 ton (T)	=	2000 pounds

Converting Measurements

In some word problems on the GED Test, and in many situations you encounter in everyday life, you need to be able to change one unit of measurement into another. For instance, if you needed a certain number of ounces of liquid, and the liquid is being sold by the quart, you would have to ***convert*** (change) the number of ounces needed into the number of quarts needed in order to figure out how much to buy.

Method 1: Using Proportion to Convert Measurements

In other chapters, you have seen how useful proportion is. You can also use proportion to change (or convert) from one unit of measurement to another.

> **TIP:** Remember from your work with proportions that each ratio must have the same label on top and the same label on the bottom.

EXAMPLE 1: Change 5 pounds to ounces.

(1) $\frac{\text{pounds}}{\text{ounces}} \quad \frac{1\text{ lb}}{16\text{ oz}} = \frac{5\text{ lb}}{x}$ **(2)** $\begin{array}{r} 16 \\ \times 5 \\ \hline 80 \end{array}$ **(3)** $1\overline{)80}$ = 80 oz

STEP 1. To start, you must know how many ounces are in one pound. Show this relationship in the ratio of pounds to ounces. On the left is the ratio of 1 lb to 16 oz. On the right put 5 lb in the pounds' position. Let x stand for the number of ounces you need to find.

STEP 2. Find the cross product of 16 and 5.

STEP 3. Divide 80 by the remaining number in the proportion, 1.
5 pounds is equal to **80 ounces**.

Rules for Changing Units of Measurement

1. Figure out the equivalency between the two units. Show this relationship in a ratio on the left. On the right side of the proportion, make a ratio using the unit of mesurement you do know. Use the letter x to represent the missing unit.
2. Find the cross product.
3. Divide by the remaining number in the proportion.

EXAMPLE 2: Dahlia is making soup that calls for 10 quarts of liquid. She needs to know how many gallons this is so that she can choose a large enough pot. How many gallons is 10 quarts?

(1) $\frac{\text{gallons}}{\text{quarts}} \quad \frac{1\text{ gal}}{4\text{ qt}} = \frac{x}{10\text{ qt}}$ **(2)** $\begin{array}{r} 10 \\ \times 1 \\ \hline 10 \end{array}$ **(3)** $4\overline{)10} = 2\frac{2}{4} = 2\frac{1}{2}\text{ gal}$

STEP 1. Make a proportion. On the left is the ratio of 1 gallon to 4 quarts. On the right, put x in the gallons' place and 10 in the quarts' place.

STEP 2. Find the cross product of 1 and 10.

STEP 3. Divide 10 by the other number in the proportion, 4.
10 quarts is equal to **$2\frac{1}{2}$ gallons.**

As you can see, you cannot set up the correct proportion if you do not know equivalent measurements. Make sure you know the equivalent units on pages 158–59 before you go any further in this chapter.

TIP: When you are converting a larger unit into smaller units, (feet into inches, for example) the number of units obtained will be greater than the number of units you started with. When you are converting smaller units into larger units, (inches into feet, for example) there will be a smaller number of units than what you started with.

Method 2: Multiplying and Dividing to Convert Measurements

The proportion method of converting measurement is very useful in most cases. However, there are some times when you just "know" to multiply or divide based on your everyday use of measurement.

EXAMPLE 1: Michelle bought 3 feet of tape. How many inches of tape is this?

SOLUTION. You know that there are 12 inches in 1 foot. To find the number of inches in 3 feet, you can simply multiply 3 by 12.
$12 \times 3 =$ **36 inches**

> **TIP:** *Multiply* when changing from a larger to a smaller unit (e.g., from feet to inches). This is because you will be getting *more* of the smaller unit.

EXAMPLE 2: Ralph worked on an assignment for 150 minutes. How many hours was this?

SOLUTION. You know that there are 60 minutes in an hour. To find the number of hours in 150 minutes, divide 150 by 60.
$150 \div 60 = \mathbf{2\frac{1}{2}}$ **hours**

> **TIP:** *Divide* when changing from a smaller unit to a larger unit (e.g., minutes to hours). This is because you will be getting *fewer* of the larger unit.

Measurement Exercise 1

Directions: Change each measurement to the new unit indicated. Use whatever method you find easiest.

1. 40 ounces = ________ pounds

2. 3 tons = ________ pounds

3. 6 feet = ________ yards

4. 23 days = ________ weeks

5. 14 ounces = ________ pound

6. 6 gallons = ________ quarts

7. 2640 feet = ________ mile

8. 100 inches = ________ yards

9. 20 ounces = ________ cups

10. 30 quarts = ________ pints

11. 3 miles = ________ feet

12. 2 yards = ________ feet

13. A group of neighbors cooked 130 quarts of tomatoes. They wanted to can them in gallon jars. How many gallon jars did they need for canning?

14. Pieter and his son went camping for three whole days. Altogether how many hours did they camp?

15. Deborah's car was double parked for 45 minutes. For what fraction of an hour was her car double parked?

16. Paul climbed a mountain which is 10,560 feet high. Find the height of the mountain in miles.

17. On Thursday, Pat typed steadily for 200 minutes, and she is paid by the hour. For how many hours should Pat be paid for this typing job?

18. Frank has to carry 3 tons of cement. If his truck carries a maximum of 500 pounds at a time, how many trips will Frank have to make?

ANSWERS AND SOLUTIONS START ON PAGE 352.

Metric Measurements

The ***metric system*** is used in most countries outside the U.S. In many ways the metric system is much easier to use than the standard system. Metric units are based on tens, hundreds, thousands, and so on.

- The basic unit of weight in the metric system is the ***gram***. A gram is about $\frac{1}{450}$ of a pound.
- The basic unit of length in the metric system is the ***meter***. A meter is just a little longer than a yard.
- The basic unit of liquid measure in the metric system is the ***liter***. A liter is a little greater than a quart.

Following are four prefixes used in metric measurements. Learn their meanings before you go on.

Prefix	**Meaning**
milli-	$\frac{1}{1000}$ or .001
centi-	$\frac{1}{100}$ or .01
deci-	$\frac{1}{10}$ or .1
kilo-	1000

EXAMPLE 1: One kilometer = ________ meters.
SOLUTION. "Kilo" means 1000. One kilometer = 1000 meters.

EXAMPLE 2: One millimeter = ________ meter.
SOLUTION. "Milli" means $\frac{1}{1000}$. One millimeter = $\frac{1}{1000}$ meter.

Below are the most common metric measurements. Take the time now to learn these units before you go on.

Measures of Length

1 meter (m)	=	1000 millimeters (mm)
1 meter	=	100 centimeters (cm)
1 kilometer (km)	=	1000 meters

Measures of Weight

1 gram (g)	=	1000 milligrams (mg)
1 gram	=	100 centigrams (cg)
1 kilogram (kg)	=	1000 grams

Liquid Measurement

1 liter (L)	=	1000 milliliters (ml)
1 liter	=	100 centiliters (cl)
1 liter	=	10 deciliters (dl)

Method 1: Moving the Decimal Point to Convert Measurements

You can change metric measurements by simply moving the decimal point to the right or the left. Since the metric system is based on the decimal system, all conversion can be completed without messy computations.

Keep the following chart in mind as you convert metric measurements.

kilo	hecto	deka	unit	deci	centi	milli

For example, how can you change 5 meters to centimeters? Look at the chart. A meter is the basic unit of length in the metric system, so start at *unit* on the chart above. Next, find *centi* on the chart. As you can see, it is two spaces to the right. To convert meters to centimeters, therefore, move the decimal point two places to the right.

To change 6400 grams to kilograms, move the decimal point three decimal places to the left. On the chart, *kilo* is three spaces to the left of the unit space.

> **TIP:** To go from a larger unit to a smaller unit, move the decimal point to the *right*.
>
> To go from a smaller unit to a larger unit, move the decimal point to the *left*.

Method 2: Using Proportion to Convert Metric Measurements

To change from one metric unit to another, you can also use a proportion as you did with standard measurements.

EXAMPLE 1: Change 5 meters into centimeters.

(1) $\dfrac{\text{meters}}{\text{centimeters}} \quad \dfrac{1\text{ m}}{100\text{ cm}} = \dfrac{5\text{ m}}{x}$ (2) $\begin{array}{r} 100 \\ \times\ 5 \\ \hline 500 \end{array}$ (3) $1\overline{)500}$ = **500**

STEP 1. Make a proportion. On the left is the ratio of 1 meter to 100 centimeters. On the right put 5 in the meters' place and x in the centimeters' place.

STEP 2. Find the cross product of 100 and 5.

STEP 3. Divide 500 by the remaining number in the proportion, 1.
5 meters is equal to **500 centimeters**.

EXAMPLE 2: Silas Chemical Company needs to send 6400 grams of a certain chemical to a pharmacy. For shipping purposes, the weight must be recorded in kilograms. How many kilograms does Silas need to send?

(1) $\frac{1\text{ kg}}{1000\text{ g}} = \frac{x}{6400}$ **(2)** $\begin{array}{r} 6400 \\ \times\ 1 \\ \hline 6400 \end{array}$ **(3)** $1000\overline{)6400.0}$ = **6.4**

STEP 1. Set up a proportion. On the left is the ratio of 1 kilogram to 1000 grams. On the right put 6400 in the grams' place and x in the kilograms' place.

STEP 2. Find the cross product of 1 and 6400.

STEP 3. Divide 6400 by the remaining number in the proportion, 1000. 6400 grams equal **6.4 kilograms**.

Measurement Exercise 2

Directions: Change each measurement to the new unit indicated. Use the proportion method if it is helpful, or try moving the decimal point.

1. 915 milliliters = ________ liters
2. 50 centimeters = ________ meter
3. 2.4 kilometers = ________ meters
4. 10 grams = ________ milligrams
5. .5 liter = ________ centiliters
6. 200 meters = ________ kilometer
7. Express 125 centigrams of gold dust in grams.
8. Julia bought 3 kilograms of apples. Find the weight of the apples in grams.
9. During an experiment, Mark bottled 4.8 liters of solution. If he wanted to store the solution in deciliter jars, how many jars would he need?
10. Heather's garden is 8750 millimeters long. To buy the right amount of fertilizer, she needs to figure out how many meters she needs to cover. What is the length of her garden in meters?

ANSWERS AND SOLUTIONS START ON PAGE 353.

Problem Solving for the GED

Comparing and Ordering Measurements

A cook has the following amounts of ingredients to add to a mixture. He wants to combine them starting with the *smallest* amount and ending with the *largest*. The ingredients have the following measurements:

Bowl A — 2 cups

Bowl B — $1\frac{1}{2}$ pints

Bowl C — $3\frac{1}{2}$ cups

Bowl D — $2\frac{1}{2}$ pints

Bowl E — 1 quart

Which of the following sequences correctly lists the order in which the ingredients should be combined?

(1) B, A, C, D, E
(2) A, B, C, E, D
(3) C, B, A, E, D
(4) B, A, D, C, E
(5) B, C, E, D, A

1. Before you can solve this problem, you need to convert some of the measurement units so that the units are all the same. This will make the measurements easier to compare.

   For this problem, first decide whether to express all measurements in cups, pints, or quarts. Try to choose one that will involve fewer conversions. In this example, we will change all measurements to cups.

2. Perform the conversions by whatever method you find easiest.

   Bowl A: 2 cups

   Bowl B: $1\frac{1}{2}$ pints = 3 cups

   Bowl C: $3\frac{1}{2}$ cups

   Bowl D: $2\frac{1}{2}$ pints = 5 cups

   Bowl E: 1 quart = 4 cups

3. Now that all measurements are expressed in the same units, put them in the specified order.

   This problem calls for *smallest* to the *largest*.

   Bowl A: 2 cups

   Bowl B: $1\frac{1}{2}$ pints = 3 cups

   Bowl C: $3\frac{1}{2}$ cups

   Bowl E: 1 quart = 4 cups

   Bowl D: $2\frac{1}{2}$ pints = 5 cups

4. Choose the answer showing the correct order.

(2) A, B, C, E, D

★ GED PRACTICE ★

Comparing and Ordering Measurements

Lyle has five different lengths of board. He wants to stack them in order of length with the *longest* board on bottom and the *shortest* board on top. The boards are the following length:

Board A — 2 feet
Board B — 20 inches
Board C — 1 yard 2 inches
Board D — 40 inches
Board E — 1 yard 2 feet

Which of the following sequences correctly lists the order in which the boards should be stacked?

(1) A, B, E, D, C
(2) E, D, C, A, B
(3) B, D, A, C, E
(4) E, C, A, D, B
(5) E, A, B, C, D

ANSWERS AND SOLUTIONS START ON PAGE 354.

Simplifying Measurements

When you add and multiply standard measurements, you will sometimes have to **simplify** your answers.

Suppose the answer to a problem is 15 inches. Since 15 inches is more than one foot, you can change the answer to feet and inches. This is called *simplifying* the solution.

EXAMPLE 1: Change 15 inches to feet and inches.

$$\begin{array}{r} 1 \text{ ft } 3 \text{ in} \\ 12\overline{)15\phantom{ \text{ ft } 3 \text{ in}}} \\ \underline{12}\phantom{ \text{ ft } 3 \text{ in}} \\ 3\phantom{ \text{ ft } 3 \text{ in}} \end{array}$$

STEP 1. The next unit of measurement larger than inches is feet. Divide 15 inches by 12, the number of inches in one foot.

STEP 2. Write the remainder in inches. 15 inches = **1 ft 3 in**. The answer could also be expressed as $1\frac{3}{12}$ = $\mathbf{1\frac{1}{4}}$ **ft**.

Rule for Simplifying Measurements

1. Divide the number of units in your answer by the number of these units contained in the *next larger* unit of measurement.
2. Write the remainder in terms of the smaller unit.

EXAMPLE 2: Simplify 150 minutes.

$$\begin{array}{r} \text{2 hr 30 min} \\ 60\overline{)150} \\ \underline{120} \\ 30 \end{array}$$

STEP 1. The next unit of measurement larger than minutes is hours. Divide 150 minutes by 60, the number of minutes in an hour.

STEP 2. Write the remainder in minutes. 150 minutes = **2 hr 30 min.** The answer could also be expressed as $2\frac{30}{60} = 2\frac{1}{2}$ **hr**.

In some problems you will find the remainder written in terms of the smaller unit. In other problems you will find the remainder changed to a fraction or a decimal.

Measurements Exercise 3

Directions: Solve each problem. Read carefully to find out how each answer should be expressed.

1. Change 19 inches to feet and inches.

2. Change 75 seconds to minutes and seconds.

3. Change 36 ounces to pounds and ounces.

4. Change 25 feet to yards and a fraction of a yard.

5. Change 200 minutes to hours and minutes.

6. Change 4500 pounds to tons and pounds.

7. Change 15 quarts to gallons and a fraction of a gallon.

8. Change 75 inches to yards and inches.

9. Change 15,840 feet to miles and feet.

10. Change 100 ounces to pounds and a fraction of a pound.

11. Maria made 9 pints of jam. Express the amount in quarts and pints.

12. Sam hiked for 40 days. Express the time in weeks and days.

ANSWERS AND SOLUTIONS START ON PAGE 354.

Arithmetic Operations

Standard Measurements

Adding, subtracting, multiplying, and dividing standard measurements require special carrying and borrowing steps. Units of measurement must be changed into the type of unit in the next column when you borrow or carry. The following are examples of the four basic operations with standard measurements. Read through each example carefully.

EXAMPLE 1: Find the sum of 2 ft 8 in, 4 ft 11 in, and 1 ft 9 in.

(1)	(2)	(3)
2 ft 8 in	2 ft 4 in	7 ft
4 ft 11 in	12)28	+2 ft 4 in
+1 ft 9 in	24	**9 ft 4 in**
7 ft 28 in	4	

STEP 1. Line up feet under feet and inches under inches. Add the units separately. **Do not** do any carrying.
STEP 2. Since 28 inches is more than one foot, you must simplify as you have just learned to do. To change 28 inches to feet and inches, divide 28 by 12, the number of inches in one foot.
STEP 3. Add the 2 ft 4 in from Step 2 to the 7 ft you got in Step 1.

EXAMPLE 2: An athlete has walked 5 miles 575 feet of the 10 miles he plans to cover in the annual Walk for Hunger march. How much further does he have to go?

(1)	(2)	(3)
10 mi	10 mi = 9 mi 5280 ft	9 mi 5280 ft
−5 mi 575 ft		−5 mi 575 ft
		4 mi 4705 ft

STEP 1. Put 10 miles on top. Line up miles under miles.
STEP 2. Notice that there is nothing to subtract 575 feet from. Borrow 1 mile from 10 miles and rewrite the amount as 9 mi 5280 ft. (Remember that 1 mi = 5280 ft.)
STEP 3. Subtract 575 feet from 5280 feet, and subtract 5 miles from the remaining 9 miles. The athlete has **4 miles 4705 feet** left to walk.

EXAMPLE 3: For a large batch of potato salad, a recipe calls for 5 lb 10 oz of potatoes. A restaurant in town goes through 3 batches every day. What weight of potatoes do they use in one day?

(1)
$$\begin{array}{r} 5\text{ lb }10\text{ oz} \\ \times \quad 3 \\ \hline 15\text{ lb }30\text{ oz} \end{array}$$

(2)
$$\begin{array}{r} 1\text{ lb }14\text{ oz} \\ 16\overline{)30} \\ \underline{16} \\ 14 \end{array}$$

(3)
$$\begin{array}{r} 15\text{ lb} \\ +\ 1\text{ lb }14\text{ oz} \\ \hline \mathbf{16\text{ lb }14\text{ oz}} \end{array}$$

STEP 1. Multiply the pounds and ounces separately.

STEP 2. Since 30 ounces is more than a pound, you must simplify. Change 30 ounces to pounds and ounces. Divide 30 by 16, the number of ounces in a pound.

STEP 3. Add the 15 lb from Step 1 to the 1 lb 14 oz from Step 2. The restaurant uses **16 pounds 14 ounces** of potatoes.

EXAMPLE 4: Lorraine and Flo used 7 gal 2 qt of laundry detergent in one year. About how much would each woman use by herself?

(1)
$$\begin{array}{r} 3\text{ gal} \\ 2\overline{)7\text{ gal }2\text{ qt}} \\ \underline{-6} \\ 1\text{ gal} \end{array}$$

(2) 1 gal = 4 qt

(3)
$$\begin{array}{r} \mathbf{3\text{ gal} \quad 3\text{ qt}} \\ 2\overline{)7\text{ gal} \quad 2\text{ qt}} \\ \underline{-6} \\ 1\text{ gal} = \underline{4\text{ qt}} \\ 6\text{ qt} \\ \underline{6\text{ qt}} \\ 0 \end{array}$$

STEP 1. Divide 2 into 7 gal.

STEP 2. Change the remainder, 1 gal, into quarts. 1 gal = 4 qt.

STEP 3. Add the 4 qt remainder to the 2 qt in the problem. Divide 2 into the new total, 6 qt. Each woman would use **3 gal 3 qt** of detergent.

In some measurement problems, you will need to convert units of measurement before you can do any computation.

EXAMPLE 5: How many cans, each holding 1 lb 8 oz of grain, can be filled from a bag that holds 24 pounds?

(1) 1 lb 8 oz =

$1\frac{8}{16} =$

$1\frac{1}{2}$ lb

(2) $24 \div 1\frac{1}{2} =$

$\frac{24}{1} \div \frac{3}{2} =$

$\frac{24}{1} \times \frac{2}{3} =$

16

STEP 1. Change 1 lb 8 oz into pounds so that you will have the same units of measurement in the problem.

STEP 2. To find out how many cans can be filled, divide 24 pounds by $1\frac{1}{2}$ pounds. The number of cans is **16**.

Problem Solving for the GED

What More Do You Need to Know?

Eddie bought 24 feet 6 inches of wire to build a chicken coop. He also bought 36 inches of steel rod. What more do you need to know to find out how much he paid for the wire?
(1) the amount he paid for the steel
(2) the total cost of his purchases
(3) the cost per foot of the wire
(4) the distance around the chicken coop
(5) the difference in price between the steel and the wire

There may be some questions like this on the GED Test — questions that ask you what other information is needed to answer a given question. In this problem, it is impossible to figure out how much Eddie paid for the wire with just the information given in the problem. What more do you need to know?

To find the total paid for a certain length of anything, you know you must multiply the length by the cost per unit. You are given the length of the wire Eddie bought, but you do not know how much the store charges for a given unit of the wire. Therefore, what you need to know is the *cost per foot of the wire*. The answer is **(3)**.

For practice, look at each of the other answer choices and see why the information in each would *not* help you find out how much Eddie paid for the wire.

★ GED PRACTICE ★

What More Do You Need to Know?

In one week, a clerk earned $81.00 in overtime pay. She also put in her regular 40-hour work week. What more do you need to know to find out the total amount of money the clerk made that week?
(1) the number of overtime hours she worked
(2) her regular hourly wage
(3) the amount per hour she is paid for overtime
(4) her total hours that week
(5) how many days she works per week

ANSWERS AND SOLUTION ARE ON PAGE 354.

Measurements Exercise 4

Directions: Solve each problem. Be sure every answer is simplified.

1. Find the sum of 4 ft 9 in and 2 ft 4 in.

2. Subtract 1 min 35 sec from 4 min 20 sec

3. Multiply 8 yd 2 ft by 5.
4. Add 2 pints 12 ounces and 10 ounces.
5. Divide 10 min 15 sec by 3.
6. Subtract 500 feet from 2 miles.
7. Multiply 2 yd 5 ft 8 in by 3.
8. Find the sum of 2 tons 500 lb, 1 ton 900 lb, and 8 tons 1200 lb.
9. A seamstress gets paid \$7.40 for each skirt she alters. She can also earn an extra \$2.50 per skirt for ironing. What more do you need to know to find out how much money she earned if she worked all day Wednesday?
   (1) her total hours worked
   (2) the number of skirts she altered and the number she ironed
   (3) her hourly wage
   (4) just the number of skirts she altered
   (5) just the number of skirts she ironed
10. For a party punch, Herb added 4 gal 2 qt of fruit juice to 3 gal 3 qt of ginger ale. If his punch bowl holds $10\frac{1}{2}$ gallons of liquid, what is the greatest amount of lemonade he can add without the bowl overflowing?
11. Ginny has completed 1 day 12 hr and 30 min of her train trip across the western United States. If the train is due to arrive at her final destination 3 days 10 hr and 15 min after her departure, how much more time does she have on the train?
12. Mrs. Lewis bought 1 lb 9 oz of shrimp for a dinner party. If there are 5 people eating, how much shrimp will each person get?
13. Wayne had a rope that measured 8 yd 2 ft 8 in. First he cut off a piece that was 1 yd 2 ft 10 in long, then he divided the remaining rope into five pieces. How long is each of the five pieces?
14. Out of 2 gallons of her homemade bubble bath, Margaret kept 3 quarts for herself. She then packaged the rest in pint-size jars for gifts. How many jars did she fill?
15. A board that is 10 yd 2 ft long is divided into 8 boards. From each of those boards, 2 feet is cut off. How long is each of the eight boards now?
16. At a price of \$3.60 a pound, how much do 2 lb 4 oz of cheese cost?

ANSWERS AND SOLUTIONS START ON PAGE 354.

Metric Measurements

Computation is generally easier in the metric system than it is in the standard system. Because it is based on the decimal system, metric system computation does not require complicated conversions. As you saw earlier, much of the work can be done by simply moving around the decimal point.

EXAMPLE 1: Find the sum of 366 grams and 2 kilograms. Express your answer in kilograms.

(1) 366 grams = .366 kilograms **(2)**

$$\begin{array}{r} 2. \\ +\ .366 \\ \hline \mathbf{2.366\ kg} \end{array}$$

STEP 1. Change 366 grams to kilograms by either making a proportion or moving the decimal point.

STEP 2. Line up 2 kg and .366 kg with point under point, and add.

EXAMPLE 2: Find the sum of 366 grams and 2 kilograms. Express your answer in grams.

(1) 2 kilograms = 2000 grams **(2)**

$$\begin{array}{r} 2000 \\ +\ 366 \\ \hline \mathbf{2366\ grams} \end{array}$$

STEP 1. Change kilograms to grams by either making a proportion or moving the decimal point.

STEP 2. Line up 366 g and 2000 g and add.

Measurements Exercise 5

Directions: Solve each problem.

1. Find the sum of 2.75 L and .495 L.

2. Subtract .6 km from 10 km.

3. Multiply 2.45 km by 3.

4. Divide 2 kg by 5.

5. Divide 5 kg by 2.

6. Take 200 mg from 5 g. First express your answer in grams. Then express your answer in milligrams.

7. Divide 4 m by 16.

8. A baker takes a loaf of bread 2 meters long and cuts off a piece measuring 25 centimeters. He then takes the remaining piece and divides it into six equal pieces. How can you find the length of each of the six pieces?
(1) Divide 2 meters by 6.
(2) Subtract 25 centimeters from 2 meters and multiply by 6.
(3) Divide 25 centimeters by 6 and subtract it from 2 meters.
(4) Add 2 meters and 25 centimeters and divide by 6.
(5) Subtract 25 centimeters from 2 meters and divide by 6.

9. Henry has .9 L of oil and he pours out 785 mL of it for work on his motorcycle. How much oil does he have left over? First express your answer in liters. Then express your answer in milliliters.

10. A fruit stand had 20 containers of its own orange juice bottled. Each container held 2.35 L of liquid. If the fruit stand paid $.40 for each liter of juice bottled, how much did it pay in all for the 20 containers?

11. Thelma participated in the Annual Race for Cancer Research by running in the 3 km race, the 450 m race, and the 100 m race. How many meters did she run altogether? How many kilometers did she run?

12. A drugstore owner has .035 kg of aspirin to divide into four bottles. How many kilograms will each bottle contain? Round your answer off to the nearest thousandth.

ANSWERS AND SOLUTIONS START ON PAGE 355.

Changing Metric to Standard and Standard to Metric Measurements

Some questions on the GED may ask you to change a metric unit to a standard unit or a standard unit to a metric unit. Don't panic! You will *always* be given an equivalent measure. This equivalent measure is a ratio that you can use to set up a proportion and solve the problem.

EXAMPLE: One mile is equal to 1.6 kilometers. How many miles is 10 kilometers?

(1) $\frac{1}{1.6} = \frac{x}{10}$

(2)

$$\begin{array}{r} \mathbf{6.25} \\ 1.6\overline{)10.0} \\ \underline{9\,6} \\ 4\,0 \\ \underline{3\,2} \\ 80 \\ \underline{80} \\ 0 \end{array}$$

STEP 1. Set up a proportion with miles on top of the ratios and kilometers on the bottom. You are given one ratio in the problem.

STEP 2. Find one cross product and divide by the remaining term. 10 kilometers = **6.25 miles**

Measurements Exercise 6

Directions: Solve the following problems. Remember that your first step should be to set up a proportion.

1. Philip poured eight quarts of soup into a pot that is measured in liters. If one quart is equal to .946 liter, how much soup did Philip have, measured in liters?

2. One meter equals 39.37 inches. A rod of steel measures 100 meters. How many inches is this?

3. There are 28.4 grams in one ounce. Measured in ounces, how much is 2272 grams?

4. There are 2.54 centimeters in an inch. Sandra has two feet of ribbon. How many centimeters does she have?

ANSWERS AND SOLUTIONS START ON PAGE 355.

Whole Number Answers in Measurement

Sometimes you will see by the way a problem is worded that *only a whole number* answer is correct. However, when you work out the problem, you may come up with an answer that includes a fraction or a remainder. Look at the following example:

EXAMPLE: How many class sessions, each 1 hr 45 min long, fit into an 8-hour day?

(1) 3
(2) 4
(3) 5
(4) 7
(5) 8

(1) 1 hr 45 min = $1\frac{3}{4}$ hr

(2) $8 \div 1\frac{3}{4} =$

$8 \div \frac{7}{4} =$

$8 \times \frac{4}{7} =$

$\frac{32}{7} = 4\frac{4}{7}$

STEP 1. Convert 1 hr 45 min to hours.
STEP 2. Divide the 8 hour day by $1\frac{3}{4}$ hours.
STEP 3. In this case, the question asks for the number of class sessions, so the fraction of a class section ($\frac{4}{7}$) is not necessary. Only 4 class sessions can be fit in this amount of time. Therefore, choice **(2)** is correct.

Scales and Meters

Reading Rulers

You may have to read inch or centimeter rulers on the GED. Figure 1 below shows a six-inch ruler.

Figure 1

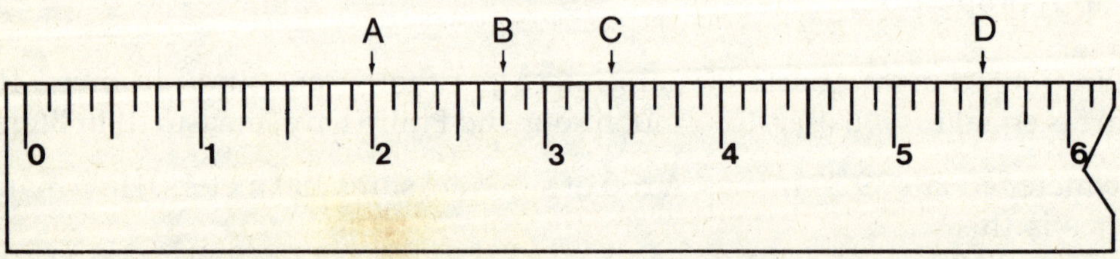

The longest lines on the ruler are inch lines. They are numbered 1, 2, 3, etc. The second longest lines are $\frac{1}{2}$ inch lines. The next longest lines are $\frac{1}{4}$ inch lines. The shortest lines are $\frac{1}{8}$ inch lines. To read a length on the ruler, decide how far to the right of zero a point is.

EXAMPLE 1: Tell how far to the right of zero the points marked A, B, C, and D on Figure 1 are.

SOLUTION. Point A — 2 inches.
Point A is at the line labeled 2.

Point B — $2\frac{3}{4}$ inches.
Point B is between 2 and 3 inches. The point is at the third $\frac{1}{4}$ inch line between 2 and 3.

Point C — $3\frac{3}{8}$ inches.
Point C is between 3 and 4 inches. The point is at the third $\frac{1}{8}$ inch line between 3 and 4.

Point D — $5\frac{1}{2}$ inches.
Point D is between 5 and 6 inches. The point is at the longest line between 5 and 6 which is the $\frac{1}{2}$ line.

Figure 2 below shows a 15-centimeter scale.

Figure 2

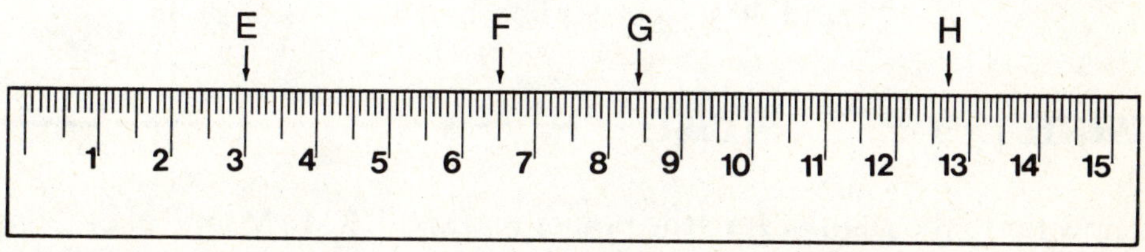

The longest lines on this scale are the centimeter lines, labeled 1, 2, 3, etc. The next longest lines are the $\frac{1}{2}$ centimeter or .5 centimeter lines. The shortest lines are millimeter lines or .1 centimeter lines. Notice in the next example that the distances on the centimeter scale are all given in decimals.

EXAMPLE 2: Tell how far to the right of zero the points marked E, F, G, and H on figure 2 are.

SOLUTION. Point E — 3 centimeters.
Point E is at the third longest line to the right of zero.

Point F — 6.5 centimeters.
Point F is between 6 and 7 centimeters. The point is at the middle line which is .5 centimeter.

Point G — 8.4 centimeters.
Point G is between 8 and 9 centimeters. The point is at the fourth millimeter or .4 centimeter line.

Point H — 12.7 centimeters.
Point H is between 12 and 13 centimeters. The point is at the seventh millimeter or .7 centimeter line.

Reading Meters

You may also have to read meters on the GED Test. A meter is a device used to measure time, distance, speed, or amount of energy. For example, a thermo*meter* measures temperature, and a speedo*meter* measures speed. You do not need to be familiar with each type of meter in order to answer questions on the test. However, you must be able to read carefully.

Figure 3 below is a meter which measures amperes. **Amperes** are a measure of an amount of electrical current.

Figure 3

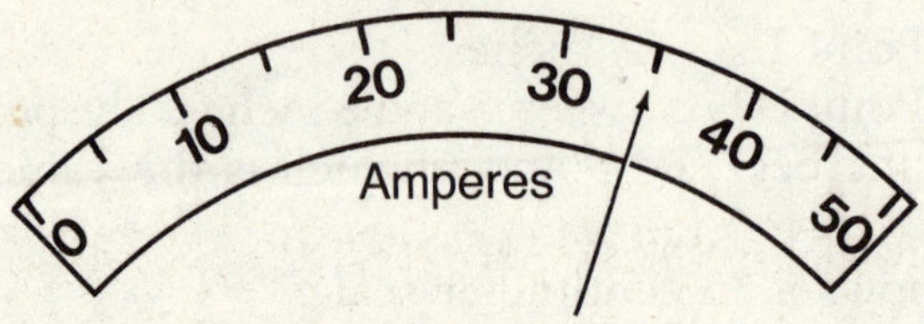

EXAMPLE: What is the reading in amperes of the meter in Figure 3?

SOLUTION. 35 amperes. The arrow in Figure 3 is exactly half way between 30 and 40 amperes. The half-way point is 35.

Measurements Exercise 7

1. Tell how far to the right of zero the points marked J, K, L, M, and N on Figure 4 are.

Figure 4

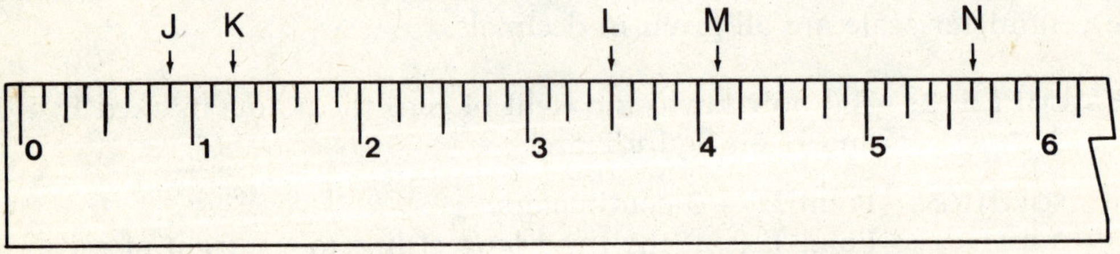

2. Tell how far to the right of zero the points marked P, Q, R, S, and T on Figure 5 are.

Figure 5

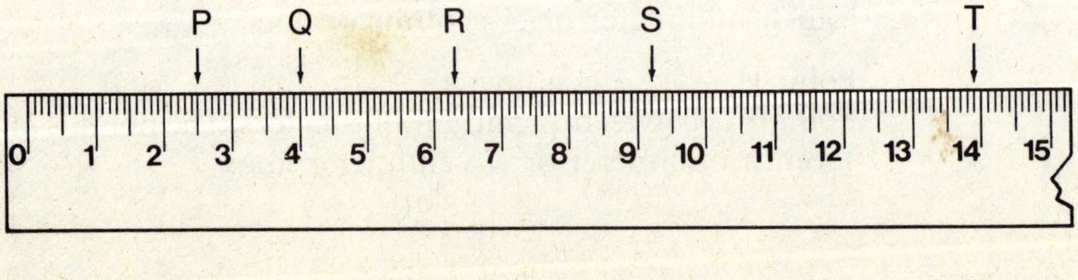

3. Approximately what is the reading in amperes of the meter in Figure 6?

Figure 6

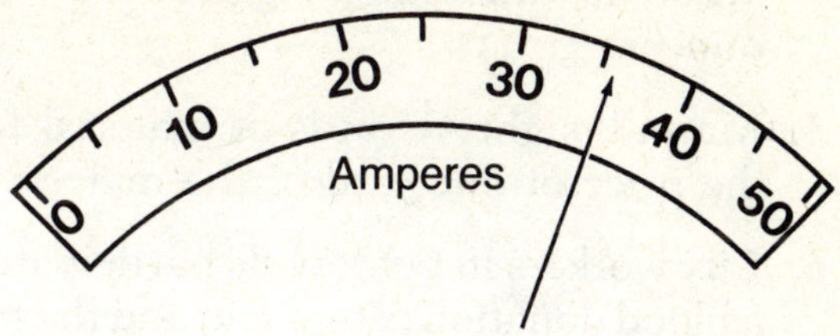

4. Figure 7 shows a volt meter. Volts are a measure of electromotive force. Approximately what is the reading in volts of the meter in Figure 7?

Figure 7

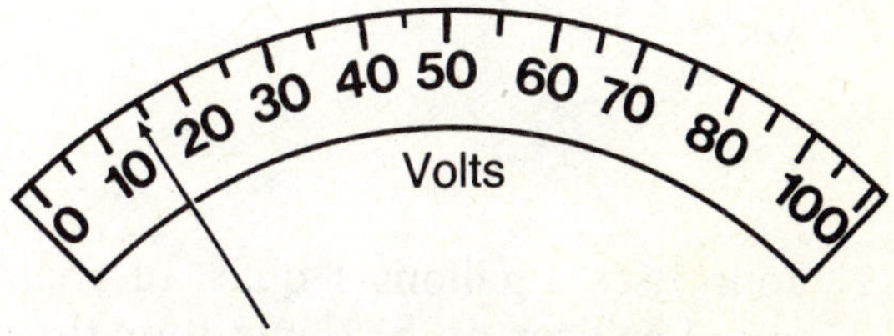

5. In Figure 8, it takes the meter $\frac{1}{10}$ of a second to rise one volt. Approximately how many seconds has it taken to rise the amount shown?

Figure 8

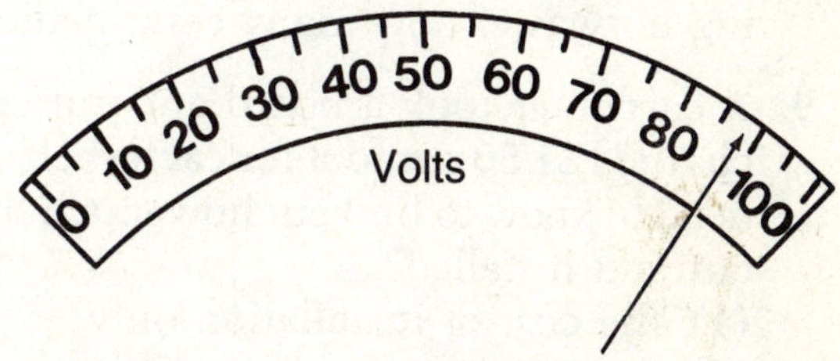

ANSWERS AND SOLUTIONS START ON PAGE 356.

Measurement Word Problems

The next exercise gives you a chance to apply your knowledge of measurements to practical situations. Read each problem carefully.

Measurements Exercise 8

Directions: Solve each problem. Remember not only what you know about measurement but also what you have learned about problem solving.

1. Mr. Asaoka mailed three packages weighing .65 kg, 1.2 kg, and 3.15 kg, respectively. What was the total weight of the packages?
2. Ann works at her brother's store. For two nights she worked 3 hours and 45 minutes each. A third night she worked 4 hours and 20 minutes. A fourth night she worked 2 hours and 30 minutes. Altogether how many hours did she work those four nights?
3. There are three doors into the Alvarados' house. The measurement around each door is 18 ft 8 in. Weatherstripping costs $.45 a foot. How much will it cost to buy weatherstripping for all three doors?

4. Fred bought 5 cans of tomato paste. Each can weighed 14 ounces and cost $.89. What was the total weight of the cans? Express your answer in pounds and ounces.

5. Maria bought $4\frac{1}{2}$ yards of material for $21.11. To the nearest cent, what was the price of one yard of the material?

6. City workers in Centerville participated in a blood donation campaign, and the results are shown at right. Find the average number of liters of blood the workers donated during the first three days of the week.

Day	Liters of Blood
Monday	29.3
Tuesday	36.5
Wednesday	42.2
Thursday	50.1
Friday	30.3

7. John uses 3 gallons 1 quart of chemicals each time he fertilizes his lawn. He puts fertilizer on his lawn four times a year. Altogether how much fertilizer does he use in a year?

8. From a cable 4 meters long, Italo cut three pieces each 1.15 meters long. Assuming no waste, how many centimeters was the remaining piece?

9. A carpenter took a board and cut some shelves, each 2 feet 6 inches long. He charged $1.50 per foot for each shelf, including installation. What more do you need to know to find out how much the carpenter was paid for the shelves he cut and installed?
   (1) the cost of installation only
   (2) the number of shelves he cut
   (3) the length of each shelf
   (4) the cost of his materials
   (5) the amount he charges per inch of shelf

10. Gianni needs 3 yards 9 inches of material to make a suit. How many suits can he make from 15 yards of material?

Use the following situation to answer questions 11 through 13.

> Flight 121 leaves Newark at 10:20 a.m. and arrives in Los Angeles at 1:05 p.m. Los Angeles time is three hours earlier than the time in Newark.

11. When Flight 121 leaves Newark, what time is it in Los Angeles?

12. When Flight 121 arrives in Los Angeles, what time is it in Newark?

13. How long is the flight from Newark to Los Angeles?

14. Following are the lengths of five pieces of pipe. Arrange them in order from longest to shortest.
    A — 19 inches
    B — 15 inches
    C — 27 inches
    D — 2 feet 2 inches
    E — 1 foot 9 inches

15. Nick needs 10 pieces of stock, each 36 inches long to frame a new kitchen counter. How many pieces of stock each 8 feet long does he need for the framing?

16. Following are the weights of five crates. Arrange them in order from heaviest to lightest.

crate A — $\frac{1}{2}$ ton
crate B — 1.1 ton
crate C — .9 ton
crate D — 1200 *pounds*
crate E — 750 *pounds*

17. On the altimeter in Figure 9, the shorter hand shows the number of thousands of feet above sea level. The longer hand shows the number of hundreds of feet above sea level. What is the altitude of a plane when its altimeter reads as shown in Figure 9?

Figure 9

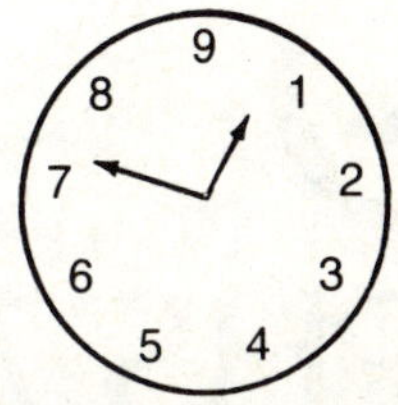

18. As shown in Figure 10, a gas meter has three dials to measure cubic feet of gas used. Dial A measures the number of 10 thousand cubic feet. Dial B measures the number of thousand cubic feet, and Dial C measures hundreds of cubic feet. Approximately how many cubic feet of gas were used?

Figure 10

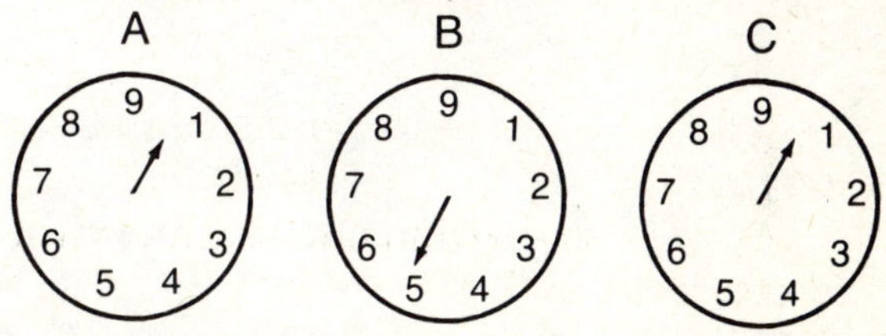

ANSWERS AND SOLUTIONS ARE ON PAGE 356.

9 GRAPHS AND TABLES

What is a Graph?

A ***graph*** is a way of organizing numbers visually. You have seen and probably used a variety of graphs in newspapers, on television, and in magazines. Basically, a graph takes a lot of numerical information (often called ***data***) and shows it in "picture" form.

Pictographs

With a graph you can make quick comparisons of numbers. Look at this information: In a recent year, the total attendance at major league baseball games was 43,746,000; at professional basketball games, 10,677,000; at professional football games, 14,092,000; at National Hockey League games, 10,534,000.

Although a lot of information is given here, you might have a hard time understanding what it all means and how different pieces of the information relate to others.

A ***pictograph*** is one way of organizing this information. A pictograph uses small pictures or symbols to represent information. Figure 1 on page 181 contains the same information as the paragraph above, but it represents it in pictorial form.

Figure 1

Yearly Attendance at Selected Professional Sports Events

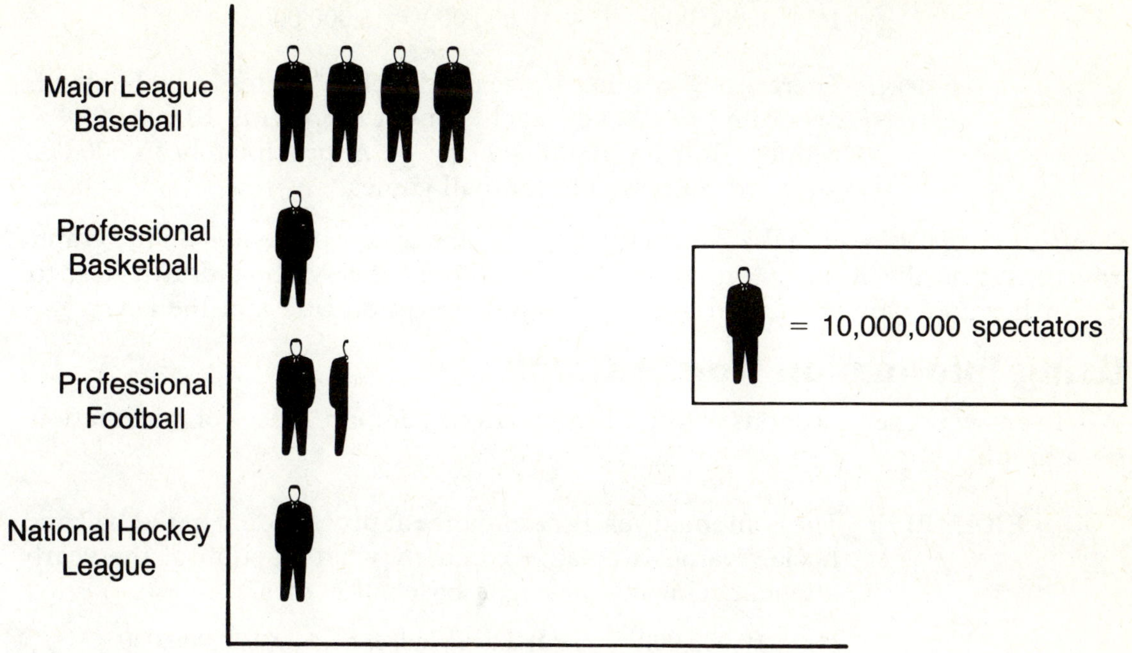

Source: Statistical Abstract of the U.S., 1984

By looking at this graph, you can easily understand the information presented — perhaps more easily than you understood the information in written form.

Graph Organization

Notice that the pictograph has a title (Yearly Attendance at Selected Professional Sports Events), a series of categories (sports events) running down the left side (or ***axis***) of the graph, and a key. According to the key, each small symbol of a person represents 10,000,000 spectators. A partial symbol represents $\frac{1}{2}$ of 10,000,000 or 5,000,000 people.

TIP: It is important that you look at the title, the key, and all labels before you attempt to examine any numerical information.

EXAMPLE: The pictograph in Figure 1 shows the yearly attendance at ________.

ANSWER. **selected professional sports events.** To answer this question, simply read the title of the graph.

EXAMPLE: What is the source of the information on the graph in Figure 1?

ANSWER. **The Statistical Abstract of the U.S., 1984**

Finding Specific Information

Once you are familiar with the way a graph is organized and what type of information it is showing you, you should be able to find specific information.

EXAMPLE: According to the pictograph in Figure 1, what was the yearly attendance at professional football games?

$$1\tfrac{1}{2} \times 10{,}000{,}000 = \frac{3}{2} \times 10{,}000{,}000 = 15{,}000{,}000$$

SOLUTION. There are $1\frac{1}{2}$ symbols in the row labeled "Professional Football." According to the key, each symbol represents 10,000,000 people. Multiply 10,000,000 by $1\frac{1}{2}$. Approximately 15,000,000 attended professional football games.

Questions like this require that you actually "read" the data represented on the graph. You must find the category on the left axis and count the symbols drawn next to it. Furthermore, you must read and understand the graph title and the key.

Using Information from a Graph

To answer other questions on the GED, you must compare pieces of information on a graph.

EXAMPLE: The combined yearly attendance at professional basketball and hockey games was approximately what fraction of the yearly attendance at major league baseball games?

(1) $\begin{array}{r} 10{,}000{,}000 \\ +10{,}000{,}000 \\ \hline 20{,}000{,}000 \end{array}$ **(2)** $\dfrac{\text{basketball \& hockey}}{\text{baseball}} \quad \dfrac{20{,}000{,}000}{40{,}000{,}000} = \dfrac{1}{2}$

STEP 1. Find the combined attendance at professional basketball and hockey games. Each has one symbol to the right, or ten million people.

STEP 2. Make a fraction with a numerator of the combined basketball and hockey attendance, 20,000,000 and a denominator of the major league baseball attendance ($4 \times 10{,}000{,}000 = 40{,}000{,}000$). Reduce the fraction. Basketball and hockey attendance was about $\frac{1}{2}$ that of baseball.

In a question like this, you need to find specific information on the graph and combine and compare it. In other words, the exact information you are looking for may not appear on the graph—you will have to do some computation to find the answer you are looking for.

Graphs Exercise 1

Directions: Use the graph in Figure 2 to answer questions 1-8.

Figure 2

New Single-Family Houses in Central County, 1981-1985

1. What does each small picture of a house represent?
2. How many new single-family houses were built in Central County in 1984?
3. How many more houses were built in 1982 than in 1983?
4. The number of houses built in 1985 was what fraction of the number built in 1982?
5. The number of houses built in 1981 was the same as the number built in which two-year period shown on the graph?
6. The number of houses built in 1983 was what percent of the number built in 1982?
7. The average cost of a new house in Central County in 1985 was $80,000. What was the total construction cost of the new single-family houses built in the county in 1985?
8. Which statement below best describes the pattern of construction of single-family houses in Central County from 1981 to 1985?
   (1) The number of new single-family houses was about the same for each year.
   (2) The number of new single-family houses decreased each year.
   (3) The number of new single-family houses increased each year.
   (4) The number of new single-family houses increased for three years and decreased for two years.
   (5) The number of new single family houses decreased for three years and increased for two years.

Directions: Use the graph in Figure 3 to answer questions 9-14.

Figure 3

Sales of Passenger Cars in the U.S., 1975–1980

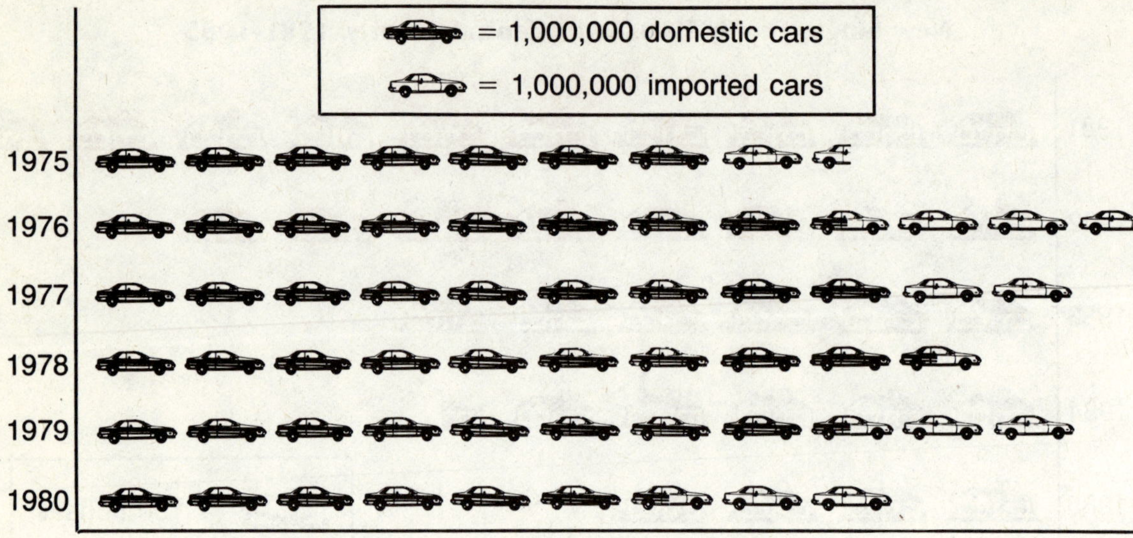

9. What does the symbol of a striped car represent?

10. The number of domestic cars sold in 1977 was how many times the number of imported cars sold that year?

11. Which year represented on this graph shows the greatest total passenger car sales?

12. By what percent did sales of imported cars increase from 1978 to 1979?

13. If car dealerships earned approximately $900 for each domestic car sold and $400 for each imported car sold, approximately how much money did United States car dealers earn in 1977?

14. Which of the following expressions shows the average of the number of domestic cars sold in 1975 and 1977?

(1) (7 + 9)/2
(2) (1 1/2 + 2)/2
(3) 1,000,000(7 + 9)/2
(4) 1,000,000(7 + 9)
(5) 2(7 + 9)/1,000,000

ANSWERS AND SOLUTIONS START ON PAGE 357.

Circle Graphs

A ***circle graph*** is another way to show data visually. A circle represents the whole, and the "pie-shaped" pieces show the different sizes of the parts. Often the parts of a circle graph are expressed in percents. When the parts of a circle graph are percents, the parts must add up to 100% or one whole.

The graph at the right shows the budget for the Johnson family. The budget is divided into five categories. Notice that the five categories added together equal 100%

Figure 4

Johnson Family Budget

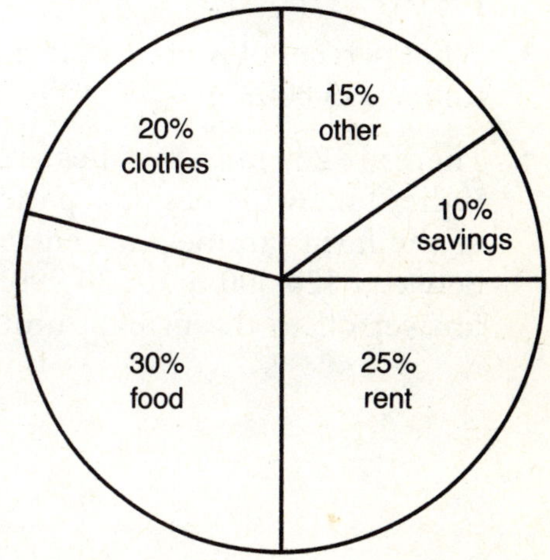

EXAMPLE 1: What percent of the Johnson family budget in Figure 4 goes for food?

SOLUTION. Find the category labeled "food" on the graph. The percent in this category is 30%.

As you saw with the pictograph on page 184, you sometimes have to apply information on a graph to information you are given in a problem.

EXAMPLE 2: The Johnsons take home \$1200 a month. According to their budget in Figure 4, how much do they pay each month for rent?

(1) $\dfrac{r}{1200} = \dfrac{25}{100}$ **(2)** $\begin{array}{r} 25 \\ \times 1200 \\ \hline 30{,}000 \end{array}$ **(3)** $\begin{array}{r} \$\ 300 \\ 100\overline{)30{,}000} \end{array}$

STEP 1. This is a simple percent problem. Fill in the proportion. 25 is the percent, and \$1200 is the whole monthly budget. Let *r* stand for the part, or rent, which is missing.

STEP 2. Find the cross product of 1200 and 25.

STEP 3. Divide by the other number in the proportion, 100. The Johnsons pay **\$300 a month** for rent.

Graphs Exercise 2

Directions: Use Figure 5 to answer the following questions.

1. How many income categories are used in Figure 5?
2. What percent of American farm families made \$10,000–\$14,999 in 1984?
3. What percent of American farm families made under \$15,000?
4. There are 200 farm families in Central County. If they follow the national pattern, how many of the farm families in Central County earn between \$20,000 and \$24,999? Round your answer off to the nearest unit.

Figure 5
Yearly Income of American Farm Families

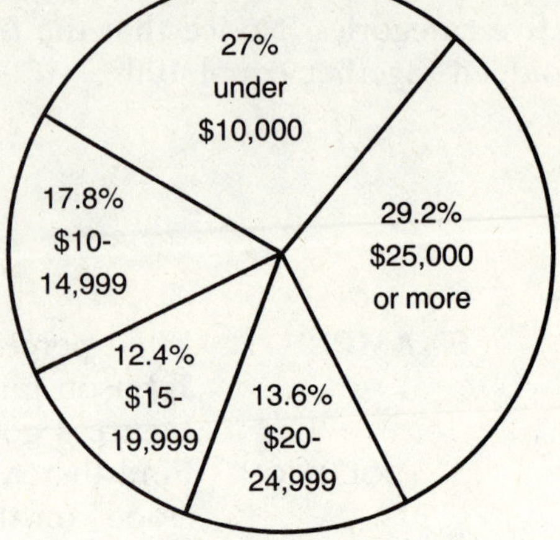

Source: Statistical Abstract of the U.S., 1984

5. According to the same census, 17% of non-farm families had a yearly income under \$10,000. The percent of farm families earning under \$10,000 is what percent greater than the percent of non-farm families earning under \$10,000?
6. Which of the following expresses approximately the fraction of farm families who earned \$25,000 or more?
   (1) $\frac{1}{4}$
   (2) $\frac{1}{5}$
   (3) $\frac{3}{10}$
   (4) $\frac{2}{5}$
   (5) $\frac{9}{10}$

ANSWERS AND SOLUTIONS START ON PAGE 357.

Other Circle Graphs

Not all circle graphs are divided into percents. For example, the parts of a circle graph may also be measured in cents. These cents are the parts of a whole dollar. Figure 6 below shows the source of a dollar of the U.S. federal budget. Notice that the sections add up to 100 cents or \$1. Each amount on the graph is the same as a percent. 11 cents out of the 100 cents in a dollar is the same as 11%.

Also notice that beneath the graph you are given some additional information — the approximate size of the federal budget. Some problems may require you to use this information in order to answer the question.

Graphs Exercise 3

Directions: Use Figure 6 to answer questions 1 through 6.

Figure 6
Source of a Dollar of the Federal Budget

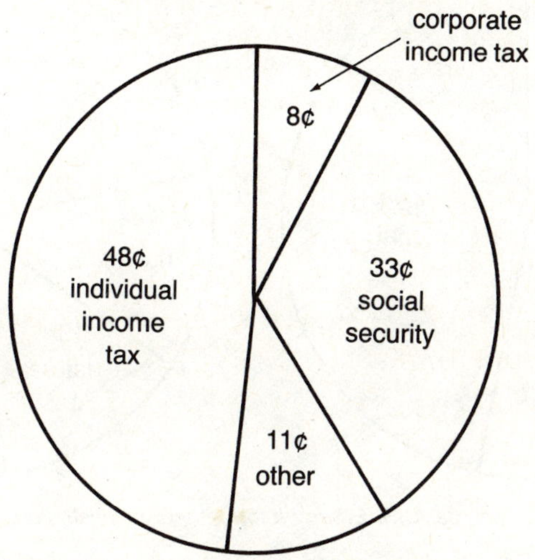

Total Federal Budget: $600 billion
Source: Statistical Abstract of the U.S., 1984

1. What percent of the federal budget comes from individual income tax?

2. Corporate income tax is what fraction of individual income tax as a source of the federal budget?

3. Social security tax is approximately what fraction of the source of the federal budget?

4. The amount of the budget that comes from social security tax is how many times the amount that comes from the source labeled "other"?

5. How many dollars of the federal budget come from corporate income tax?

6. According to the graph, which of the following statements is true?
   (1) Together individual income tax and social security tax make up over 80% of the federal budget.
   (2) Corporate income tax represents the greatest source of the federal budget.
   (3) Social security tax payments make up over half of the federal budget.
   (4) Together social security tax and corporate income tax make up 25% of the federal budget.
   (5) Individual income taxes were higher in 1984 than ever before.

Directions: Use Figure 7 to answer questions 7 through 12.

Figure 7
Sources of Energy in the U.S.

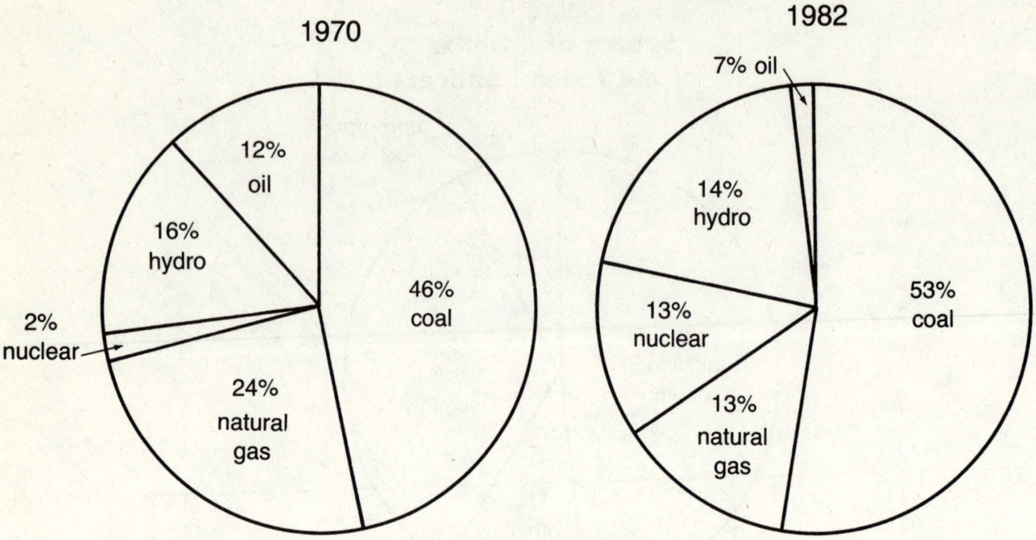

Source: U.S. Energy Information Administration

7. For what years do the graphs in Figure 7 describe energy sources?

8. For the years shown on the graphs, which energy source dropped the greatest percent?

9. For the years shown on the graphs, which energy source rose the greatest percent?

10. The percent of energy coming from nuclear power in 1982 was how many times as great as the percent coming from nuclear power in 1970?

11. Together coal and oil made up what fraction of the energy sources in the U.S. in 1982?

ANSWERS AND SOLUTIONS START ON PAGE 357.

Bar Graphs

A ***bar graph*** is more complicated than a pictograph or a circle graph. To find the values represented on a bar graph, you must read two sets of information. One set of information goes up and down along the vertical axis (side). The other set of information goes across along the horizontal axis (bottom).

Figure 8 is a bar graph that shows noon temperatures for five days. The vertical axis, running up and down along the left side, measures temperature in degreees. The horizontal axis, across the bottom of the graph, lists the days of the week.

Figure 8
Daily Noon Temperatures

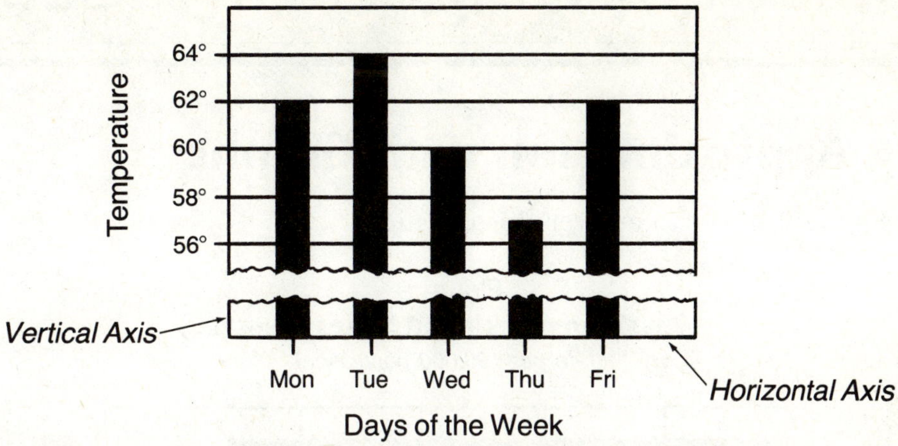

Week of April 7, 1986

EXAMPLE 1: According to Figure 8 what was the noon temperature on Wednesday, April 9, 1986?

SOLUTION. Find Wednesday on the horizontal axis. Follow the bar above "Wed" to the top. Read the temperature straight across on the axis to the left. The bar stops at the line labeled "60°." The temperature at noon on Wednesday was **60°**.

Notice also that the specific dates mentioned can be found by looking at the note below the graph itself. Important information can be found in places other than the title and the axes, so be sure to read *all* information contained on a graph.

EXAMPLE 2: On what day was the temperature 57°?

SOLUTION. Look halfway between 56° and 58° to estimate 57° on the vertical axis. Look across until you find a bar that rises to this point. The bar labeled "Thu" ends halfway between 56° and 58°. The noon temperature was **57°** on Thursday.

Problem Solving for the GED

Using Approximation with Graphs

Use Figure 9 below to answer this question:

Figure 9
Physicians by United States Region
(rate per 100,000 population)

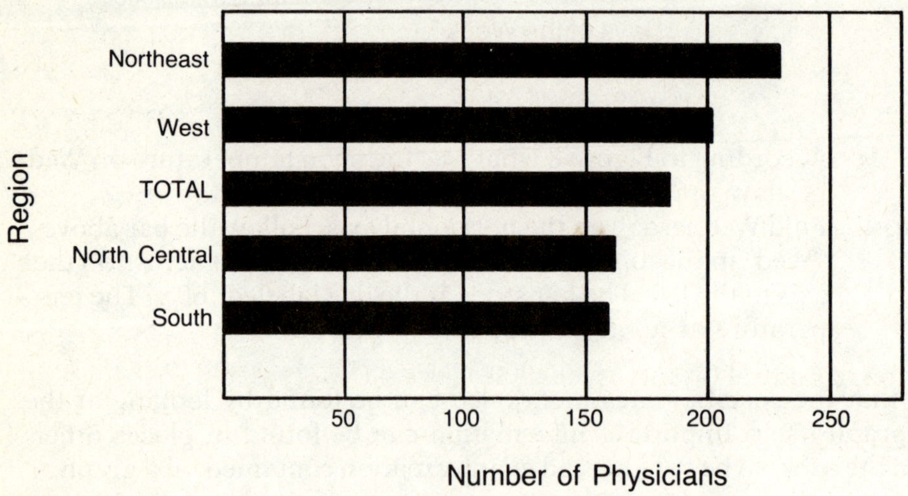

Source: U.S. Surgeon General

Approximately how many more physicians are there in the West than in the South (per 100,000 population)?

(1) 10
(2) 20
(3) 40
(4) 100
(5) 150

As you can see, it is difficult to read exact values on a bar graph. On this graph, you are shown values at intervals of 50. Unless a bar ends *exactly* on one of those intervals, you cannot tell exactly where the bar *does* end. For this reason, some questions on the GED will ask you to *approximate* values. Here are some steps to help you do this:

1. After you have identified the bar(s) you want to evaluate, locate the two horizontal or vertical lines that come before and after the end of the bar.

   In this problem, the West bar ends after the 200 line, but much before the 250 bar. This means that the value is very close to 200. The South bar ends about $\frac{1}{5}$ of the way between the 150 bar and the 200 bar. This means that the value is close to 160.

2. Next, estimate an answer using these approximate values.

   If the West has about 200 physicians and the South has about 160, then the West has approximately 40 more physicians per 100,000 people than the South.

3. Check to see if your approximation is one of the answer choices given. If it is not, is there an answer close to the one you have come up with?

   Answer choice **(3)** is correct. Both **(1)** and **(2)** are clearly too low; **(4)** and **(5)** are too high.

> **TIP:** In questions involving approximation with graphs, if two answer choices seem close, go back and check your work again. Remember that only one answer can be correct.

Graphs Exercise 4

Directions: Use Figure 9 to answer the following questions.

1. What is the approximate number of physicians per 100,000 population for the Northeast?

2. Which region has the most physicans per person?

3. The population of Central County is 300,000. Central County is in the North Central part of the U.S. If Central County is typical of the region, approximately how many physicians are there in the county?
(1) 165
(2) 350
(3) 495
(4) 600
(5) 750

4. The entire population of the U.S. is 240,000,000. According to the graph, approximately how many physicians are there in the U.S.?
(1) 185
(2) 240
(3) 2400
(4) 200,000
(5) 444,000

5. A typical town of 50,000 people in the Northeast has approximately how many more physicians than a typical town of 50,000 people in the North Central states?
(1) 15
(2) 20
(3) 25
(4) 30
(5) 35

6. The rate of physicians per population is approximately what percent greater in the West than in the South?
(1) 40%
(2) 35%
(3) 30%
(4) 25%
(5) 20%

ANSWERS AND SOLUTIONS START ON PAGE 357.

More Bar Graphs

Bar graphs sometimes use different kinds of bars to compare information. The graph in Figure 10 uses two different kinds of bars to compare dollar amounts for different years. Practice working with graphs like this one containing more information.

Graphs Exercise 5

Directions: Use Figure 10 to answer the following questions.

Figure 10
Yearly Sales of Goods and Services in Selected Types of Stores in Central County

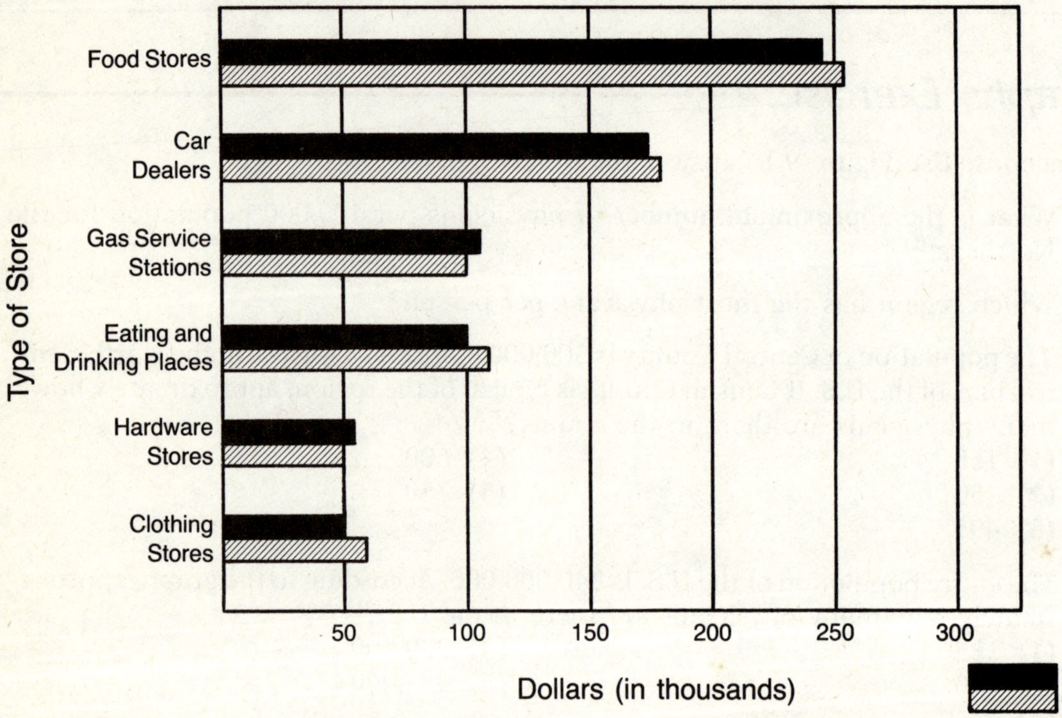

1. What was the approximate yearly total for sales of goods and services for
   a. food stores in 1985?
   b. car dealers in 1984?
   c. eating and drinking places in 1984?
2. Which establishments had lower sales in 1985 than in 1984?
3. In 1985, hardware stores did approximately what percent of the business of food stores?
4. Hardware stores do approximately what fraction of the business of gas service stations?
5. In 1985 the combined sales of clothing stores and eating and drinking places was nearly equal to the sales of what type of store?

6. There are three gas service stations in Central County. If each did approximately the same amount of business, how much did each take in for 1985? Round your answer to the nearest thousand.

ANSWERS AND SOLUTIONS START ON PAGE 358.

Line Graphs

A ***line graph***, like a bar graph, has information on both a vertical axis and a horizontal axis. The values on a line graph are displayed along a continuous line. Every point along the line has a value. To tell the value of a point, read across to the vertical axis at the side and down to the horizontal axis across the bottom.

Figure 11 is a line graph that shows the net income from farming per farm.

Figure 11

Net Income from Farming per Farm (in 1967 dollars)

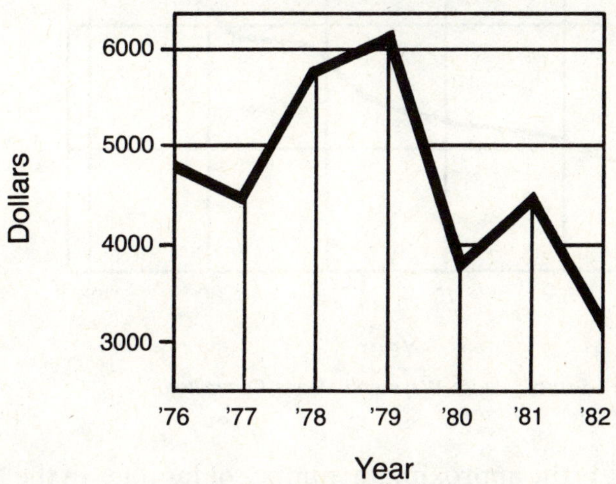

Source: Statistical Abstract of the U.S., 1984

EXAMPLE 1: What was the net income per farm in 1981?

SOLUTION. Find '81 on the horizontal axis. Follow the line labeled "81" up until it meets the continuous heavy line. Read the dollar income on the vertical axis directly to the left. The point on the vertical axis is halfway between $4000 and $5000. The net income per farm in 1981 was **$4500**.

EXAMPLE 2: In what year shown on the graph did net income per farm reach $6000?

SOLUTION. Find $6000 on the vertical axis. Follow the line labeled "$6000" directly across to the continuous heavy line. Read the year directly below on the horizontal axis. The net income per farm reached $6000 in **1979**.

Problem Solving for the GED

Reading a Graph Carefully

Directions: Use the graph in Figure 12 to answer the following questions.

Figure 12
Families in the U.S. Headed by Females

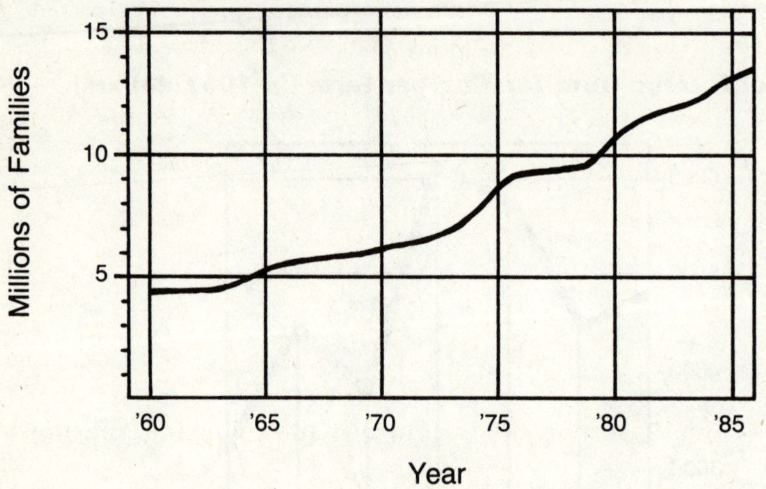

Source: House Ways and Means Committee

1. Which of the following is the approximate number of families in the U.S. headed by females in 1965?
(1) 5
(2) 5000
(3) 5 million
(4) 6 million
(5) 8 million

2. In 1959, the number of U.S. families headed by females was
(1) around 5 million
(2) below 5 million
(3) less than in 1958
(4) more than in 1960
(5) Not enough information is provided.

To answer questions like these on the GED, you need to read the graph, and all the information it contains very carefully. If you do not, you may make some very common mistakes.

> In problem 1, for example, if you look at the graph quickly, you will see that for the year 1965, the line graph crosses a horizontal line very close to the number "5." Answering quickly, you may choose answer (1).
>
> However, if you read the information on the graph more carefully, you will see that the horizontal values are given *in millions*. This means that the "5" actually represents a value of *5 million*. Answer choice (3) is correct.

You also must be careful that answers you give are based on the information provided.

> In problem 2, for example, you might be tempted to choose answer (2). The rest of the graph shows that the number of families headed by females rose each year. You might guess that in 1959, the number was below the number in 1960. However, since the year 1959 *is not included* on the graph, you cannot be sure of its value. You do not have enough information to answer the question, so answer choice (5) is correct.

★ GED PRACTICE ★

Reading a Graph Carefully

Directions: Use Figure 12 on page 194 to answer the following question.

In 1985, the number of families in North America headed by females was
(1) 14
(2) 1400
(3) 12 million
(4) 14 million
(5) Not enough information is provided.

ANSWER IS ON PAGE 358.

More Line Graphs

Line graphs sometimes use different kinds of lines to compare information. The graph in Figure 13 on page 196 uses three different kinds of lines to compare inflation rates in different parts of the world. Sometimes you will need to look at all three lines to answer a question correctly.

Graphs Exercise 6

Directions: Use the graph in Figure 13 to answer the following questions.

Figure 13
Unemployment Rates in Three World Areas

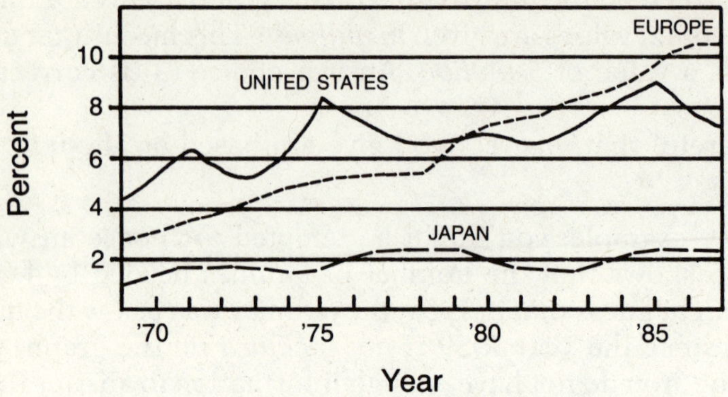

Source: International Monetary Fund

1. For the years shown on the graph, when did the unemployment rate in the U.S. first reach 8%?
(1) 1970
(2) 1972
(3) 1975
(4) 1981
(5) 1985

2. Approximately what was the unemployment rate in Japan in 1968?
(1) 5%
(2) 4%
(3) 2%
(4) 1%
(5) Not enough information is provided.

3. In what year was the unemployment rate of Europe the same as that of the U.S.?
(1) 1979
(2) 1975
(3) 1970
(4) 1985
(5) 1973

4. In 1971 the unemployment rate in the U.S. was about how many times that of Europe?
(1) 5
(2) 4
(3) 3
(4) 2
(5) $\frac{1}{2}$

ANSWERS AND SOLUTIONS START ON PAGE 358.

Tables

Another graphic representation of data that is often found in newspapers and books is a table. A ***table*** is a list of exact numbers displayed in columns and rows. Although graphs make quick comparisons of numbers easy, tables allow for greater accuracy. The reader does not have to approximate values because they are actually written out for him. The table in Figure 14 compares the populations of five large metropolitan areas.

Figure 14

Population of Five Largest Metropolitan Areas in the U.S.

	1970	1980
New York City	18,193,000	17,539,000
Los Angeles	9,981,000	11,498,000
Chicago	7,779,000	7,937,000
Philadelphia	5,749,000	5,681,000
San Francisco	4,754,000	5,368,000

Source: U.S. Census

EXAMPLE 1: What was the population of metropolitan Chicago in 1980?

SOLUTION. Find Chicago in the list of cities. Look to the right in the column labeled "1980." The population of metropolitan Chicago in 1980 was 7,937,000.

EXAMPLE 2: For which metropolitan areas on the table was the population less in 1980 than in 1970?

SOLUTION. To answer this question you must compare every row of numbers. Look for the rows where the number under 1980 is less than the number under 1970. The two metropolitan areas which had less population in 1980 than in 1970 were **New York City** and **Philadelphia.**

Graphs Exercise 7

Directions: Use the table in Figure 15 to answer the following questions.

Figure 15

Union Membership (in thousands)

	1955	1965	1975	1985
United Steel Workers	980	876	1062	572
International Ladies Garment Workers Union	383	363	363	210
International Typographical Union	78	87	81	38
Communication Workers of America	249	288	476	524
United Automobile Workers	1260	1150	n.a.*	974
Service Employees International Union	205	305	480	688
American Federation of State, County, and Municipal Employees	99	237	647	997

Source: A.F.L.-C.I.O.

**The letters* n.a. *mean that information was not available for this year.*

1. Tell the membership for each of the following.
   a. United Steel Workers in 1985.
   b. International Typographical Union in 1965.
   c. Communication Workers of America in 1955.

2. Which three unions represented on the table increased in membership for each year shown?

3. Which union lost the greatest number of members from 1975 to 1985?

4. Which union lost the greatest percent of members from 1975 to 1985?

5. The 1985 membership of the American Federation of State, County, and Municipal Employees was approximately how many times as great as the 1955 membership?

6. Which of the following expressions gives the average number of members in the United Auto Workers for the years 1955 and 1965?
   (1) 980 + 876
   (2) (980 + 876)/2
   (3) (1260 + 1150)/2
   (4) 1260 + 1150
   (5) 1000 (1260 + 1150)/2

ANSWERS AND SOLUTIONS START ON PAGE 358.

Graphs and Tables Summary

The next exercise will serve as a review of your knowledge of graphs and tables. Keep the following ideas in mind.

- Graphs give a visual picture of numerical information or data. They are not particularly useful in providing *exact* information, but they provide an easy way to compare different values without doing a lot of reading or computing.
- Tables are made up of numerical information organized in rows and columns. They usually give more exact values than graphs do.
- Graphs and tables contain a lot of information in a small area. It is always important to read *all* the information provided, including titles, subtitles, notes, and labels, as well as the numerical data presented.
- Approximation skills are important in reading graphs. Although you will not have to find *exact* values for most graphs on the GED, you will often be required to choose the correct *approximate* value.
- Try not to "read too much" into a graph or table. Although you may *think* a certain answer seems sensible, make sure it is supported by the information in front of you. Watch out especially when one of the answer choices is "*not enough information*."

★ GED PRACTICE ★

Exercise 8: Graphs and Tables Review

Directions: Use the table in Figure 16 to answer the following questions.

Figure 16

Personnel on Active Duty in U.S. Armed Services (in thousands)

	Number	Percent
Officers	273.7	100
White	246.3	90
Black	15.4	6
Hispanic origin	3.6	1
Other	8.4	3
Enlisted	1804.2	100
White	1258.0	70
Black	397.5	22
Hispanic origin	72.9	4
Other	75.8	4

Source: U.S. Department of Defense

1. What is the number of personnel in each of the following categories?
a. Total officers
b. Total enlisted personnel
c. Black officers
d. Enlisted personnel of Hispanic origin

2. Blacks make up what percent of enlisted personnel?

3. The percent of white officers in the armed services is how many times that of black officers?

4. Altogether, blacks, personnel of Hispanic origin, and those labeled "other" make up what percent of officers in the armed services?

5. According to the U.S. Department of Defense, how many enlisted people on active duty in the U.S. armed services are of Hispanic origin?
   (1) 4
   (2) 72.9
   (3) 4000
   (4) 72,900
   (5) Not enough information is provided.

6. According to the U.S. Department of Defense, there are how many more white enlisted personnel than white officers?
   (1) 20
   (2) 10,117
   (3) 1,011,700
   (4) 1,530,500
   (5) Not enough information is provided.

Directions: Questions 7–10 are based on the following chart and graph.

Maynard Family February Income

	Salary	Overtime	Odd Jobs
Sally	$845	$32.50	—
Doug	$680	—	$170

Maynard Family February Expenditures
(percent of total take-home pay)

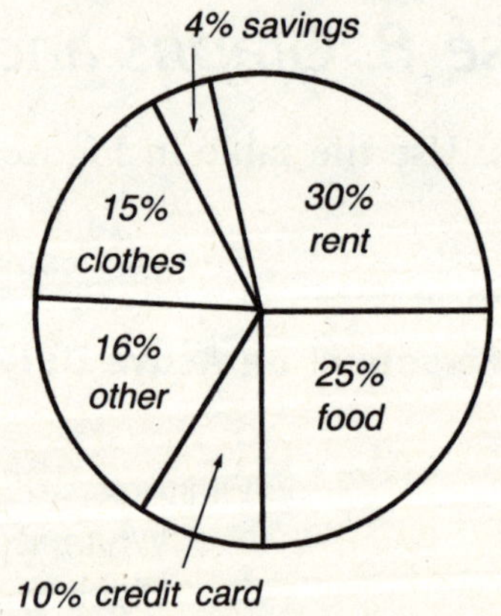

7. How much money did the Maynards spend on credit card payments in February?
   (1) $10
   (2) $15.25
   (3) $17.27
   (4) $172.75
   (5) $340.00

8. The money that Doug made in salary was how many times as much as he made doing odd jobs?

(1) $\frac{1}{4}$

(2) $\frac{1}{3}$

(3) 4

(4) 170

(5) 115,600

9. Not including odd jobs, Doug works 34 hours per week. Assuming that his monthly salary is based on 4 weeks per month, how much money did Doug make per hour?
(1) $4.00
(2) $5.00
(3) $20.00
(4) $36.00
(5) $136.00

10. How much more money did Doug and Sally save than they made in overtime work?
(1) $12.50
(2) $32.50
(3) $36.60
(4) $69.10
(5) $101.60

ANSWERS AND SOLUTIONS START ON PAGE 358.

10 THE BASICS OF ALGEBRA

Algebra, like all the math you have studied so far, is a tool you can use to solve many problems. In this chapter, you will become familiar with some of algebra's most important and powerful features:

- First, you will learn to use letters to represent unknown numbers, sometimes called **variables**.
- Second, you will learn about **equations**. An equation is a statement that two amounts are equal.
- Third, you will work with specific types of equations called **formulas**. Formulas will be extremely helpful in your work with geometry and algebra.

Once you have mastered these basic algebra concepts, you are on your way to solving many different kinds of word problems found on the GED.

Writing Algebraic Expressions

In algebra, letters are often used to stand for numbers that we want to find. We call these letters ***variables*** or ***unknowns***. To use algebra, you must learn to translate words into algebraic expressions. An algebraic expression always contains a variable or unknown. Any letter can be used to represent an unknown, but x and n are most commonly used.

Below are several examples of algebraic expressions using the four basic arithmetic operations of addition, subtraction, multiplication, and division. Read each example carefully. The letters you see represent unknown numbers, or variables. Watch for the words that suggest which mathematical operations to write. Any letter can be used to represent an unknown number.

Example	Expression	Notes
A number increased by four	$x + 4$ or $4 + x$	*Increased by* means to add.
Nine more than a number	$r + 9$ or $9 + r$	*More than* means to add.
Eight less than a number	$y - 8$	*Less than* means to subtract.
Eight decreased by a number	$8 - y$	*Decreased by* means to subtract. (Compare the order of this example to the last one.)
Four times a number	$4e$	*Times* means to multiply.
The product of ten and a number	$10b$	A *product* is the answer to multiplication. (Notice that there is no sign between the number and the unknown in multiplication expressions.)
A number divided by 5	$m/5$ or $\frac{m}{5}$	Notice that the divisor goes on the bottom.
Five divided by a number	$5/m$ or $\frac{5}{m}$	Notice that the unknown is the divisor here.
One-fourth of a number	$\frac{1}{4}m$ or $\frac{m}{4}$	Remember that $\frac{1}{4}$ of a number is equal to that number over four.

Notice that with subtraction and division of unknowns, the order of the letters and numbers is very important. For example, the expression $y - 8$ is *not* equal to the expression $8 - y$. The first means "take 8 away from a number." The second means "take a number away from 8." Similarly, the expression $\frac{m}{5}$ is *not* the same as the expression $\frac{5}{m}$.

Basics of Algebra Exercise 1

Directions: Write an algebraic expression for each of the following. Use the letter x to represent each unknown.

1. A number increased by nine.
2. The product of seven and a number.
3. Fives times a number.
4. The sum of a number and eight.

5. A number minus ten.
6. One less than a number.
7. Three divided by a number.
8. A number decreased by twenty.
9. A number divided by fifteen.
10. A number plus one-half.
11. The quotient of a number divided by two.
12. Four more than a number.

ANSWERS AND SOLUTIONS START ON PAGE 358.

Using Variables in Word Problems

As a first step in solving some word problems, it is helpful to express number relationships with letters, called variables. In the next examples, watch again for the words that suggest which mathematical operations to write.

EXAMPLE 1: Let r represent the amount of rent that Max pays in a month. Next year Max will have to pay $20 more each month for rent. Write an expression that tells his monthly rent next year.

$r + 20$

SOLUTION. The expression "$20 more" means to add. Max's monthly rent this year is r. His monthly rent next year will be $\boldsymbol{r + 20}$.

EXAMPLE 2: Four friends jointly won w dollars in a raffle. They decided to share the money equally. Write an expression for the amount each person received.

$\frac{w}{4}$

SOLUTION. To share equally means to divide. The total amount they won was w. The amount each received was w divided by 4, or $\frac{w}{4}$.

Basics of Algebra Exercise 2

Directions: Write an expression for the situation described in each problem.

1. Jack makes w dollars an hour. He will get a $2-an-hour raise. Write an expression for his new hourly wage.
2. Normally there are s students in John's GED class. Because of bad weather, five students were absent one night. Write an expression for the number of students who came to class that night.
3. Meat costs c per pound. Write an expression for the cost of three pounds of meat.
4. Let i stand for Ellen's gross income. Of her income, $\frac{1}{5}$ goes for taxes and social security. Write an expression for the amount of Ellen's income that is withheld for taxes and social security.

5. The total town budget for summer programs is x. Four programs share the money equally. Write an expression for the amount each program gets.

6. The Smiths take home t dollars every month. They spend $\frac{5}{6}$ of their take-home pay for basic expenses. Write an expression for the amount of their basic expenses each month.

7. The length of a rectangle is l. The width is two inches less than the length. Write an expression for the width of the rectangle.

8. The sales tax in Bill's state is 6% (.06). Write an expression for the amount of tax Bill has to pay on an item that costs m dollars.

9. Frank is y years old. Write an expression for his age in 10 years.

10. Beverly weighs p pounds. Write an expression for her weight after she loses 15 pounds.

ANSWERS START ON PAGE 359.

Equations

What Is an Equation?

An ***equation*** is a statement that two amounts are equal. You can write an equation with words or with symbols. Here is an equation written with symbols:

$$9 + 6 = 15$$

In words this equation says, "Nine plus six is equal to fifteen." As you can see, the amount on each side of the equal sign is the same.

The equations you will use most often in algebra will include a letter that is called the ***unknown***. The value of an unknown that makes an equation true is called the ***solution*** to an equation. In this section you will learn to solve equations using the basic arithmetic operations of addition, subtraction, multiplication, and division.

Following are examples of equations with the four basic operations.

Addition equation: $m + 43 = 74$

Subtraction equation: $y - 15 = 15$

Multiplication equation: $6x = 132$

Division equation: $\frac{w}{7} = 10$

Solving One-Step Equations

Look at this equation: $m + 43 = 74$. This equation is asking you "What number plus 43 will give me a sum of 74?" You may be able to just figure out the answer by trying different values for m. Your knowledge of mathematics may be enough to solve a problem like this. However, if you have trouble seeing a solution right away, algebra can help you handle this type of problem easily. Also, as problems with unknowns become more complex, you will find algebra more and more useful.

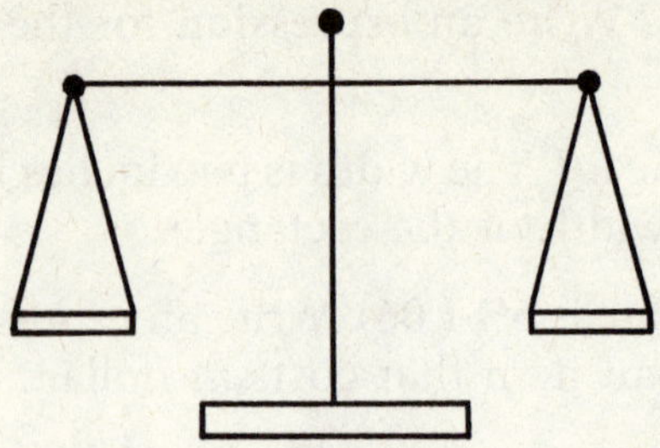

The equal sign (=) makes an equation like an old-fashioned balance scale. Imagine a scale with weights on one side and apples on the other side. If you remove an apple from one side, you must remove a corresponding weight from the other side to keep the scale balanced.

You can "balance" an equation like the one above by performing the *same* operation on both sides of the equal sign.

When you solve an equation, your goal is to get the unknown to stand alone. In other words, to solve the equation $m + 43 = 74$, you want to get the *m alone* on one side of the equation. As you can see, m is not alone in the equation $m + 43 = 74$ because it has $+ 43$ right next to it.

How can you get the unknown (m in this case) to stand alone? In an equation with one operation (such as addition), you can get the unknown to stand alone by performing the ***inverse*** (opposite) operation to both sides of the equation.

- The inverse of addition is subtraction.
- The inverse of subtraction is addition.
- The inverse of multiplication is division.
- The inverse of division is multiplication.

Rule for Solving Equations with One Operation

Identify the operation in the equation. Then perform the inverse operation on **both** sides of the equation in order to get a statement that means, "Unknown = value," or "Value = unknown."

Remember: Whatever you do to one side of an equation you *must* do the other side.

EXAMPLE 1: Solve for the unknown in $m + 43 = 74$.

$$\begin{array}{rcr} m + 43 & = & 74 \\ -\ 43 & & -\ 43 \\ \hline m & = & 31 \end{array}$$

SOLUTION. On the left side, 43 is added to the unknown. Since the inverse of addition is subtraction, subtract 43 from both sides. The solution for m is 31. $\boldsymbol{m = 31}$

To check the solution for an equation, substitute the value of the unknown into the equation. For the example on page 206, replace m with 31.

$$\text{Check:}\quad \begin{aligned} m + 43 &= 74 \\ 31 + 43 &= 74 \\ 74 &= 74 \end{aligned}$$

The correct solution to an equation makes both sides of the equation equal. When you substitute 31 for m in the equation in Example 1, both sides of the equation equal 74. This means that 31 is the correct solution.

EXAMPLE 2: Solve and check the equation $39 = y - 15$.

(1) $$\begin{aligned} 39 &= y - 15 \\ +15 &\quad +15 \\ \hline 54 &= y \end{aligned}$$

(2) $$\begin{aligned} 39 &= 54 - 15 \\ 39 &= 39 \end{aligned}$$

STEP 1. On the right side of the equation, 15 is subtracted from the unknown. Since the inverse of subtraction is addition, add 15 to both sides. The solution for y is 54. **y = 54**

STEP 2. To check, substitute 54 for y in the original equation.

Basics of Algebra Exercise 3

Directions: Solve and check each equation.

1. $f + 20 = 57$

2. $42 = t + 7$

3. $n + 36 = 60$

4. $41 = c + 18$

5. $b - 19 = 28$

6. $33 = k - 8$

7. $c - 4 = 27$

8. $18 = d - 6$

ANSWERS AND SOLUTIONS START ON PAGE 359.

Solving Equations with Multiplication and Division

In the next examples, you will see equations with multiplication and division. Remember that the inverse of multiplication is division. The inverse of division is multiplication.

EXAMPLE 1: Solve and check the equation $6x = 132$.

(1) $$\begin{aligned} 6x &= 132 \\ \frac{6x}{6} &= \frac{132}{6} \\ \mathbf{x} &= \mathbf{22} \end{aligned}$$

(2) $$\begin{aligned} 6x &= 132 \\ 6(22) &= 132 \\ 132 &= 132 \end{aligned}$$

STEP 1. On the left side, the unknown is being multiplied by 6. Since the inverse of multiplication is division, divide both sides of the equation by 6. On the left side you get $1x$. Write this simply as x. On the right side you get the solution, 22.

STEP 2. To check, substitute 22 for x in the original equation.

> **TIP:** When the number 1 appears before the unknown, as in $1x$, the unknown is written by itself. This is because any value multiplied by 1 is the same value.

EXAMPLE 2: Solve and check the equation $10 = \frac{w}{7}$.

(1)

$$10 = \frac{w}{7}$$

$$10 \times 7 = \frac{w}{7} \times \frac{7}{1}$$

$$\mathbf{70 = w}$$

(2)

$$10 = \frac{w}{7}$$

$$10 = \frac{70}{7}$$

$$10 = 10$$

STEP 1. On the right side of the equals sign, the unknown is being divided by 7. Since multiplication is the inverse of division, multiply both sides of the equation by 7. The solution for w is **70.**

STEP 2. To check, substitute 70 for w.

Basics of Algebra Exercise 4

Directions: Solve and check each equation.

1. $8y = 96$

2. $11 = 2d$

3. $15p = 75$

4. $25z = 100$

5. $\frac{x}{3} = 9$

6. $9 = \frac{m}{4}$

7. $\frac{a}{5} = 8$

8. $\frac{c}{9} = 2$

ANSWERS AND SOLUTIONS START ON PAGE 359.

Writing One-Step Equations to Solve Word Problems

In the previous section, you learned to solve one-step equations. On the GED Test, you may first have to read a word problem and *write an equation* based on it before you can solve the problem.

At the beginning of this chapter, you learned to write algebraic expressions from words. To write equations from words, watch for verbs such as *is* or *equals*. These verbs tell you where to put the equal sign.

EXAMPLE 1: A number increased by twelve is twenty. Find the number.

(1) $x + 12 = 20$

(2)

$$\begin{aligned} x + 12 &= 20 \\ -12 &\quad -12 \\ \hline x &= 8 \end{aligned}$$

(3)

$$x + 12 = 20$$

$$8 + 12 = 20$$

$$20 = 20$$

STEP 1. Use a letter to represent the unknown in the expression, "a number increased by twelve." The verb *is* tells you to put the equal sign before 20.

STEP 2. Solve the equation by subtracting 12 from both sides. The solution is 8.

STEP 3. Check by substituting 8 for x in the original equation.

EXAMPLE 2: Andy's gross pay minus the $45 his employer withholds each week is his net pay. Andy's net pay is $325 a week. What is Andy's gross pay?

(1) $x - 45 = 325$

(2)
$$\begin{array}{rcr} x - 45 & = & 325 \\ +\ 45 & & +\ 45 \\ \hline x & = & 370 \end{array}$$

STEP 1. Use a letter to represent Andy's gross pay. Then write an equation that expresses the gross pay minus the amount withheld. The verb *is* in the first sentence tells where to put the equal sign.

STEP 2. Solve the equation by adding 45 to both sides. The solution is 370. Andy's gross pay is **$370** a week.

You may have found the two problems above easy to solve without writing an equation. This is OK. You should use whatever method is successful for you. However, as word problems get more challenging, you will find that equations can make your work a lot easier. Therefore, it's a good idea to practice writing and solving them.

Basics of Algebra Exercise 5

Directions: For questions 1 through 6, choose the letter of the algebraic equation at the right that corresponds to the verbal equation described on the left.

1. ______ A number increased by nine is fifteen.

2. ______ Two-thirds of a number equals twelve.

3. ______ A number divided by three is twenty.

4. ______ When you subtract a number from sixteen, the result is ten.

5. ______ The product of a number and three is twenty.

6. ______ When a number is divided by seven, the result is four.

a. $\frac{3}{n} = 20$

b. $3s = 20$

c. $t + 15 = 9$

d. $7a = 4$

e. $c + 9 = 15$

f. $(\frac{2}{3})x = 12$

g. $16 - y = 10$

h. $\frac{a}{7} = 4$

i. $\frac{d}{3} = 20$

j. $\frac{2}{3} \times 12 = x$

Directions: For questions 7 through 12, write an equation for each statement. Then solve for the unknown number.

7. Nine less than a number is fifteen.

8. The product of six and a number is 27.

9. The quotient of a number divided by five equals fifty.

10. Three is equal to a number decreased by five.

11. Sixty is equal to three-fourths of a number.

12. The product of seven and a number is 84.

Directions: For questions 13 through 16, write and solve an equation for each problem.

13. The original price minus a $40 discount is the price Max paid for a new color TV. He paid $290 for the TV. Find the original price.

14. The total payroll at Acme, Inc., divided by 15 is the amount each employee takes home every week. Every worker takes home $195 a week. Find the total weekly payroll.

15. Sandy pays one-fourth of her monthly income for rent. Her monthly rent is $520. How much is her income each month?

16. Of the members of a block association, 75% (.75) came to the last meeting. Altogether, 90 people attended the meeting. How many people belong to the association?

ANSWERS AND SOLUTIONS START ON PAGE 359.

Solving Longer Equations

Writing Longer Algebraic Expressions

At the beginning of this chapter, you learned to write algebraic expressions with the four basic operations of addition, subtraction, multiplication, and division.

Algebra is a useful way to write mathematical relationships that contain more than one of these basic operations. Read the next examples carefully. Watch for the words that suggest which mathematical operations to write.

Example	Expression	Notes
The sum of a number and one-third of the same number	$c + \frac{1}{3}c$	*Sum* means to add, and *one-third of* means to multiply.
Twice a number decreased by seven	$2k - 7$	*Twice* means to multiply, and *decreased by* means to subtract.
Twice the quantity of a number decreased by seven	$2(k - 7)$	*Twice* means to multiply, and *decreased by* means to subtract. *Quantity* groups the expression *a number decreased by seven.* (Compare this example to the last one to see how parentheses are used.)
Three more than a half number	$\frac{1}{2}r + 3$ OR $\frac{r}{2} + 3$	*More than* means to add, and *half* can mean to multiply by $\frac{1}{2}$ or to divide by 2.

Example	Expression	Notes
The sum of six and a number, all multiplied by three	$3(p + 6)$	The parentheses group the sum together.
Five divided into the sum of three and a number	$\frac{x + 3}{5}$ OR $(x + 3)/5$	The fraction bar groups the $x + 3$ together. First find the sum. Then divide by 5.

Basics of Algebra Exercise 6

Directions: For questions 1 through 8, choose the letter of the algebraic expression at the right that corresponds to the verbal expression described on the left.

1. ______	Two less than seven times a number.	**a.** $7c - 2c$
2. ______	The quantity m minus five, divided by four.	**b.** $16 - 3p$
3. ______	A number increased by eight times itself.	**c.** $2a + 6$
4. ______	Twice the quantity of a number increased by six.	**d.** $y + 8y$
5. ______	Three more than one-half of a number.	**e.** $7c - 2$
6. ______	Twice a number subtracted from seven times the same number.	**f.** $2(g + 6)$
7. ______	Six more than twice a number.	**g.** $\frac{m - 5}{4}$
8. ______	Three times a number subtracted from sixteen.	**h.** $\frac{1}{2}s + 3$
		i. $3p - 16$

Directions: Write an algebraic expression for each of the following. Use x to represent the unknown number.

9. Twice a number increased by six.

10. Five less than the product of eight and a number.

11. The sum of two times a number and four times the same number.

12. Eight more than half a number.

13. Subtract one from one-third of a number.

14. Take three times a number from fifteen.

15. Four divided into the quantity of a number increased by seven.

16. The sum of twelve and a number, all multiplied by ten.

17. The quantity of four less than a number, all multiplied by nine.

ANSWERS AND SOLUTIONS START ON PAGE 360.

Solving Equations with More than One Operation

Earlier in this chapter, you learned to solve one-step equations. But look at this equation:

$$4m + 2 = 26$$

As you can see, both multiplication and addition are involved. To solve equations with more than one operation, you still want to get the unknown value *by itself* on one side. To do this, use the inverse operation for every operation in the equation.

> **TIP:** In equations with more than one operation, remember that you *must* take care of addition and subtraction first. Then take care of multiplication and division.

EXAMPLE 1: Solve for m in $4m + 2 = 26$.

(1) $\begin{array}{rcr} 4m + 2 & = & 26 \\ -2 & & -2 \\ \hline 4m & = & 24 \end{array}$ (2) $\dfrac{4m}{4} = \dfrac{24}{4}$, $\mathbf{m = 6}$

STEP 1. Since 2 is being added on the left side of the equals sign, use the inverse operation of addition. Subtract 2 from both sides of the equation. You are now left with the equation $4m = 24$.

STEP 2. Since m is being multiplied by 4, use the inverse operation, division. Divide both sides of the equation by 4. The solution is 6. $\mathbf{m = 6}$

EXAMPLE 2: Solve for s in $5 = \dfrac{s}{3} - 7$.

(1) $\begin{array}{rcl} 5 & = & \frac{s}{3} - 7 \\ +7 & & \quad +7 \\ \hline 12 & = & \frac{s}{3} \end{array}$ (2) $3 \cdot 12 = \left(\frac{s}{3}\right) \cdot 3$, $\mathbf{36 = s}$

STEP 1. Add 7 to both sides.

STEP 2. Multiply both sides by 3. The solution is 36. $\mathbf{36 = s}$

EXAMPLE 3: Find the value of r in $\frac{2}{3}r + 5 = 23$.

(1) $\begin{array}{rcr} \frac{2}{3}r + 5 & = & 23 \\ -5 & & -5 \\ \hline \frac{2}{3}r & = & 18 \end{array}$ (2) $\frac{3}{2} \cdot \frac{2}{3}r = \frac{18}{1} \cdot \frac{3}{2}$, $\mathbf{r = 27}$

STEP 1. Because 5 is being added and subtraction is the inverse of addition, subtract 5 from both sides.

STEP 2. Since r is multiplied by $\frac{2}{3}$, divide both sides of the equation by $\frac{2}{3}$. Remember that to divide by a fraction means to multiply by the reciprocal of the fraction. Multiply both sides of the equation by $\frac{3}{2}$. The solution is 27. $\mathbf{r = 27}$

Basics of Algebra Exercise 7

Directions: Solve and check each equation.

1. $7m - 2 = 54$
2. $\frac{a}{3} + 5 = 9$
3. $7 = \frac{c}{2} + 3$
4. $82 = 9d + 10$
5. $25c - 17 = 183$
6. $\frac{w}{2} - 7 = 3$
7. $2 = 6x - 10$
8. $\frac{1}{3}p + 8 = 11$
9. $40 = 13z + 14$
10. $\frac{n}{2} + 3 = 7$
11. $\frac{3}{4}y - 3 = 12$
12. $39 = 16k - 9$
13. $10 = 6a + 7$
14. $9r + 15 = 18$

ANSWERS AND SOLUTIONS START ON PAGE 360.

Solving Equations with Parentheses

Part of the information in some equations may be contained in parentheses. A set of parentheses groups together the numbers and letters that are to be multiplied or divided by some other amount. In the expression $5(m + 6)$, both m and 6 are to be multiplied by 5.

Before you can solve an equation that has parentheses, you must get rid of the parentheses. To do this, multiply each number and letter inside the parentheses by the number on the outside. This is an application of the distributive law, which you learned in Chapter 2. See page 40 if you need a review of this law.

EXAMPLE 1: $5(m + 6)$
$5 \cdot m + 5 \cdot 6$
$5m + 30$

EXAMPLE 2: $3(2m - 1)$
$3 \cdot 2m - 3 \cdot 1$
$6m - 3$

To solve an equation that contains parentheses, first multiply the terms inside the parentheses by the number on the outside. Then follow the rules for solving equations. Remember that a dot (·) between two numbers or letters means multiplication.

EXAMPLE 3: Solve for c in $4(c - 6) = 12$.

STEP 1. Multiply both c and 6 by 4.

$$4(c - 6) = 12$$
$$4c - 24 = 12$$

STEP 2. Add 24 to both sides.

$$\begin{array}{rcr} 4c - 24 & = & 12 \\ +24 & & +24 \\ \hline 4c & = & 36 \end{array}$$

STEP 3. Divide both sides by 4. The solution is 9. $\mathbf{c = 9}$

$$\frac{4c}{4} = \frac{36}{4}$$
$$\mathbf{c = 9}$$

Basics of Algebra Exercise 8

Directions: Solve each equation.

1. $3(x + 4) = 27$

2. $10 = 5(a - 3)$

3. $36 = 4(c + 4)$

4. $28 = 7m - 32$

5. $3(y - 2) = 3$

6. $80 = 8(s + 7)$

7. $3(t + 10) = 90$

8. $7(m - 2) = 28$

ANSWERS AND SOLUTIONS START ON PAGE 361.

Writing and Solving Longer Equations

You learned earlier in this chapter to write one-step equations from words. To write longer equations, follow the same procedure. Watch for verbs such as *is* and *equals*. These verbs tell you where to put the equal sign.

EXAMPLE 1: Eight more than three times a number is twenty. Find the number.

STEP 1. Use a letter to represent the unknown in the expression, "Eight more than three times a number." The verb *is* tells you to put the equal sign before 20.

$$3x + 8 = 20$$

STEP 2. Solve the equation. Your first step is to subtract 8 from both sides.

$$\begin{aligned} 3x + 8 &= 20 \\ -8 &\quad -8 \\ 3x &= 12 \end{aligned}$$

STEP 3. To get the unknown alone, divide both sides by 3. $\mathbf{x = 4}$

$$\frac{3x}{3} = \frac{12}{3}$$

$$\mathbf{x = 4}$$

EXAMPLE 2: One-third of a number decreased by seven equals four. Find the number.

STEP 1. Use a letter to represent the unknown, and write the equation. The verb *equals* tells you where to put the equal sign.

$$\tfrac{1}{3}x - 7 = 4$$

STEP 2. Solve the equation. Your first step is to add 7 to both sides.

$$\begin{aligned} \tfrac{1}{3}x - 7 &= 4 \\ +7 &\quad +7 \\ \tfrac{1}{3}x &= 11 \end{aligned}$$

STEP 3. To get the unknown alone, multiply both sides by 3. $\mathbf{x = 33}$

$$3 \cdot \tfrac{1}{3}x = 11 \times 3$$

$$\mathbf{x = 33}$$

Solving Equations with Separated Unknowns

Sometimes the unknowns are separated in an equation. You must combine the unknowns before you can solve the equation. The numbers next to letters can be added *only* if the letters are the same. Look at these examples carefully.

$$\begin{array}{r} 5x \\ +\ 2x \\ \hline 7x \end{array} \qquad \begin{array}{r} 4a \\ +\ \ a \\ \hline 5a \end{array} \qquad \begin{array}{r} 6c \\ -\ 5c \\ \hline c \end{array}$$

Notice that in the second example, a is understood to be $1a$. In the third example, the answer c is the same as $1c$.

EXAMPLE 1: Solve for x in $5x - 2x + 8 = 26$.

STEP 1. Combine the unknowns (x's): $5x - 2x = 3x$.

$$5x - 2x + 8 = 26$$
$$3x + 8 = 26$$

STEP 2. To get the unknowns alone on one side, first perform the inverse of addition. Subtract 8 from both sides.

$$\begin{array}{rcr} 3x + 8 & = & 26 \\ -\ 8 & & -\ 8 \\ \hline 3x & = & 18 \end{array}$$

STEP 3. Divide both sides by 3. The solution is 6.
$\boldsymbol{x = 6}$

$$\frac{3x}{3} = \frac{18}{3}$$
$$\boldsymbol{x = 6}$$

EXAMPLE 2: Solve for a in $9a - 3 = 2a + 11$.

STEP 1. Since the a's are on different sides of the equal sign, use inverse operations to combine them. Subtract $2a$ from both sides.

$$\begin{array}{rcr} 9a - 3 & = & 2a + 11 \\ -\ 2a & & -\ 2a \\ \hline 7a - 3 & = & 11 \end{array}$$

STEP 2. Use inverse operations again to isolate the $7a$ on the left side. Add 3 to both sides.

$$\begin{array}{rcr} 7a - 3 & = & 11 \\ +\ 3 & & +\ 3 \\ \hline 7a & = & 14 \end{array}$$

STEP 3. Divide both sides by 7. The solution is 2.
$\boldsymbol{a = 2}$

$$\frac{7a}{7} = \frac{14}{7}$$
$$\boldsymbol{a = 2}$$

Rules for Solving Equations with Separated Unknowns

1. Combine the unknowns.
   a. If the unknowns are on the *same* side of the equal sign, combine them according to the rules for adding numbers.
   b. If the unknowns are on *different* sides of the equal sign, combine them by using inverse operations.
2. Use inverse operations to solve the equation.

Notice that in Example 2, the unknowns are combined on the side with the greater value ($9a$ is more than $2a$). Later on, when you learn about positive and negative numbers, you will see that you could also subtract $9a$ from $2a$ on the right side. For now, however, always combine unknowns on the side that has the greatest unknown value.

EXAMPLE 3: Solve for y in $20 - 2y = 3y$.

STEP 1. The unknowns are on different sides of the equal sign. Combine them on the side with the greater unknown value by adding $2y$ to both sides.

$$\begin{array}{rcr} 20 - 2y & = & 3y \\ + 2y & & + 2y \\ \hline 20 & = & 5y \end{array}$$

STEP 2. Divide both sides by 5. The solution is 4. $\mathbf{y = 4}$

$$\frac{20}{5} = \frac{5y}{5}$$
$$\mathbf{4 = y}$$

Basics of Algebra Exercise 9

Directions: Solve and check each equation.

1. $5y - y = 19 + 9$
2. $6t + 8 + 4t = 58$
3. $9c = 44 - 2c$
4. $8m = 2m + 30$
5. $4a + 55 = 9a$
6. $4p = p + 18$
7. $6f = 14 - f$
8. $3 = y + 8y$
9. $8r + 17 = 5r + 32$
10. $7n - 9 = 3n + 7$
11. $6z + 11 = 5z + 20$
12. $5y - 4 = 2y + 77$

ANSWERS AND SOLUTIONS START ON PAGE 361.

Problem Solving for the GED

Using Algebra to Solve Word Problems

The Millers, the Rigbys, and the Smiths went on a camping trip. The Millers spent \$100 more than the Rigbys, and the Smiths spent twice as much as the Millers. If the cost of the trip was \$580 altogether, how much did the Smith family spend?

(1) \$70
(2) \$100
(3) \$170
(4) \$195
(5) \$340

When you use algebra to solve a word problem, your goal is to *write and solve an equation* based on the problem. There are several steps that will help you do this.

1. Use the variable x to represent a value that you do not know in the problem.

   In the problem above, use x to represent the amount of money the Rigbys spent.
   x = \$ spent by Rigbys

2. Express the other information given in the problem by using algebraic expressions as you learned earlier.

   You are told that the Millers spent \$100 more than the Rigbys, which we have represented with x.
   $x + 100$ = \$ spent by Millers

   You are also told that the Smiths spent twice as much as the Millers. Write an algebraic expression for this amount.
   $2(x + 100)$ = \$ spent by Smiths

3. Once you have written algebraic expressions for the information in the problem, write an equation that is based on this information.

   In this problem, you are told that the amount spent by all three families was \$580. You have written expressions for the amounts spent by each of the three. Your equation should look like this:
   $x + (x + 100) + 2(x + 100) = 580$

4. Solve the equation.

$$
\begin{aligned}
x + (x + 100) + 2(x + 100) &= 580 \\
2x + 100 + 2x + 200 &= 580 \\
4x + 300 &= 580 \\
-300 &\quad -300 \\
\hline
4x &= 280 \\
\frac{4x}{4} &= \frac{280}{4} \\
\boldsymbol{x} &= \mathbf{\$70}
\end{aligned}
$$

5. Now that you know the value of x, you can find the amount you are looking for.

   This problem asks for the amount the Smiths spent. Substitute \$70 for x in the expression $2(x + 100)$.

   $2(70 + 100) = 340$
   answer: **(5)** \$340

★ GED PRACTICE ★

Using Algebra to Solve Word Problems

Stan's age is one more than twice his brother Bill's age. If Stan is seventeen years old, how old is Bill?

(1) 5
(2) 8
(3) 9
(4) 16
(5) 18

ANSWERS AND SOLUTIONS START ON PAGE 362.

Basics of Algebra Exercise 10

Directions: Solve each problem. Your first step should be to use a variable to stand for a value you do not know. Follow the steps described above if you need help.

1. The sum of two consecutive whole numbers is 27. Find the numbers. (The word *consecutive* means one following another.)

2. The sum of three numbers is 60. The second number is two more than the first. The third number is two more than the second. Find the three numbers.

3. There are six times as many union workers as there are nonunion workers at the Acme Dye Plant. Eight less than the number of union workers is equal to four times the number of nonunion workers. How many union workers are there at the plant?

4. There are three times as many women in a GED class as there are men. When the number of women is decreased by two, the result is the same as when the number of men is increased by ten. Find the number of men and women in the class.

5. Juan and his boss Felipe are house painters. For every dollar that Juan gets, Felipe gets \$3. They made \$360 for painting a house. How much did Felipe make on the job?

6. Mr. Migliaccio makes twice as much money as his wife does. Together the Migliaccios make \$294 a week. Which of the following expressions gives the amount of money Mrs. Migliaccio makes per week?

   (1) $(294 \times 2) - 294$
   (2) $\frac{294}{3}$
   (3) $\frac{294}{2}$
   (4) $294 \div \frac{1}{2}$
   (5) $294 \div \frac{1}{3}$

7. Paul, Jeff, and Jerry worked together fixing Paul's car. Jeff worked twice as many hours as Jerry, and Paul worked six hours more than Jeff. Altogether they worked 51 hours. How many hours did Paul work?

8. In the last local election in Centerville, three voters failed to vote for every voter who went to the polls. There are 22,000 registered voters in Centerville. How many of them failed to vote during the last election?

9. David is four times as old as Laila. Nancy is two years older than David. If Laila is eighteen years old, which of the following expressions gives you Nancy's age?

   (1) $\frac{18}{4} + 2$
   (2) $18 + 4 + 2$
   (3) $18 + 2$
   (4) $(4 \times 18) + 2$
   (5) $4(18 + 2)$

10. Unextended, the ladder Audrey uses to reach the roof of her house is one half the height of the house. The ladder can then extend 5 more feet. If Audrey's house is forty feet tall, which of the following expressions gives the length of the ladder, fully extended?

    (1) $(\frac{1}{2} \times 40) + 5$
    (2) $(\frac{1}{2} \times 40) - 5$
    (3) $\frac{1}{2}(40 + 5)$
    (4) $(2 \times 40) + 5$
    (5) $(2 \times 40) - 5$

ANSWERS AND SOLUTIONS START ON PAGE 362.

Formulas

What Is a Formula?

A ***formula*** is an algebra equation much like those you have been working with. A formula is different from an equation in one way: It is a rule that is always true. For example, when you learned that "Interest = Principal × Rate × Time" in Chapter 6, you were really using a formula: $i = prt$. The equations you have been working with so far have only *one* letter. Formulas have more than one letter. In this chapter and in Geometry, you will learn to use many different formulas to help you solve problems.

To use any formula, you substitute values you *do* know for certain letters in the formula. When you ***substitute***, you replace a letter with a number. For example, how can you solve the problem below?

> Sam put $2,000 in a savings account that pays 5% interest per year. He left the money in for one year. How much interest did he earn?

As you probably remember from your work in Chapter 6, you need to use the formula $i = prt$. You are looking for i (interest), so you substitute the values given in the problem for the other letters of the formula. The principal (p) is $2,000, the rate ($r$) is 5%, and the time ($t$) is 1.

$$i = p \times r \times t$$
$$i = \$2{,}000 \times 5\% \times 1$$
$$i = \$100$$

In the example below, notice how your work with equations in this chapter can help you with formulas.

EXAMPLE: Margaret received $360 in interest payments from the $3,000 she put in her savings account. Her rate of interest was 6%. How long was her money in this account in order to earn this interest?

(1) $i = prt$

(2) $\$360 = \$3{,}000 \times 6\% \times t$

(3) $\$360 = \$180 \times t$

(4) $\dfrac{\$360}{\$180} = \dfrac{\$180 \times t}{\$180}$

$\mathbf{2 = t}$

STEP 1. Since this is an interest problem, write down the formula for simple interest.

STEP 2. Substitute values you do know for the letters in the formula. You know that the interest is $360, the principal is $3,000, and the rate if 6%. Time (t) is the value you are trying to find.

STEP 3. As a first step in solving the equation, multiply $3,000 by 6%. Then substitute the result in the equation.

STEP 4. As with any equation, your goal is to get the unknown value *alone* on one side. Since t is being multiplied by 180, use the inverse operation, division. Divide *both sides* of the equation by 180. t = **2 years.**

Rules for Using Formulas to Solve Word Problems

1. Choose the formula you need, and *write it down*.
2. Substitute values you *do* know for the letters in the formula.
3. Solve the equation using inverse operations where necessary.

In the next chapter, you will learn more about different formulas and when they can be used. You will not need to memorize any formulas because you will be provided with a "formulas page" with the GED Math Test. On this page, you will find all the formulas you will need to solve problems on the GED. You will, however, need to be able to choose and use the correct formula. This section and the following chapter will help you learn to do this effectively.

The Distance Formula

One formula that will appear on the GED formula page is the ***distance formula***.

$d = rt$

In this formula, d = distance, r = rate, and t = time. In other words, you must multiply the rate by the time in order to find distance. Remember that when two numbers or letters appear next to each other with no operation symbol between them, you multiply them together. Let's take a look at a word problem in which this formula would be helpful:

Linda drove 180 miles at a consistent speed of sixty miles per hour. How many hours did she drive?

Follow the rules for using formulas to solve word problems:

1. Choose a formula and write it down.

   $d = rt$

2. Substitute the values you know for the letters.

   $180 = 60t$

3. Solve the equation by using inverse operations. You want to get the unknown (t) alone on a side.

   $$\frac{180}{60} = \frac{60t}{60}$$
   $$3 = t$$

   Linda was driving for **3 hours.**

Total Cost Formula

Another formula on the GED formula page tells you how total cost relates to unit cost and number of units:

$c = nr$

In this formula, c = total cost, n = number of units, and r = rate, or cost per unit. Use this formula to solve the following problem.

Kate bought some towels for $7.90 each. Her total bill came to $31.60. How many towels did she buy?

1. Choose a formula and write it down.

   $c = nr$

2. Substitute the values you know for the letters.

   $\$31.60 = n(\$7.90)$

3. Solve the equation using inverse operations.

   $$\frac{\$31.60}{\$7.90} = \frac{n(\$7.90)}{\$7.90}$$
   $$4 = n$$

   Kate bought 4 towels.

You may have found that you did not need a formula to solve either the distance problem or the total cost problem. This is fine—you should use whatever method you find successful. But formulas can be useful as problems get more difficult, so you should become familiar with working with them.

Basics of Algebra Exercise 11

Directions: Try using formulas to solve the following problems.

1. A plane travels 1,676 miles in 4 hours. What is the plane's rate of travel?

2. A manufacturer sold $111.25 worth of hammers to a hardware store. The cost per unit of the hammers was $1.25. How many hammers did the hardware store buy?

3. Temperature can be read in two different ways—Celsius and Fahrenheit. To change from one system to another, you use the following formula: $F = (\frac{9}{5})C + 32$, where F = degrees in Fahrenheit and C = degrees in Celsius.

If the temperature was 20 degrees Celsius on Tuesday, what was the temperature in Fahrenheit?

4. Anne's savings account earned $32.50 interest in $\frac{1}{2}$ year. If the bank's interest rate was 5%, how much money did Anne have in the bank to begin with?

ANSWERS AND SOLUTIONS START ON PAGE 363.

Basics of Algebra Exercise 12: Review

Directions: Write out the following expressions.

1. Write an algebraic expression for a number increased by eleven.
2. Write an expression for the sum of three times a number and one-third of the same number.
3. Cheese costs c per pound. Write an expression for the cost of four pounds of cheese.
4. Andrew's gross salary is s. He takes home $\frac{4}{5}$ of his gross salary. Write an expression for Andrew's take-home salary.

Directions: For questions 5 through 12, solve each equation.

5. $9x = 108$
6. $\frac{c}{4} = 8$
7. $14 + x = 3x - 8$
8. $\frac{21}{p} + 1 = 8$
9. $\frac{1}{8}y = 24$
10. $12.5 = a + 1.5$
11. $96 + 2x = x + 101$
12. $20 - 4z = 5z - 35$

13. A number decreased by eighteen is one. Write an equation and solve to find the number.
14. Four-fifths of a number is forty. Write an equation and solve to find the number.
15. Nine times a number decreased by five is the same as six times the same number increased by seven. Write an equation and solve to find the number.

★ GED PRACTICE ★

16. When Ron's truck was full, it weighed 7.5 tons. The weight he carried at that time was 3.25 tons. How many tons does Ron's truck weigh when it is empty?
    (1) 3.25
    (2) 4.25
    (3) 10.75
    (4) 21.75
    (5) 42.50

17. 10% (.10) of Sal's take-home pay goes for car payments each month. His car payments are $84 a month, and his rent is $260 per month. How much is Sal's take-home pay each month?
    (1) $8.40
    (2) $34.40
    (3) $344
    (4) $840
    (5) $3440

18. There are 25 more women than men in the night school Louise attends. Altogether there are 115 students in the school. How many women attend the night school?
    (1) 25
    (2) 45
    (3) 70
    (4) 90
    (5) 140

19. Sam and Joe are painters. Sam makes \$2 an hour less than Joe. On a job which took them both 40 hours to complete they made \$440. How much does Sam make an hour?
(1) \$3.50
(2) \$4.50
(3) \$6.50
(4) \$45
(5) \$65

Directions: Questions 20 through 23 refer to the following situation.

During the annual Walk for Peace, Nina walked a total of six miles more than her husband Ed. Their combined mileage was 46. Ed's company pledged to donate \$1.00 for each mile that Ed and Nina walked.

20. How many miles did Nina walk?
(1) 6
(2) 14
(3) 20
(4) 26
(5) 40

21. What is the total amount that Ed's company must donate for Ed's and Nina's participation in the walk?
(1) \$6
(2) \$20
(3) \$40
(4) \$46
(5) Not enough information is provided.

22. What more do you need to know to find out the total number of hours that Ed walked?
(1) the hours that Nina walked
(2) Ed's rate of walking
(3) the distance traveled by Ed
(4) the time that Ed stopped walking
(5) the time that Ed started walking

23. Three of Nina's friends each donated \$.50 per mile that Nina and her husband walked. Which of the following expressions gives the total amount of money donated by Nina's three friends?
(1) $3 \times \$.50$
(2) $46 \times \$.50$
(3) $3(46 \times \$.50)$
(4) $\frac{46 \times \$.50}{3}$
(5) $(3 + \$.50)46$

24. Mel Webster makes \$280 a month more than his wife Kathleen. Together they make \$1760 per month. How much money does Mel make in one month?
(1) \$360
(2) \$740
(3) \$840
(4) \$1020
(5) \$1480

25. Ten times a number decreased by seven equals 101 plus that number. Find the number.
(1) 12
(2) 13
(3) 18
(4) 20
(5) 23

ANSWERS AND SOLUTIONS START ON PAGE 363.

11 GEOMETRY

Common Geometric Figures

Geometry is the study of lines, angles, flat figures, and three-dimensional shapes. Geometry is useful in many practical situations you may encounter in everyday life. For example, when you need to figure out how much fencing you should buy to surround a garden or how much liquid you'll need to fill a water tank, knowledge of geometry can help. Many problems in this chapter will show examples of these uses. Geometry also makes up 20 percent of the GED Math Test, and the problems in this chapter will help you prepare for this as well.

To solve geometry problems, you must know some special terms. Take the time now to memorize the terms you do not already know.

An ***angle*** is a figure made of two lines extending from the same point. Figures 1, 2, and 3 are examples of angles. The size of an angle has nothing to do with the length of its lines. The angle in Figure 2 is larger than the other two angles because it is more open.

Figure 1

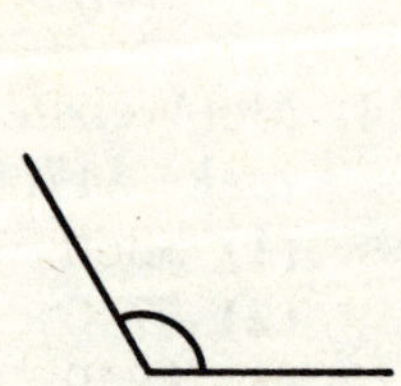
Figure 2

A ***right angle*** looks like a square corner. Figure 3 is a right angle. A small square inside the angle is often used to indicate a right angle. Many everyday objects contain examples of right angles. The top and side of a door form a right angle. The corners of this page are right angles. You will learn the names of other types of angles later in this section.

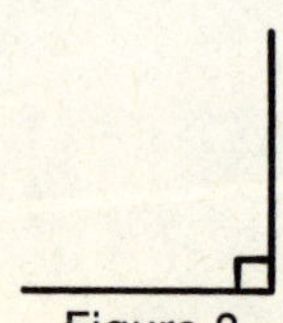
Figure 3

Parallel lines are lines that run in the same direction. Parallel lines never cross no matter how far they are extended. Train tracks are examples of parallel lines. Figures 4, 5, and 6 each show pairs of parallel lines. In Figure 4 the two parallel lines are vertical. ***Vertical*** means running straight up and down. The parallel lines in Figure 5 are horizontal. ***Horizontal*** means running straight across. The parallel lines in Figure 6 are neither vertical nor horizontal.

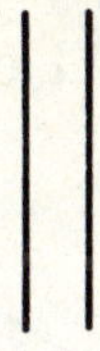

Figure 4

Figure 5

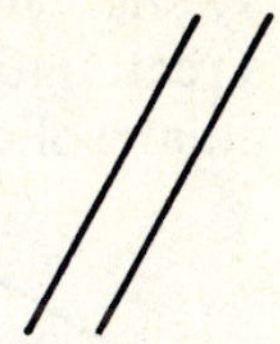

Figure 6

Perpendicular lines are lines that cross or meet to form right angles. Figure 7 is an example of perpendicular lines. The wood or metal supports of a window are usually perpendicular lines.

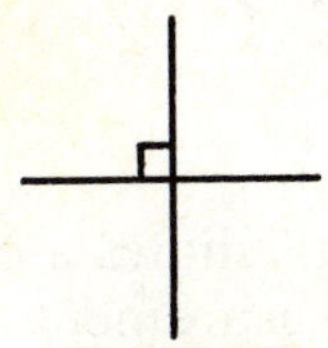

Figure 7

Three common flat shapes are the rectangle, the square, and the triangle. These three shapes are examples of plane figures. A ***plane figure*** is completely flat.

A ***rectangle*** is a four-sided figure with four right angles. The sides across from each other are parallel and equal in length. Figure 8 is a rectangle. A short side and a long side of a rectangle are perpendicular to each other. Rectangles are very common figures. Most walls, floors, windows, and doors are rectangles.

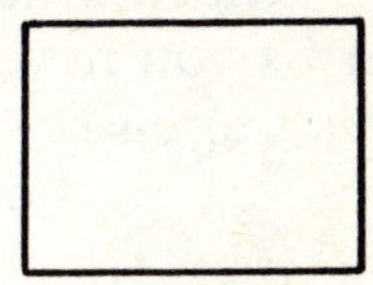

Figure 8

A ***square*** is a four-sided figure with four right angles and four equal sides. The sides across from each other are parallel. The only difference between a square and a rectangle is that the sides of a square all have the same measurement. Figure 9 is a square.

Figure 9

A ***triangle*** is a three-sided figure. Figures 10, 11, and 12 are all triangles. In Figure 10 each side has a different measurement. In Figure 11 each side has the same measurement. In Figure 12 two of the sides are the same, and the third side is longer. You will learn the names of different types of triangles later.

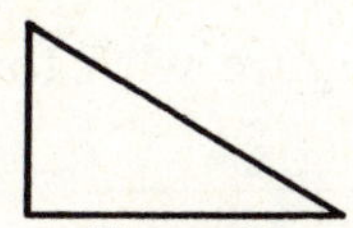

Figure 10

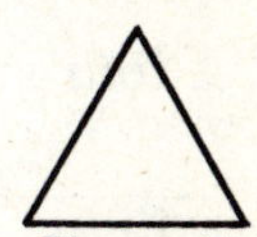

Figure 11

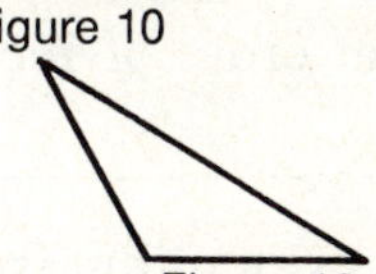

Figure 12

Another common plane figure is the ***circle***. A circle is a flat figure made up of a curve. Every point of the curve is the same distance from the center. Figure 13 shows a circle. The ***diameter*** of a circle is a line that crosses the circle through its center from one side to the other. It also measures the distance across the circle. A ***radius*** of a circle is a line from the center of the circle to any point on the curve of the circle. A radius is half the distance across a circle. In other words a radius is half of the diameter of a circle.

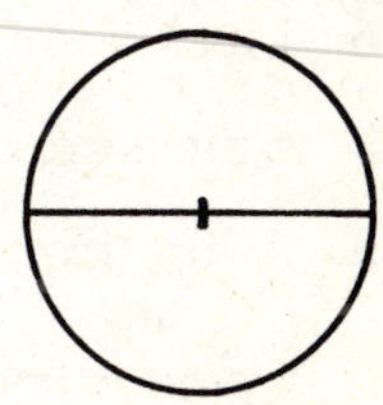

Figure 13

Three-dimensional figures are sometimes called ***solids***. Figures 14, 15, 16, and 17 show four common solids. Figure 14 shows a ***rectangular solid***. The corners of a rectangular solid are all right angles. A cardboard carton is the shape of a rectangular solid.

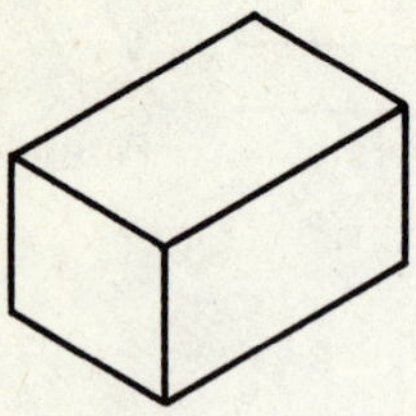

Figure 14

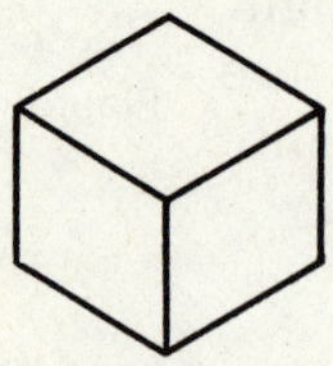

Figure 15

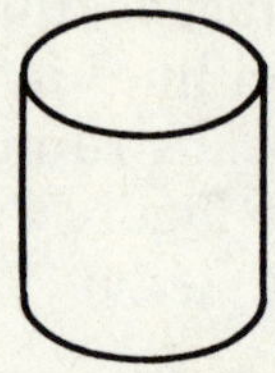

Figure 16

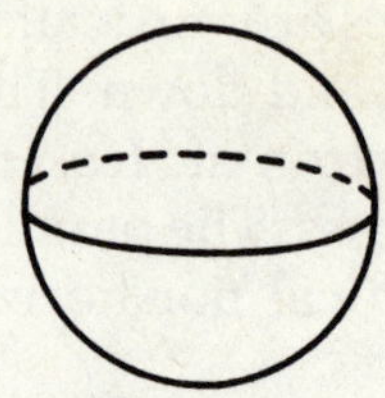

Figure 17

Figure 15 shows a cube. A ***cube*** is a rectangular solid in which each side has the same measurement. A child's letter blocks are usually in the shape of cubes.

Figure 16 shows a ***cylinder***. The top and bottom of a cylinder are circles. The sides of a cylinder are at right angles to the top and bottom. A tin can is in the shape of a cylinder.

Figure 17 shows a sphere. ***Sphere*** is the geometrical term for the shape of a ball.

The next exercise gives you a chance to test your knowledge of the geometrical terms you have learned so far. Be sure you know the terms above before you try this exercise.

Geometry Exercise 1

Directions: For questions 1 through 9, fill in each blank.

1. Lines that run in the same direction and do not cross each other are called ____________________ lines.
2. A four-sided figure with four right angles and four equal sides is called a ____________________.
3. Lines that cross or meet each other to form right angles are called ____________________ lines.
4. Two lines extending from the same point form an ____________________.
5. Lines running straight up and down are called ____________________ lines.
6. Lines running straight across are called ____________________ lines..
7. A measure of the distance across a circle is called the ____________________.
8. A line from the center of a circle to the curve that forms the circle is called the ____________________.
9. The geometrical term for the shape of a basketball is a ____________________.

Directions: Identify the following figures.

10. 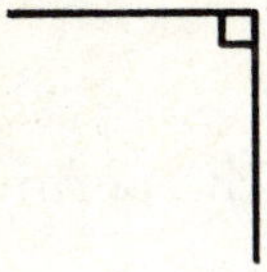

12. 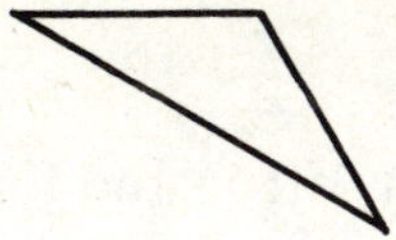

14. 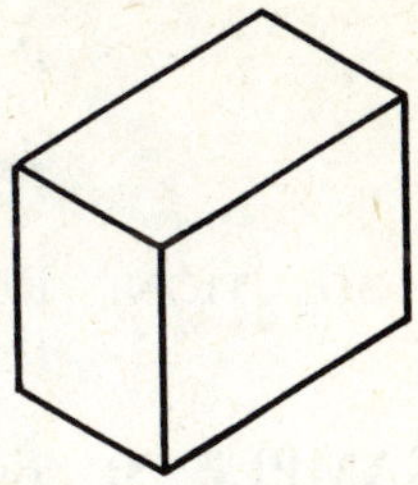

11. 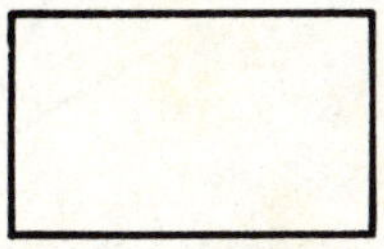

13. 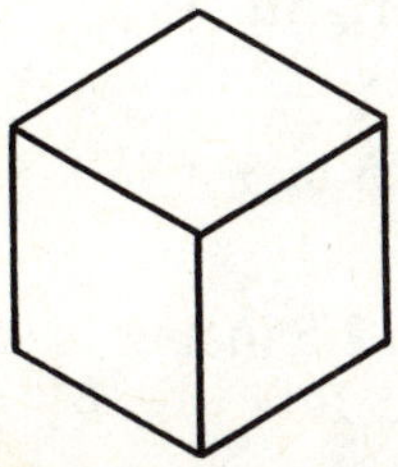

15. 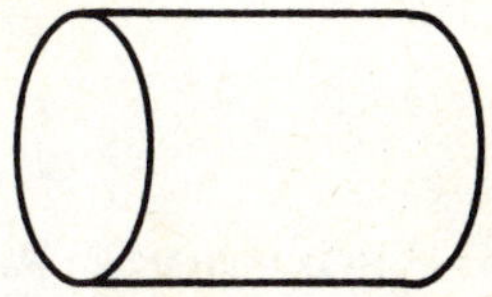

ANSWERS AND SOLUTIONS START ON PAGE 364.

Perimeter and Circumference

Perimeter is the measurement of the distance around a flat figure. The following are the formulas for finding the perimeter of a rectangle, a square, and a triangle. As you know, these formulas will be provided for you on the GED Test, but it is useful to memorize them.

Perimeter of a rectangle: $P = 2l + 2w$; where P = perimeter, l = length, w = width

Perimeter of a square: $P = 4s$; where P = perimeter, s = side

Perimeter of a triangle: $P = a + b + c$; where P = perimeter, and a, b, c are the sides of a triangle

As you recall, to use a formula, start by substituting letters with any values that you know.

EXAMPLE 1: Find the perimeter of the rectangle pictured at the right.

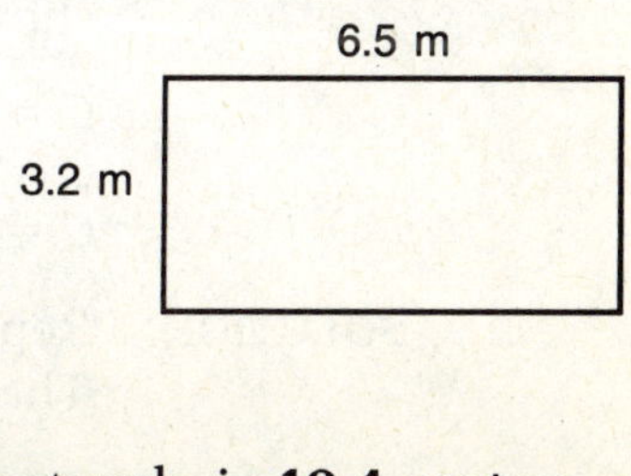

$P = 2l + 2w$
$P = 2(6.5 \text{ m}) + 2(3.2 \text{ m})$
$P = 13 \text{ m} + 6.4 \text{ m}$
$P = \mathbf{19.4 \text{ m}}$

SOLUTION. Replace l with 6.5 m and w with 3.2 m in the formula $P = 2l + 2w$. The perimeter of the rectangle is **19.4 meters.**

EXAMPLE 2: Find the perimeter of the square shown at the right.

$P = 4s$
$P = 4(3\text{ ft})$
$P = 12\text{ ft}$

SOLUTION. Replace s with 3 ft in the formula $P = 4s$. The perimeter of the square is **12 feet**.

EXAMPLE 3: Find the perimeter of the triangle shown at the right.

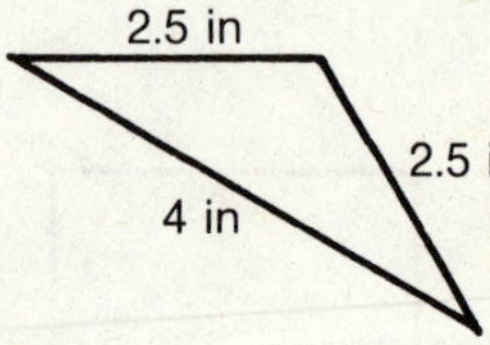

$P = a + b + c$
$P = 2.5\text{ in} + 2.5\text{ in} + 4\text{ in}$
$P = \mathbf{9\text{ in}}$

SOLUTION. Replace a with 2.5 inches, b with 2.5 inches, and c with 4 inches in the formula $P = a + b + c$. The perimeter of the triangle is **9 inches.**

> **TIP:** In problems like these examples, you may find it just as easy to add up the length of each side instead of using a formula. In more challenging problems, however, you may find that using a formula is a big help.

The distance around a circle is called the ***circumference***, not the perimeter. The formula for the circumference of a circle includes the Greek letter π (pi). The letter π stands for the ratio of the circumference of a circle to the diameter.

The value of π is approximately 3.14 or $\frac{22}{7}$. The symbol that means "approximately equal to" is $\cong$. We write $\pi \cong 3.14$, or $\pi \cong \frac{22}{7}$.

> **TIP:** It is useful to be familiar with *both* values of π, because in some problems a fraction is easier to use, while in others a decimal will make the computation easier.

The formula for the circumference of a circle is:

$$C = \pi d;\text{ where } c = \text{circumference},\ \pi = 3.14,\text{ and } d = \text{diameter}$$

On the GED Test, you will be told when to use the value $\frac{22}{7}$ for π. Use the value 3.14 for π unless a problem states otherwise.

EXAMPLE 4: Find the circumference of the circle shown at the right.

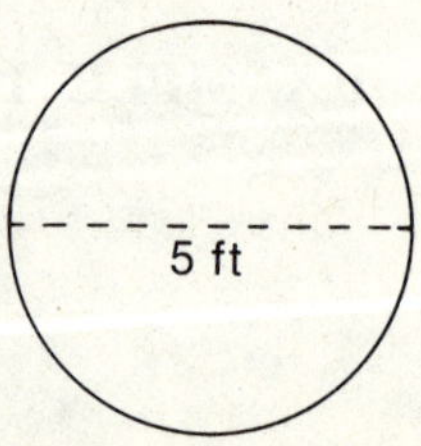

$C = \pi d$
$C = 3.14 \times 5\text{ ft}$
$C = 15.70 = 15.7\text{ ft}$

SOLUTION. Replace π with 3.14 and d with 5 ft in the formula $C = \pi d$. The circumference of the circle is **15.7 feet.**

EXAMPLE 5: Find the circumference of the circle shown at the right. Use $\frac{22}{7}$ for π.

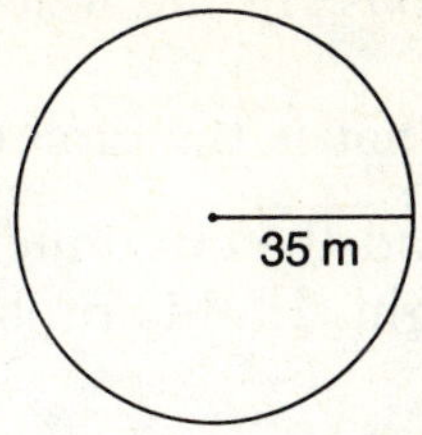

(1) $d = 2 \times 35 = 70$ m

(2) $C = \pi d$
$C = \frac{22}{7} \times 70$
$C = \mathbf{220\ m}$

STEP 1. Notice that the picture shows the radius of the circle. To find the diameter, multiply the radius by 2.

STEP 2. Replace π with $\frac{22}{7}$ and d with 70 m in the formula $C = \pi d$. The circumference of the circle is **220 m.**

Geometry Exercise 2

Directions: Solve each problem.

1. What is the perimeter of the rectangle shown at the right?

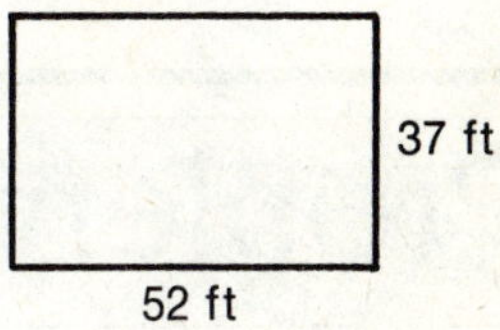

2. Find the perimeter of the triangle shown at the right.

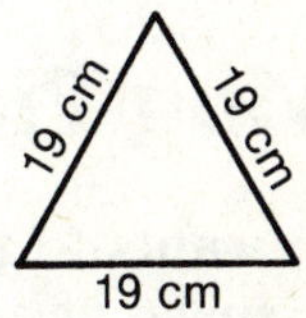

3. Find the perimeter of the square shown at the right.

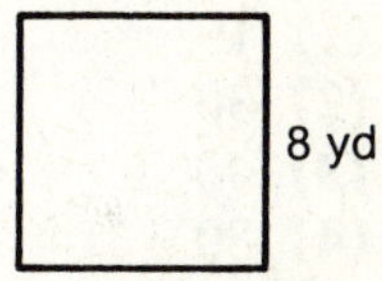

4. Find the perimeter of a rectangle with a length of $2\frac{3}{4}$ inches and a width of $1\frac{1}{2}$ inches.
5. Find the perimeter of a square with a side that measures $\frac{1}{2}$ inch.
6. Find the perimeter of a triangle whose sides measure $7\frac{3}{4}$ inches, $9\frac{1}{4}$ inches, and $10\frac{1}{2}$ inches, respectively.
7. What is the perimeter of a rectangle that is 6.3 miles long and 5.2 miles wide?
8. Each side of a square measures 1.75 meters. Find the perimeter of the square.
9. What is the perimeter of the triangle shown at the right?

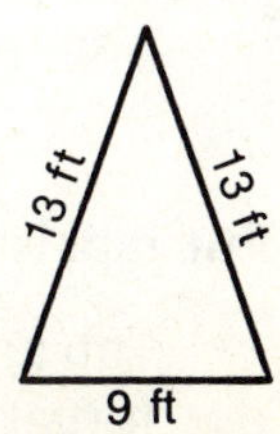

Directions: In the following problems use $\pi = 3.14$ unless told to do otherwise.

10. What is the circumference of a circle with a diameter of 10 yards?

11. Find the circumference of the circle pictured at the right. In this problem, use $\pi = \frac{22}{7}$.

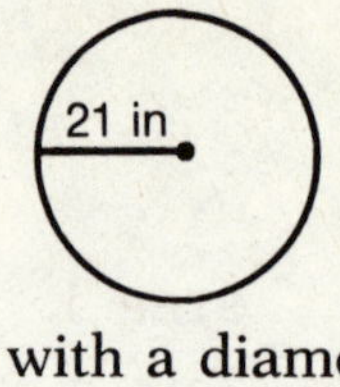

12. Find the circumference of a circle (to the nearest tenth) with a diameter of 1.5 cm.

13. Find the circumference of a circle with a diameter of 14 yards. In this problem, use $\pi = \frac{22}{7}$.

ANSWERS AND SOLUTIONS START ON PAGE 364.

Problem Solving for the GED

Drawing Pictures

Pete wants to put a fence around a rectangular section of his backyard. The space he wants to enclose measures 30 feet in length, and it is $\frac{1}{3}$ as wide. How many feet of fencing does he need?

(1) 10
(2) 30
(3) 40
(4) 80
(5) 120

It may be difficult to decide just how to put numbers together in a problem like this. You may be trying to picture the fencing the same time you are figuring out numbers. Sometimes it can help to actually *draw* a rough picture of the figure being described. Once you have drawn the figure, you can jot down measurements where they belong. Watch how the steps below help to solve this problem neatly and logically.

1. Decide what kind of figure is being described (circle, triangle, rectangle, square), and draw it.

   This problem mentions a *rectangular* space.

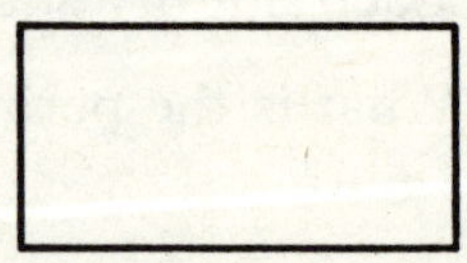

2. Put labels on the dimensions you are given.

   You are told that the length is 30 feet. Don't forget to write 30 next to *both* lengths. You are also told that the width is $\frac{1}{3}$ the length, or 10 feet. Write 10 next to *both* widths.

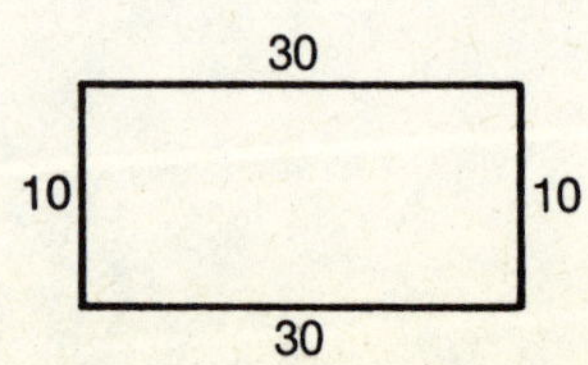

3. Use a formula or your own method to find the distance around the figure, the perimeter.

$P = 2l + 2w$
$P = (2 \times 30) + (2 \times 10)$
$P = 60 + 20 =$ **80 feet** of fencing.
Answer: **(4)**

★ **GED PRACTICE** ★

Drawing Pictures

Joanna's job is to sew strips of lace around four cushions. Two square cushions have sides that measure 10 inches. Two rectangular cushions are each 12 inches in length and $\frac{1}{2}$ as wide. How many inches of lace does Joanna need to sew around these cushions?

(1) 22
(2) 76
(3) 112
(4) 124
(5) 152

ANSWER AND SOLUTION ARE ON PAGE 364.

Perimeter and Circumference Word Problems

Word problems give you a chance to apply your knowledge of perimeter and circumference to practical situations. The words *circumference* and *perimeter* will not always appear when you need to use these skills. Sometimes you will have to figure out for yourself that you need to find one of these values.

Remember that perimeter is a measure of the distance around a flat figure. The phrase "find the distance around" means to find the perimeter. Fencing, weatherstripping, and framing are some practical applications of perimeter.

Problems with Different Units of Measurement

Watch out for problems where the units of measurement change. It is important to include labels with any numbers so that you can keep track of what type of measurement you are using — for example, feet or inches, pounds or ounces. You may find it helpful to review what you learned about changing units of measurement in Chapter 8.

EXAMPLE: How many *feet* of picture framing are required to go around a square picture with a side of 30 inches?

(1) $P = 4s$
$P = 4 \times 30$ in

(2) $P = 120$ in

(3) $12\overline{)120}$ = 10 ft

STEP 1. Write the formula for the perimeter of a square. Substitute in the length of one side.
STEP 2. Find P, the total number of inches in the perimeter.
STEP 3. Notice that the question asks for the number of *feet*, and you have found the number of *inches*. To find the number of feet in 120 inches, either set up a proportion or divide as shown above. **10 feet** of framing are required to go around the picture.

"Backward" Perimeter Problems

So far you have been working on problems where you need to find perimeter. In some problems, you may be *given* the perimeter and one other dimension of a figure. You will then be asked to find the other dimension. Your work with equations in The Basics of Algebra will help you a great deal with this type of problem.

EXAMPLE: Sam has a rectangular garden surrounded by 50 feet of fencing. The fencing is 10 feet wide. How long is the fencing?

(1) $P = 2l + 2w$

(2) $50 = 2l + (2 \times 10)$

(3) $50 = 2l + 20$

(4)
$$\begin{array}{rcl} 50 & = & 2l + 20 \\ -20 & & \quad -20 \\ \hline 30 & = & 2l \end{array}$$

(5) $\frac{30}{2} = \frac{2l}{2}$

$\mathbf{15 = l}$

STEP 1. Write down the formula for perimeter.
STEP 2. Substitute in the values you *do* know: $P = 50$ and $w = 10$.
STEP 3. Do the computation on the right side of the equation.
STEP 4. To get the unknown value alone, use inverse operations and first subtract 20 from each side.
STEP 5. Using inverse operations again, divide both sides by 2.
l = **15 feet**

Geometry Exercise 3

Directions: Solve each problem. Draw a figure if it will help you to visualize a problem.

1. Mark wants to put a frame around an 8-inch by 10-inch photograph. How many inches of framing does he need?

2. The garden in Sylvia's yard is 26 meters long and half as wide. How many meters of fencing does she need to enclose the garden?

3. Find the measurement of the side of the largest square frame that can be made from 100 inches of framing. Express your answer in feet and inches.

4. Find the measurement in feet and inches around a circular mirror with a diameter of 35 inches. Let $\pi = \frac{22}{7}$.

Directions: Questions 5 through 7 refer to the following situation. You may want to try drawing a picture as shown in the problem-solving section on page 230.

There are six windows on the second floor of Mrs. Jackson's house. Each window has a base of 36 inches and a height of 42 inches. Around each window, Mrs. Jackson wants to put weatherstripping that costs $.60 a foot.

5. What is the distance in inches around one window on the second floor of Mrs. Jackson's house?

6. Find the measurement in feet around all the windows on the second floor.

7. What is the total cost of weatherstripping for the six windows?

8. The figure at the right shows the dimensions of the triangular garden in the corner of Robert's yard. How many yards of fencing are required to enclose the garden?

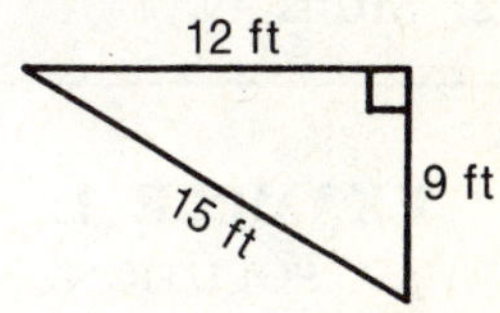

9. Workers are putting brick trim around the edge of a circular pool in the park in Central City. The diameter of the pool is 50 feet. How many 9-inch-long bricks are needed to go around the edge of the pool? Your answer must be a whole number.

10. Paul has 19 feet of oak trim to put around a rectangular countertop that he is building in his kitchen. The width of counter is 30 inches. How long can the countertop be if Paul uses all of the oak trim?

11. Which of the following represents the perimeter of the figure at the right?
(1) 2.3×1.4
(2) $2 \times 2.3 \times 1.4$
(3) $(2 \times 2.3) + (2 \times 1.4)$
(4) $(2.3 + 1.4)/2$

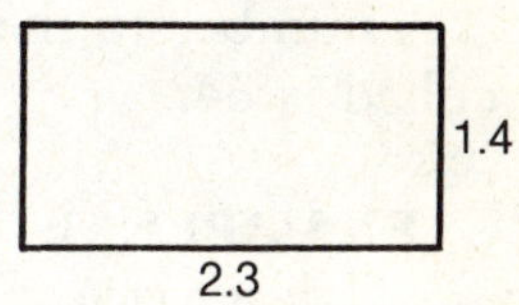

12. Which of the following represents the perimeter of the figure at the right?
(1) $2(1.6) + 2.4$
(2) $1.6 + 2.4$
(3) $1.6 + 2(2.4)$
(4) $2(1.6) + 2(2.4)$

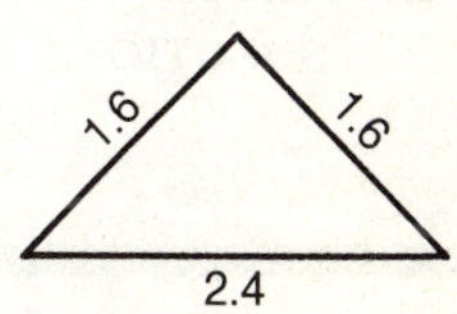

ANSWERS AND SOLUTIONS START ON PAGE 365.

Powers and Roots

Powers

Before you go on to study area, you need to learn two skills that you will use in both geometry and algebra. These skills — finding powers and roots — are an extension of the arithmetic skills you have already developed.

The figure 5^2 is read as "five to the second power." The 5 is called the ***base***. The 2 is called the ***exponent***. The exponent tells how many times to write the base in a multiplication problem.

Rule for Finding Powers

1. Write the base as many times as the exponent indicates.
2. Multiply.

EXAMPLE 1: What is the value of 5^2?

SOLUTION. Write 5 two times and multiply.
$5^2 = 5 \times 5 = 25$

A number to the second power is called the ***square*** of a number. We can say that 5 squared is 25.

EXAMPLE 2: Evaluate 4^3.

SOLUTION. Write 4 three times and multiply. The first product is $4 \times 4 = 16$. The next product is $16 \times 4 = 64$.
$4^3 = 4 \times 4 \times 4 = 64$

A number to the third power is called the ***cube*** of a number. We can say that 4 cubed is 64.

EXAMPLE 3: Find the value of $.03^2$.

SOLUTION. Write .03 two times and multiply. Notice that the answer has four decimal places.
$.03^2 = .03 \times .03 = .0009$

EXAMPLE 4: Evaluate $(\frac{2}{3})^3$.

SOLUTION. Write $\frac{2}{3}$ three times and multiply.
$(\frac{2}{3})^3 = \frac{2}{3} \times \frac{2}{3} \times \frac{2}{3} = \frac{8}{27}$

The number 1 is a special case with powers.
1 to any power equals 1.

EXAMPLE 5: What is 1^6?

SOLUTION. Write 1 six times and multiply. Notice that no matter how many times you multiply 1 by 1, you will get 1 as the answer.
$1^6 = 1 \times 1 \times 1 \times 1 \times 1 \times 1$

> Any number to the first power is that number.

EXAMPLE 6: What is 7^1?

SOLUTION. There is no multiplication in this problem. 7 to the first power means simply to write 7 one time.
$7^1 = 7$

> Any number to the 0 power is 1.

EXAMPLE 7: What is 8^0?

SOLUTION. The zero power actually means a number divided by itself,
$8^0 = \frac{8}{8} = 1$

Reviewing the Order of Operations

You can use powers in combination with other operations. As you learned in Chapter 2, there is a specific order in which to perform operations. The list below shows you where powers fit into this order:

1. Do operations that are grouped in parentheses.
2. Find powers (and roots, which you will learn about in this chapter).
3. Do multiplication or division.
4. Do addition or subtraction.

EXAMPLE 1: Simplify $3^2 + 2^4$.

(1) $3^2 = 3 \times 3 = 9$
$2^4 = 2 \times 2 \times 2 \times 2 = 16$

(2) $9 + 16 = \mathbf{25}$

STEP 1. Find the value of each power.
STEP 2. Add the results.

EXAMPLE 2: Find the value of xy^2 when $x = 9$ and $y = 3$.

$$\begin{aligned} xy^2 &= 9 \cdot 3^2 \\ &= 9 \cdot 9 \\ &= \mathbf{81} \end{aligned}$$

STEP 1. Substitute 9 for x and 3 for y.
STEP 2. Find 3 to the second power.
STEP 3. Multiply 9 by 9.

Geometry Exercise 4

Directions: Find the value of each of the following.

1. 6^2
2. 1^3
3. 5^3
4. $(\frac{3}{5})^2$
5. 3^5
6. $5^2 + 3^2$
7. $5^3 - 10^2$
8. $(.06)^2$
9. 12^0
10. 25^2
11. $11^2 - 4^2$
12. $(.005)^2$
13. $10^0 + 10^1 + 10^2$
14. 20^3
15. 13^1
16. $(\frac{1}{4})^3$
17. 10^4
18. $8^2 - 2^4$
19. $3^3 + 7^2$
20. $(0.1)^3$

ANSWERS AND SOLUTIONS START ON PAGE 365.

Square Roots

In addition to finding powers, another important operation is finding ***square roots***. This operation is indicated by the symbol $\sqrt{\ }$. The square root of a number is found by asking, "What number multiplied by itself will give me this number?"

EXAMPLE 1: Find $\sqrt{25}$.
SOLUTION. 5 multiplied by itself gives 25.
$\sqrt{25} = \mathbf{5}$

EXAMPLE 2: Find $\sqrt{144}$.
SOLUTION. 12 multiplied by itself gives 144.
$\sqrt{144} = \mathbf{12}$

Following is a list of common square roots. Take the time now to memorize this list.

$\sqrt{1} = 1$	$\sqrt{36} = 6$	$\sqrt{121} = 11$	$\sqrt{400} = 20$
$\sqrt{4} = 2$	$\sqrt{49} = 7$	$\sqrt{144} = 12$	$\sqrt{900} = 30$
$\sqrt{9} = 3$	$\sqrt{64} = 8$	$\sqrt{169} = 13$	$\sqrt{1600} = 40$
$\sqrt{16} = 4$	$\sqrt{81} = 9$	$\sqrt{196} = 14$	$\sqrt{2500} = 50$
$\sqrt{25} = 5$	$\sqrt{100} = 10$	$\sqrt{225} = 15$	$\sqrt{3600} = 60$

There are several methods for finding the square roots of larger numbers. One method is to average. The averaging method for finding a square root is a good way to improve your estimating skills. Study the next examples carefully. Think about the first "guess" in each solution.

EXAMPLE 3: Find $\sqrt{484}$.

(1) 20?

(2) $\begin{array}{r} 24 \\ 20\overline{)484} \\ \underline{40} \\ 84 \\ \underline{80} \\ 4 \end{array}$

(3) $\begin{array}{r} 24 \\ \underline{+20} \\ 44 \end{array}$ $\qquad \begin{array}{r} 22 \\ 2\overline{)44} \end{array}$

(4) $\begin{array}{r} 22 \\ \underline{\times 22} \\ 44 \\ \underline{44} \\ 484 \end{array}$

STEP 1. Guess a round number that you think may be fairly close to the correct answer. For example, 50 is a bad guess because $50 \times 50 = 2{,}500$. But 20 is a good guess because $20 \times 20 = 400$, which is much closer to 484.

STEP 2. Divide 20 into 484. Ignore the remainder.

STEP 3. Average the division answer, 24, with the guess, 20.

STEP 4. Check by multiplying 22 by itself. Since $22 \times 22 = 484$, **22** is the square root of 484.

Rules for Finding Square Roots

1. Guess. Choose a number that you think is fairly close to the correct square root.
2. Divide the number you guessed into the number for which you are finding the square root.
3. Add your guess and the number you get when you divide. Divide by 2 to get the average.
4. To check your answer, multiply the average by itself.

EXAMPLE 4: Find $\sqrt{841}$.

(1) 30?

(2) $\begin{array}{r} 28 \\ 30\overline{)841} \\ \underline{60} \\ 241 \\ \underline{240} \\ 1 \end{array}$

(3) $\begin{array}{r} 30 \\ \underline{+28} \\ 58 \end{array}$ $\qquad \begin{array}{r} 29 \\ 2\overline{)58} \end{array}$

(4) $\begin{array}{r} 29 \\ \underline{\times 29} \\ 261 \\ \underline{58} \\ 841 \end{array}$

STEP 1. Guess 30 because $30 \times 30 = 900$, which is close to 841. Notice that the guess can be larger than the final answer.

STEP 2. Divide 841 by 30. Ignore the remainder.

STEP 3. Average 30 and 28.

STEP 4. Check by multiplying 29 by itself.

Many students find it difficult to guess an answer. This takes practice and a willingness to experiment. Sometimes the average of the two numbers does not turn out to be the square root. In this case, use the average as a *second* guess. Example 5 on the next page shows you how this works.

EXAMPLE 5: Find $\sqrt{1,444}$.

(1) 30?

(2)
$$\begin{array}{r} 48 \\ 30\overline{)1,444} \\ \underline{1\,20} \\ 244 \\ \underline{240} \\ 4 \end{array}$$

(3)
$$\begin{array}{r} 30 \\ \underline{+\,48} \\ 78 \end{array} \qquad \begin{array}{r} 39 \\ 2\overline{)78} \end{array}$$

(4)
$$\begin{array}{r} 39 \\ \underline{\times\,39} \\ 351 \\ \underline{1\,17} \\ 1,521 \end{array}$$

STEP 1. Suppose you guess 30 because $30 \times 30 = 900$.
STEP 2. Divide 1,444 by 30.
STEP 3. Average 30 and 48.
STEP 4. Check by multiplying 39 by itself. In fact, 39 is *not* the square root of 1,444. Use 39 as your next guess and repeat steps 2, 3, and 4.

(2)
$$\begin{array}{r} 37 \\ 39\overline{)1,444} \\ \underline{1\,17} \\ 274 \\ \underline{273} \\ 1 \end{array}$$

(3)
$$\begin{array}{r} 39 \\ \underline{+\,37} \\ 76 \end{array} \qquad \begin{array}{r} 38 \\ 2\overline{)76} \end{array}$$

(4)
$$\begin{array}{r} 38 \\ \underline{\times\,38} \\ 304 \\ \underline{1\,14} \\ 1,444 \end{array}$$

STEP 2. Divide 1,444 by 39.
STEP 3. Average 39 and 37.
STEP 4. Check by multiplying 38 by itself. Yes, **38** is the square root of 1,444.

Geometry Exercise 5

Directions: Find the following square roots.

1. $\sqrt{196}$

2. $\sqrt{441}$

3. $\sqrt{1,024}$

4. $\sqrt{1,521}$

5. $\sqrt{361}$

6. $\sqrt{676}$

7. $\sqrt{1,849}$

8. $\sqrt{3,364}$

9. $\sqrt{4,624}$

10. $\sqrt{6,241}$

11. $\sqrt{7,396}$

12. $\sqrt{9,604}$

ANSWERS AND SOLUTIONS START ON PAGE 366.

Area

Area is a measure of the amount of surface within the perimeter of a flat figure. Area is measured in square units — for example, square inches, square feet, or square meters.

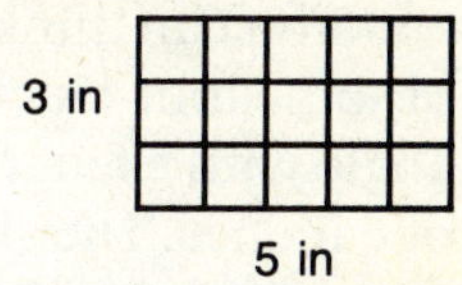

The figure at the right represents a rectangle that has a length of 5 inches and a width of 3 inches. The surface of the rectangle is covered with 15 small squares. Each square is one square inch. The surface, or area, of the rectangle is 15 square inches.

Following are the formulas for area that you will use in this section. The formulas will be given on the GED, but it is also useful to memorize them.

Area of a rectangle: $A = lw$; where A = area, l = length, and w = width

Area of a square: $A = s^2$; where A = area, and s = side

Area of a triangle: $A = \frac{1}{2}bh$; where A = area, b = base, and h = height

Area of a circle: $A = \pi r^2$; where A = area, $\pi = 3.14$ or $\frac{22}{7}$, and r = radius

EXAMPLE 1: Find the area of the rectangle pictured at the right.

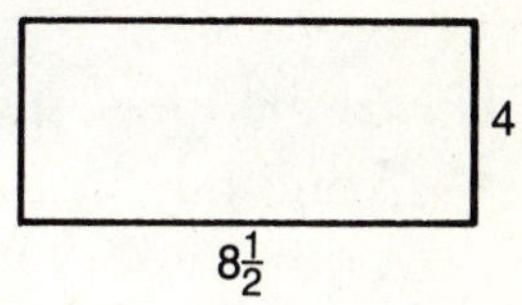

$A = lw$

$A = 8\frac{1}{2} \times 4$

$\frac{17}{2} \times 4 = 34$ square inches or 34 in^2

SOLUTION. Replace l with $8\frac{1}{2}$ and w with 4 in the formula $A = lw$, and multiply.

Notice that the answer is labeled **square inches**. This label is often abbreviated **sq in.** It can also be written **in^2**, which means *inches to the second power.*

EXAMPLE 2: Find the area of a square whose side measures 7 feet.

$A = s^2$
$A = 7^2$
$A = 7 \times 7 = 49$ sq ft, or 49 ft^2

SOLUTION. Replace s with 7 in the formula $A = s^2$.

The length and width of a rectangle are perpendicular to each other. The base and height of a triangle are also at right angles to each other. Look at the location of the base and height for each triangle below.

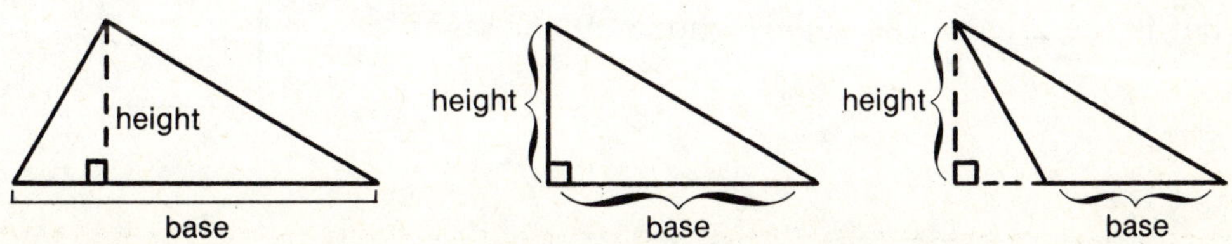

In the triangle at the left, the height is inside the triangle. In the middle triangle, the height is the left-hand side. For the triangle at the right, the height is an imaginary line from the top of the triangle to an extension of the base.

You can think of the area of a triangle as half the area of some four-sided figure. The figure at the right is a rectangle with a length (base) of 9 inches and a width (height) of 6 inches. The shaded part of the rectangle is a triangle with a base of 9 inches and a height of 6 inches. You can use this idea to remember that the area of a triangle ($\frac{1}{2}bh$) is half the area of a rectangle.

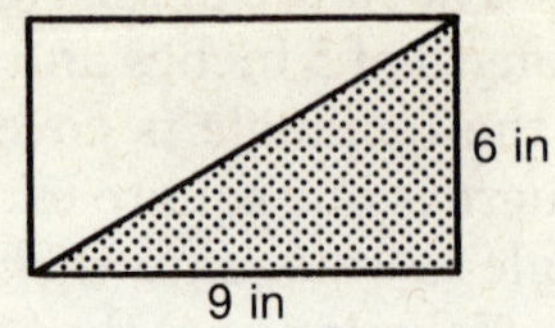

EXAMPLE 3: Find the area of a triangle with a base of 9 inches and a height of 6 inches.

$$A = \frac{1}{2}bh$$
$$= \frac{9 \times 6}{2} = 27 \text{ in}^2$$

SOLUTION. Replace b with 9 and h with 6 in the formula $A = \frac{1}{2}bh$. The area of the triangle is 27 in^2, which is exactly half the area of the rectangle.

TIP: Sometimes, the area of a triangle may be represented as $\frac{bh}{2}$, which, as you know from your study of fractions, is the same as $\frac{1}{2}bh$. You should be familiar with both forms.

EXAMPLE 4: Find the area of a circle with a radius of 3 inches.

$$A = \pi r^2$$
$$A = 3.14 \times 3^2$$
$$A = 3.14 \times 9 = 28.26 \text{ sq in or in}^2$$

SOLUTION. Replace π with 3.14 and r with 3 in the formula $A = \pi r^2$. The area of the circle is 28.26 in^2

Geometry Exercise 6

Directions: Solve each problem.

1. What is the area of the rectangle pictured at the right?

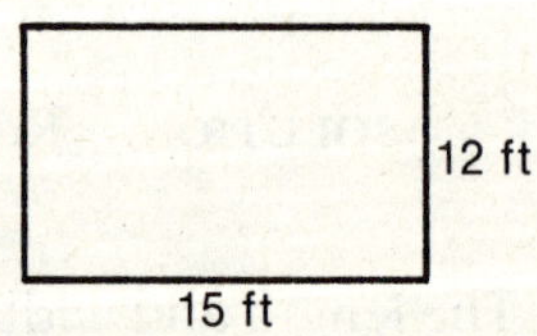

2. What is the area of the square shown at the right?

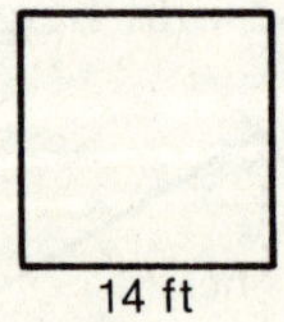

3. Find the area of the triangle shown at the right.

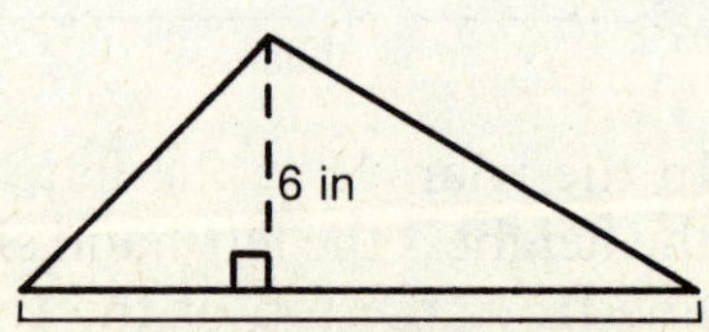

4. Find the area of a rectangle with a length of 8 inches and a width of $3\frac{1}{2}$ inches.

5. What is the area of a square whose side measures 6.5 meters?

6. Find the area of the triangle pictured at the right.

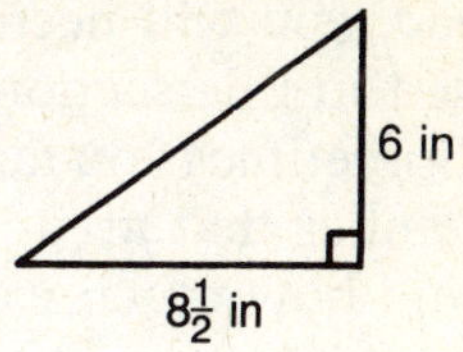

7. A rectangle has a length of 12.6 cm and a width of 5.4 cm. Find the area of the rectangle to the nearest square centimeter.

8. Find the area of a square with a side that measures $4\frac{1}{2}$ inches.

9. Which of the following expressions represents the area (in miles) of the shaded portion of the picture at the right?

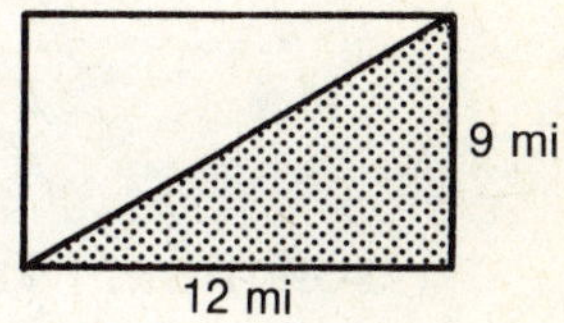

(1) 9×12

(2) $12 + 9$

(3) $\frac{1}{2}(12 + 9)$

(4) $\frac{1}{2}(12 \times 9)$

(5) $\frac{1}{2}(12 + 12 + 9 + 9)$

10. A rectangle has a length of 0.15 m and a width of 0.06 m. Find the area of the rectangle.

11. What is the area of a square if the side measures $\frac{3}{8}$ inch?

12. Find the area of the triangle pictured at the right.

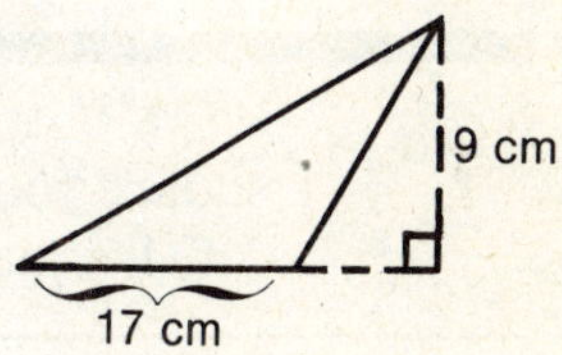

13. Find the area of the circle pictured at the right.

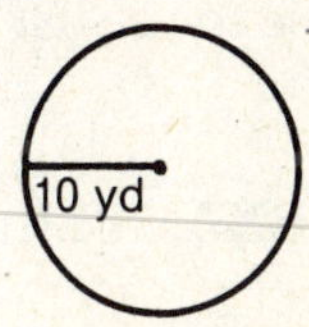

14. A circle has a radius of 7 feet. Find the area of the circle. In this problem use $\pi = \frac{22}{7}$.

15. A circle has a diameter of 1 meter.

a. What is the radius of the circle?

b. What is the area of the circle?

16. Find the area of a circle with a radius of 1.2 meters. Find the area to the nearest tenth of a square meter.

ANSWERS AND SOLUTIONS START ON PAGE 366.

Area Word Problems

This section gives you a chance to apply your knowledge of area to practical situations. The word *area* will not always appear where you need to use this skill. Instead, you will need to recognize when to use the formulas for area you have learned in this section.

Sometimes you may wonder whether you should be finding perimeter or area. Remember that area is a measurement of the amount of surface on a flat figure. Finding how much carpet to put on a floor, how much paint to put on walls, and how much material to buy for sewing all involve finding area. Perimeter measures *distance around* something — fencing and framing are two common situations where the perimeter formula is useful.

EXAMPLE 1: Luba wants to put carpet on her bedroom floor. Her bedroom measures 9 feet by 10 feet. How many square feet of carpet does she need?

$A = lw$
$A = 9 \times 10 = 90$ sq ft

SOLUTION. There are two clues that tell you to find the area in this problem. One is the word *carpet*. Carpet covers a surface. The other clue is "square feet." Only area is measured in square feet. Replace w with 9 and l with 10 in the formula $A = lw$.

Problems with Different Units of Measurement

Watch out for area problems where you need to change units of measurement. When finding area, you cannot multiply feet by inches or yards by feet. First you must convert, or change, one unit to another. Before you go on, learn the following equivalents:

1 square foot = 12 inches × 12 inches = 144 square inches
1 square yard = 3 feet × 3 feet = 9 square feet

Add these to the list of measurement equivalents you saw on pages 158–59 of Chapter 8.

EXAMPLE: How many square *yards* of fabric does Ernie need to cover the figure below?

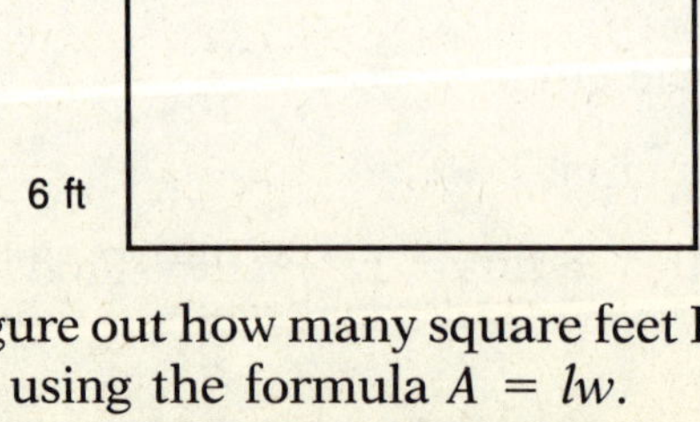

STEP 1. Figure out how many square feet Ernie needs by using the formula $A = lw$.

$A = lw$
$A = 10\frac{1}{2} \times 6$
$A = 63$ sq ft

STEP 2. Since the problem asks for square yards, convert the square feet into square yards by setting up a proportion. On the left is the ratio of 1 square yard to 9 square feet. On the right put the area of the figure in square feet on the bottom. Put x in the place of square yards.

$$\frac{1 \text{ sq yd}}{9 \text{ sq ft}} = \frac{x}{63}$$

STEP 3. Find the cross product of 1 and 63.

$$\begin{array}{r} 63 \\ \times\ 1 \\ \hline 63 \end{array}$$

STEP 4. Divide 63 by the remaining number in the proportion, 9. Ernie needs **7 square yards** of fabric.

$$9\overline{)63} = 7 \text{ sq yd}$$

"Backward" Area Problems

In some area problems, you will be given the area and one dimension for a rectangle. Then you will have to find the other dimension. In the example below, notice how your work with equations helps you to solve "backward" problems easily.

EXAMPLE: A gallon of paint covers 300 square feet of walls. The walls in Carla's house are 8 feet high. What is the total length of wall that a gallon of paint will cover in Carla's house?

(1) $A = lw$ **(2)** $300 = l \times 8$ **(3)** $\frac{300}{8} = \frac{l \times 8}{8}$ $8\overline{)300.0} = 37.5$ ft

STEP 1. Decide what formula to use and write it down. You are given area in this problem, so use the area formula.

STEP 2. Substitute values you know for letters in the formula.

STEP 3. Use inverse operations to get the unknown, l, alone on one side. Division is the inverse of multiplication, so divide both sides by 8. The length of the wall is **37.5 feet**.

Geometry Exercise 7

Directions: Solve each problem.

1. Colin is putting new wood flooring in the spare room of his house. The room is 9 feet wide and 12 feet long. How much flooring will Colin need for the house?

2. The two figures at right have the same area. How long is one side of the square?

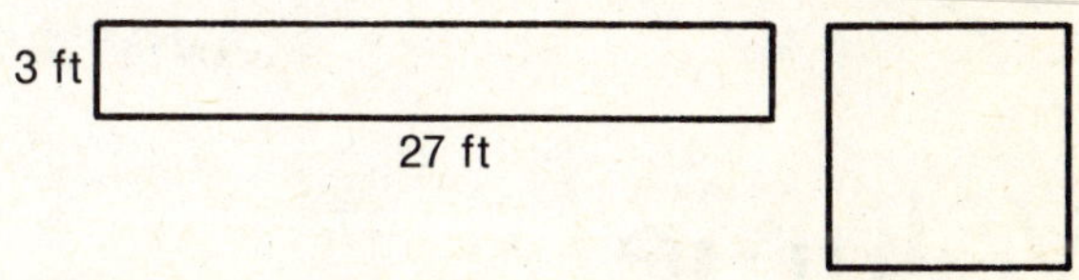

3. How many square inches of glass does Mark need to cover a photograph that measures 8 inches by 10 inches?

4. Bill wants to put 2-foot by 2-foot carpet tiles on the floor of his living room, which is shown at the right. Find the smallest number of carpet tiles Bill needs to completely cover the floor.

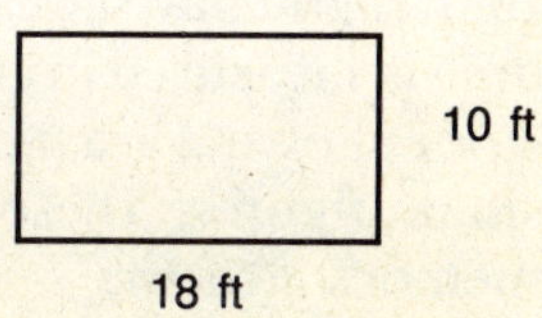

5. A gallon of floor paint will cover about 200 square feet of concrete. The floor of Harold's basement is 25 feet wide and 40 feet long. How many gallons of paint does Harold need to put one coat of paint on the basement floor?

6. The picture at the right shows the garden in Robert's yard. He wants to cover the garden with plastic because of a sudden drop in temperature. How many square yards of plastic does he need?

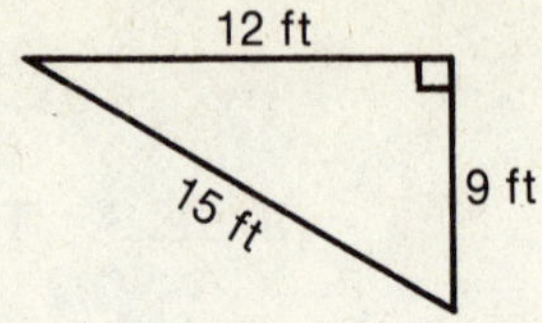

7. Mary wants to make curtains for the three windows in her living room. For each pair of curtains, she needs material measuring 6 feet by 8 feet. Altogether how many square yards of material does she need?

8. A round carpet that measures ten feet across covers how many square feet of floor?

9. Roman wants to repave a circular patio. The radius of the patio is 20 feet. How large is the surface that he wants to repave?

10. One package of grass seed contains enough to grow 400 square feet of lawn. Find the measurement of the side of the largest square that can be covered in grass from the seed in the package.

11. Phil used the package of grass seed in problem 10 to cover a side yard that is only 10 feet wide. Find the length of the stretch of yard that he can cover with the seed from one package.

12. Which of the following represents the area of the figure at the right?
(1) $(2 \times 8) + (2 \times 12.5)$
(2) 8×12.5
(3) $2 \times 8 \times 12.5$
(4) $8 + 12.5$
(5) Not enough information is provided.

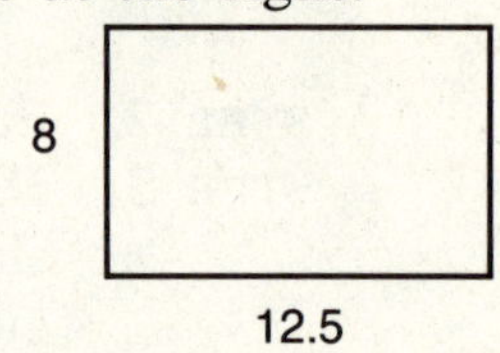

13. Which of the following represents the area of the figure at the right?
(1) $2(5.8) + 6$
(2) 5.8×6
(3) $5.8 \times 6 \times 2$
(4) $(5.8 \times 6)/2$
(5) Not enough information is provided.

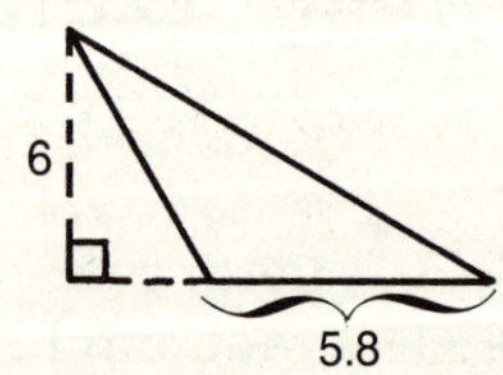

ANSWERS AND SOLUTIONS START ON PAGE 367.

Volume

Volume is a measure of the amount of space inside a three-dimensional or solid figure. You have already learned the names *rectangular solid, cube* and *cylinder*. Volume is measured in cubic units such as cubic inches, cubic feet, and cubic meters.

Following are the formulas for finding the volume of three common three-dimensional figures. These formulas will be given on the GED, but it may be useful to memorize them.

Volume of a rectangular solid or container:	$V = lwh$; where V = volume, l = length, w = width, and h = height
Volume of a cube:	$V = s^3$; where V = volume, and s = side
Volume of a cylinder:	$V = \pi r^2h$; where V = volume, $\pi = 3.14$, r = radius, and h = height

EXAMPLE 1: Find the volume of the rectangular solid pictured at the right.

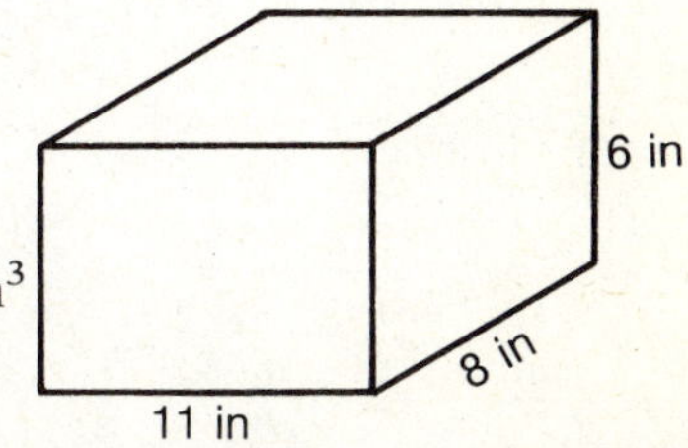

$V = lwh$
$V = 11 \times 8 \times 6 = 528$ cu in, or 528 in^3

SOLUTION. Replace l with 11, w with 8, and h with 6 in the formula $V = lwh$.

EXAMPLE 2: Find the volume of the cube pictured at the right.

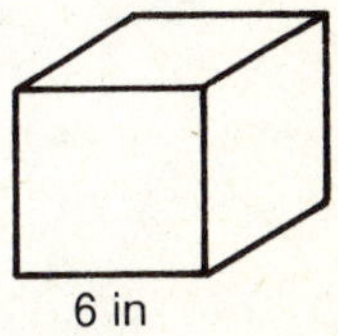

$V = s^3$
$V = 6^3$
$V = 6 \times 6 \times 6 = 216$ cu in or 216 in^3

SOLUTION. Replace s with 6 in the formula $V = s^3$.

EXAMPLE 3: Find the volume of the cylinder pictured at the right.

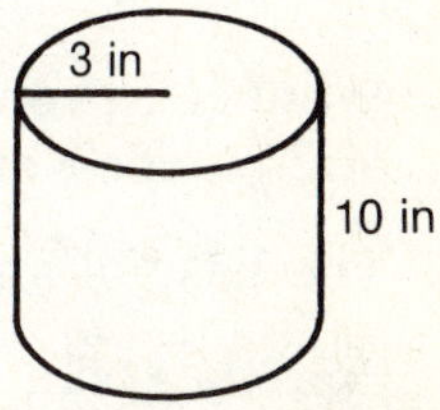

$V = \pi r^2h$
$V = 3.14 \times 3^2 \times 10$
$V = 3.14 \times 9 \times 10 = 282.6$ cu in or in^3

SOLUTION. Replace π with 3.14, r with 3, and h with 10 in the formula $V = \pi r^2h$. First find 3^2. Then multiply across.

Remember that a ***sphere*** is a three-dimensional shape like a basketball. The radius of a sphere is the distance from the center to the outside. The formula for a sphere is not given on the formula page on the GED, but it *will* be given to you in any problem where it is needed.

TIP: In some perimeter, area, and volume problems on the GED test you will have to recognize what shape the figure is. In other words, you will not be told if the figure is a rectangle, a cylinder, or another shape. In problems like these, make sure you take some time to determine what kind of figure you are looking at so that you choose the correct formula.

Geometry Exercise 8

Directions: Solve each problem, and give each answer the correct label.

1. Find the volume of the figure shown on the right.

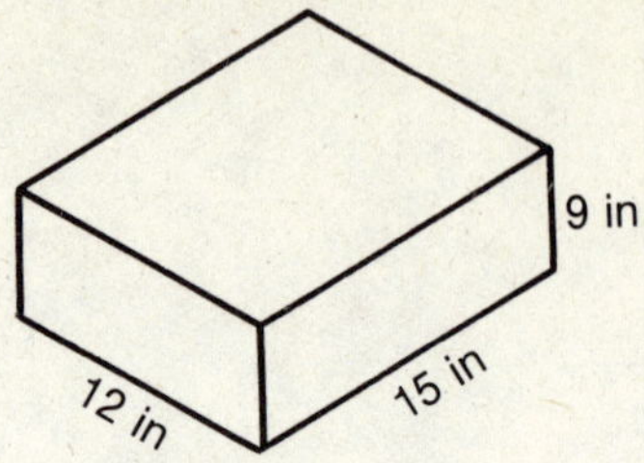

2. Find the volume of the cube shown on the right.

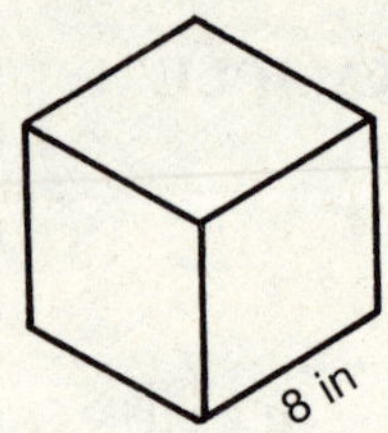

3. Find the volume of grain that the silo at right can hold if it is filled up to the heavy black line. Use $\frac{22}{7}$ for π.

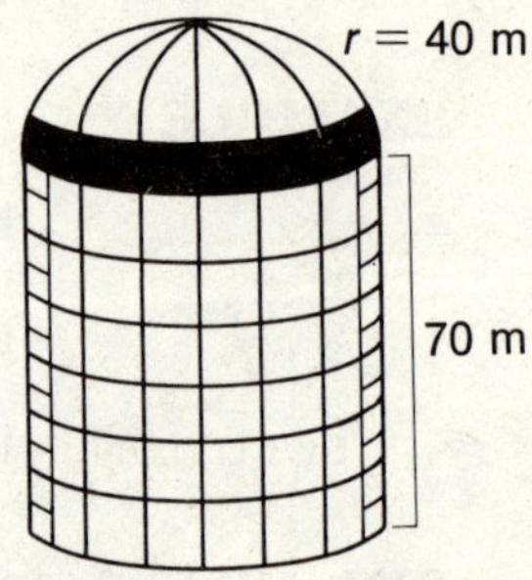

4. What is the volume of a rectangular box that has a base of $5\frac{1}{4}$ inches, a height of $2\frac{2}{3}$ inches, and a width of $3\frac{1}{2}$ inches?

5. The base of a cube measures 1.2 centimeters. Find the volume.

6. Find the volume of a rectangular solid with a base of 30 feet, a height of $\frac{1}{2}$ foot, and a width of 5 feet.

7. What is the volume of a cylinder with a radius of 10 feet and a height of 35 feet?

8. Find the volume of a cube if the base measures $\frac{1}{2}$ inch.

9. What is the volume of soil that the truck at right can carry if it is filled to the top?

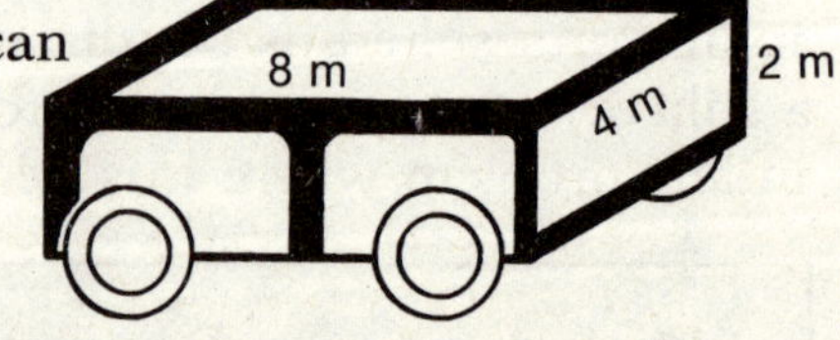

10. Find the volume of the cylinder pictured at the right. Round your answer to the nearest tenth of a cubic meter.

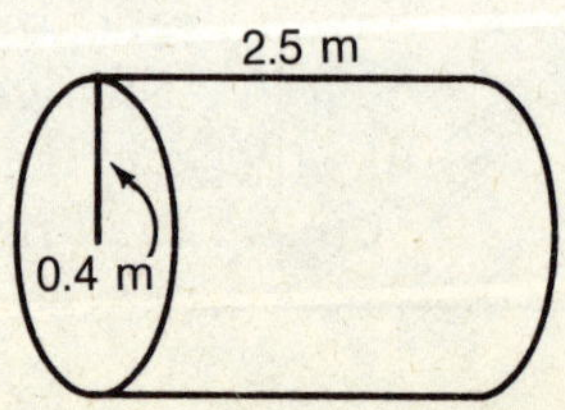

11. What is the volume of a cube with a base that measures $1\frac{1}{4}$ inches?

12. The formula for the volume of a sphere is $V = \frac{4}{3}\pi r^3$, where V is the volume, $\pi = 3.14$, and r is the radius. Find the volume of a sphere with a radius of 3 inches.

Questions 13 and 14 refer to the figure below.

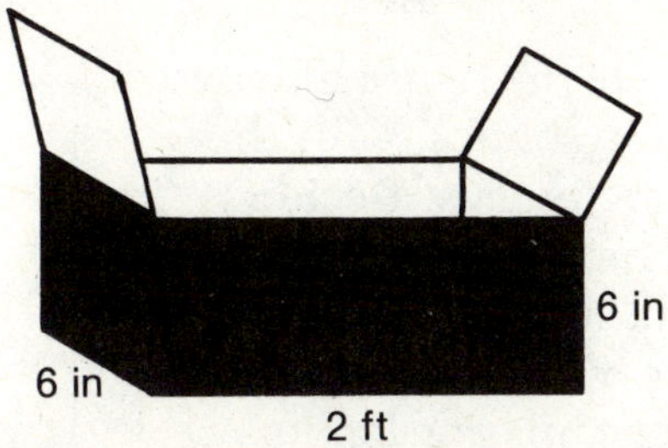

13. What is the volume in square inches of the carton shown?

14. How many boxes that measure 1 foot by 6 inches by 1 inch can you fit into the carton shown?

ANSWERS AND SOLUTIONS START ON PAGE 368.

More Perimeter, Area, and Volume Word Problems

On the GED, you may have to find the area of figures other than rectangles, squares, triangles, and circles. For these figures, you will be given the formulas that you need.

EXAMPLE 1: The figure at the right is a trapezoid. In this figure the top and the bottom are parallel. The left and right sides are not. The formula for the area of a trapezoid is $A = \frac{h(b_1 + b_2)}{2}$; where A is the area, h is the height, and b_1 and b_2 are the parallel sides. Find the area of the trapezoid pictured at the right.

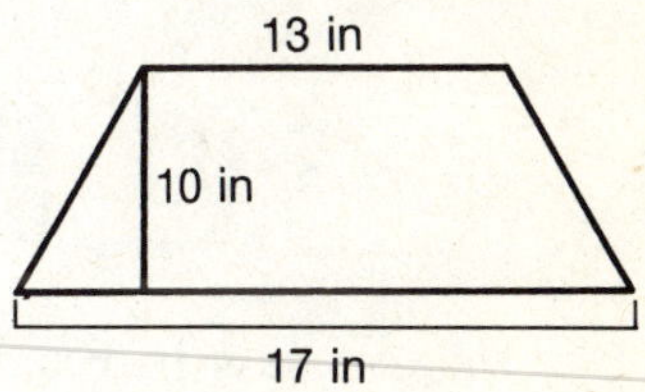

$$A = \frac{h(b_1 + b_2)}{2}$$

$$A = \frac{10(13 + 17)}{2}$$

$$A = \frac{10 \times 30}{2} = 150$$

SOLUTION. Replace h with 10, b_1 with 13, and b_2 with 17 in the formula given in the problem. Remember to do the operation in the parentheses first.

Note: In some problems like the one above, you will find small numbers written below and to the right of a variable (b_1). These numbers are used for identification only. Do not confuse them with numbers involved in computation.

In some problems you may be asked to find the area of complicated figures. These figures are often made up of two or more simple figures.

EXAMPLE 2: Find the area of the figure shown at the right.

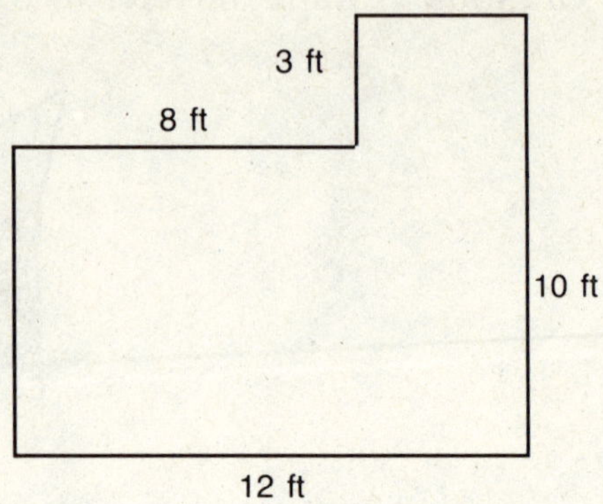

STEP 1. Separate the figure into two familiar shapes — in this case rectangles. Decide what measurements you are missing and label them (x and y in this figure).

STEP 2. Find the missing lengths of the sides by subtracting values you *do* know.
Side x is 12 ft − 8 ft = 4 ft
Side y is 10 ft − 2 ft = 7 ft

STEP 3. Find the area of each rectangle by using the correct formula. First, replace l with 4 and w with 3 in the formula $A = lw$. Then, replace l with 12 and w with 7 in the formula $A = lw$.
Area I $= lw$
$= 4 \times 3$
$= 12$ sq ft
Area II $= lw$
$= 12 \times 7$
$= 84$ sq ft

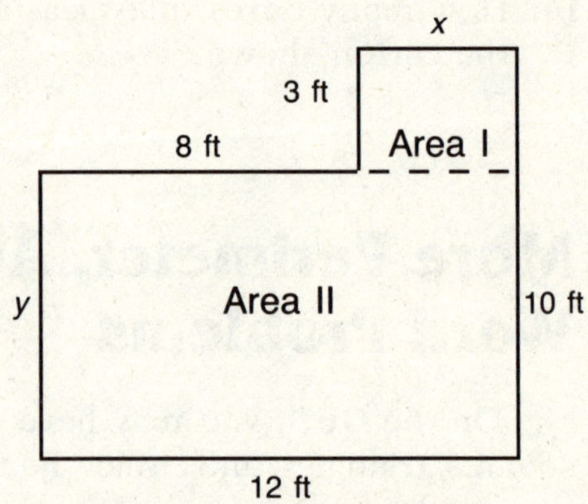

STEP 4. Add the areas of the two rectangles. The total area of the figure is **96 sq ft.**
Area I + Area II =
12 + 84 = 96 sq ft

As you saw in Example 2, you can often figure out missing measurements by looking at the figures and subtracting measurements that you *do* know. However, watch out for problems where not enough information is given to allow you to do this.

EXAMPLE 3: How many square feet are in the figure at the right?

(1) 60
(2) 84
(3) 100
(4) 140
(5) Not enough information is provided.

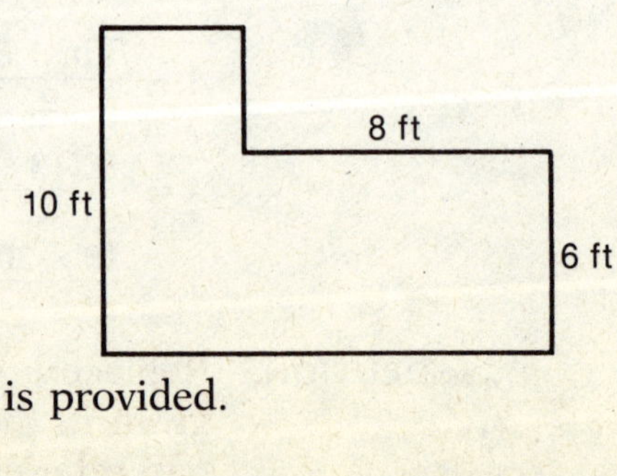

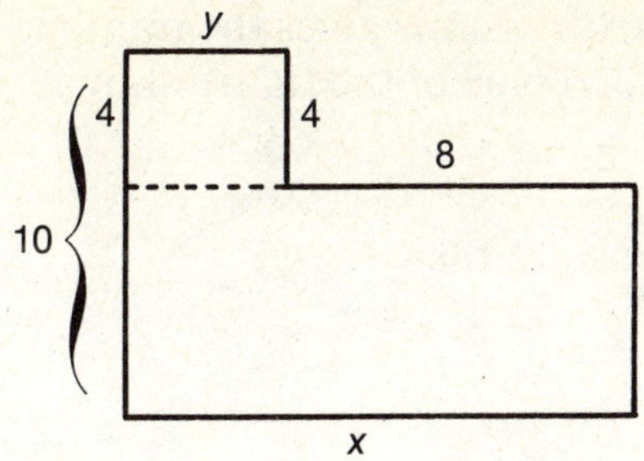

STEP 1. Divide the figure into two familiar shapes and decide what measurements you are missing. Label them with letters.

STEP 2. Figure out the missing lengths by subtracting values you know. As you can see, without knowing the length of side x, you cannot figure out the length of side y either. There is not enough information to solve this problem. Answer choice (5) is correct.

Geometry Exercise 9

Directions: Solve each problem.

1. A parallelogram is a four-sided figure with two pairs of parallel sides. The sides across from each other are equal in length. The formula for the area of a parallelogram is $A = bh$ where A is the area, b is the base, and h is the height. Find the area of the parallelogram pictured at the right.

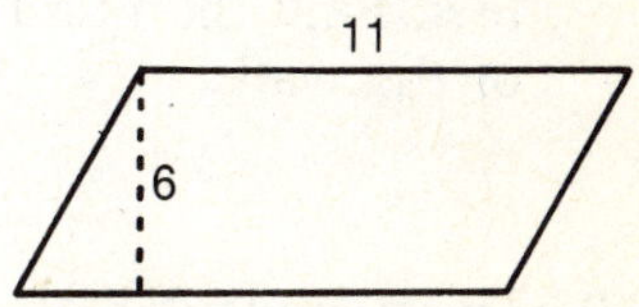

2. The formula for the surface area of a sphere is $A = 4\pi r^2$, where A is the area, π is 3.14, and r is the radius. Find the surface area of a sphere with a radius of 10 inches.

3. The formula for the area of a trapezoid is $A = h(b_1 + b_2)/2$ where A is the area, h is the height, and b_1 and b_2 are the parallel sides. Find the area of the trapezoid pictured at the right.

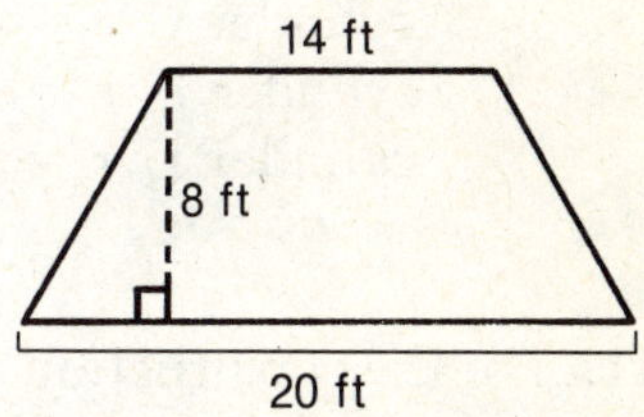

4. Find the total number of inches around the figure shown at the right.

5. Find the number of square inches on the surface of the figure in problem 4.

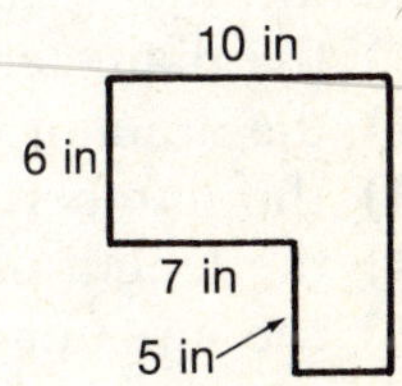

6. Find the area of the figure shown at the right.

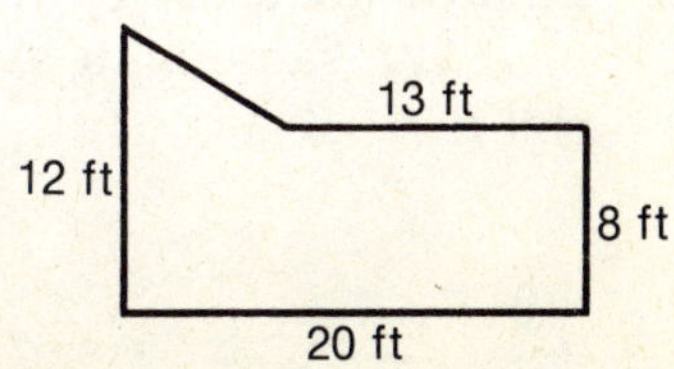

7. Which of the following expressions gives the number of square feet in the shaded part of the figure shown at the right?
(1) $12^2 + \pi(8^2)$
(2) $(12 + 1) + 8$
(3) $12^2 - 8^2$
(4) $12^2 - \pi(8^2)$
(5) $12^2 - \pi(4^2)$

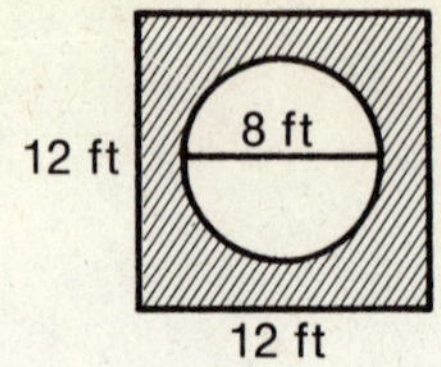

8. The diagram at the right shows the measurements of the floor of the Porters' living room and dining room. How many square yards of carpet do they need to cover the floors of both rooms?

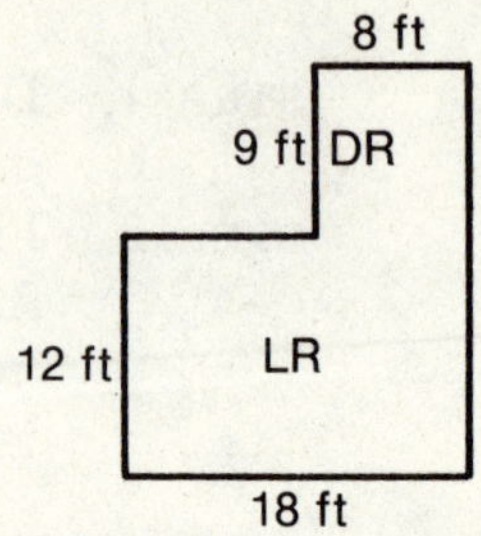

9. The shaded part of the figure at the right shows the walk around the pool at the Local Community Center. The pool is a rectangle that measures 34 feet by 18 feet. Find the number of square feet on the surface of the walk.

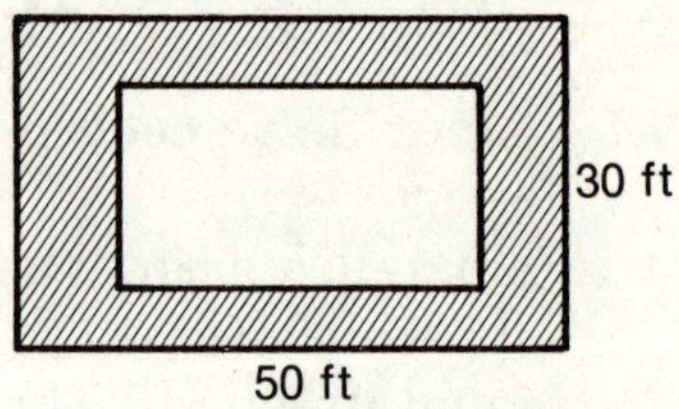

10. The rectangular container pictured at the right has inside measurements of 6 inches by 6 inches by 10 inches. Following are the measurements of five cylinders. Which would fit into the rectangular container?

cylinder A $r = 6$ inches and $h = 10$ inches
cylinder B $r = 5$ inches and $h = 9$ inches
cylinder C $r = 4$ inches and $h = 10$ inches
cylinder D $r = 3$ inches and $h = 10$ inches
cylinder E $r = 2$ inches and $h = 16$ inches

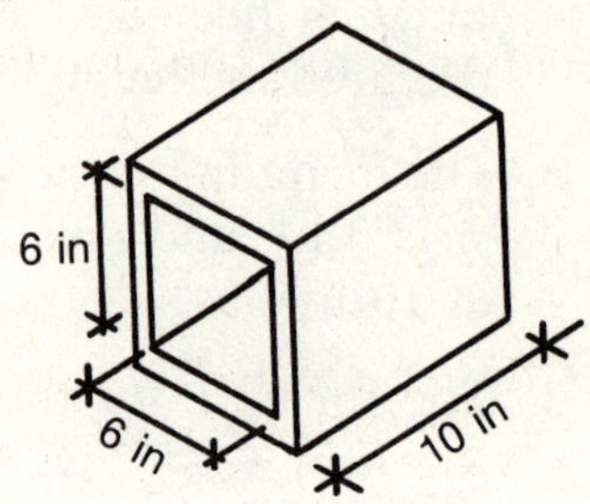

11. For the construction of a new building, a hole was dug 4 yards deep, 20 yards wide, and 30 yards long. Which of the following do you need to know to find out how many truckloads were needed to carry away all the dirt from the hole?
(1) the diameter of the hole
(2) the amount of dirt in each truckload
(3) the number of hours it took
(4) the length of the truck
(5) the amount of dirt left over at the site

12. Find the measurement of the side of a square that will have the same area as the rectangle pictured at the right.

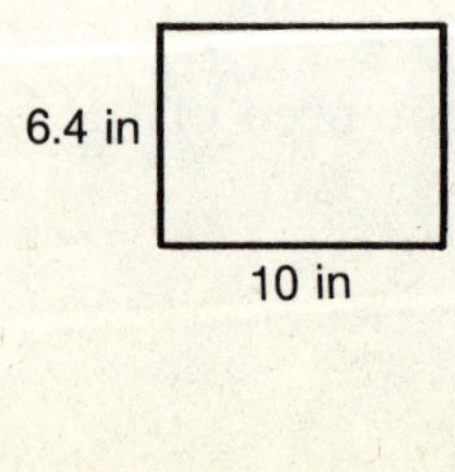

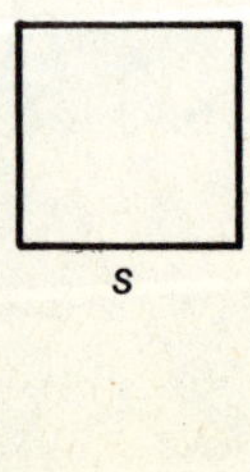

13. Find the volume of the mailbox at right. Think of it as a rectangular solid on bottom and a half cylinder on top.

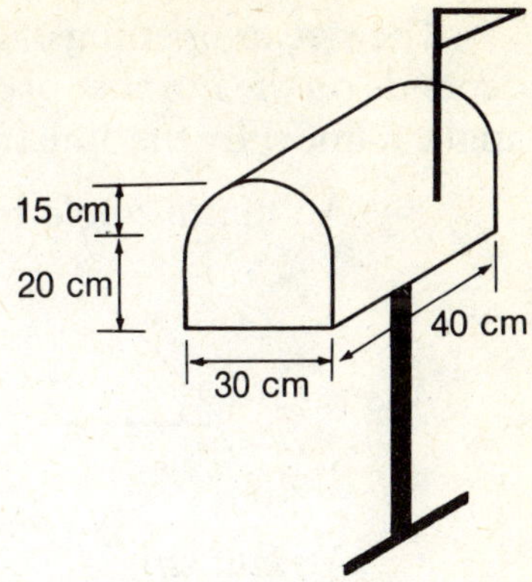

14. Each figure at the right has the same area. Find the base of the rectangle at the right.

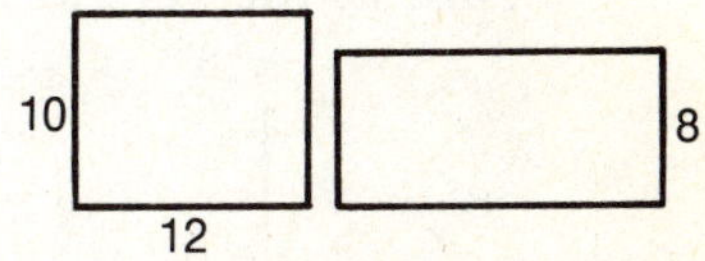

15. Which of the following represents the area of the figure shown at the right?
(1) 19
(2) 38
(3) 73
(4) 130
(5) Not enough information is provided.

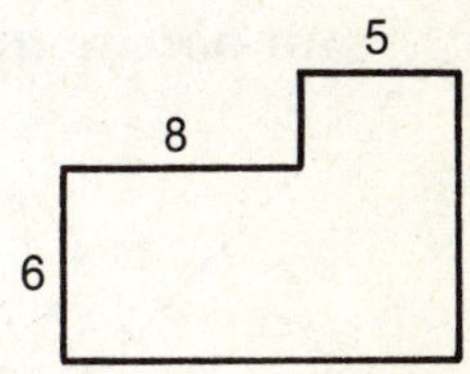

16. Which of the following represents the area of the figure at the right?
(1) 120
(2) 124
(3) 140
(4) 160
(5) Not enough information is provided.

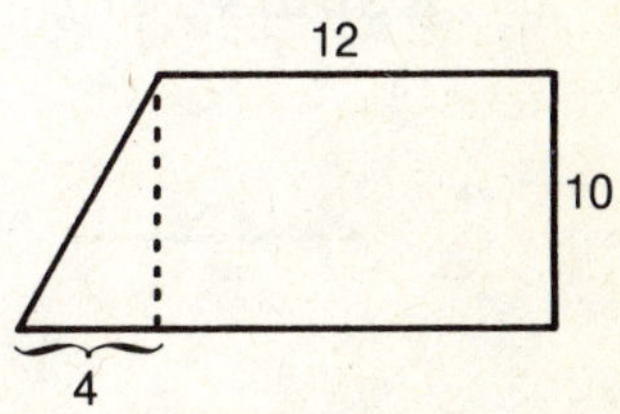

ANSWERS AND SOLUTIONS START ON PAGE 368.

Angles

You learned on page 224 that an angle is formed by two lines extending from the same point. The point is called the **vertex** of an angle. The symbol $\angle$ stands for the word *angle*.

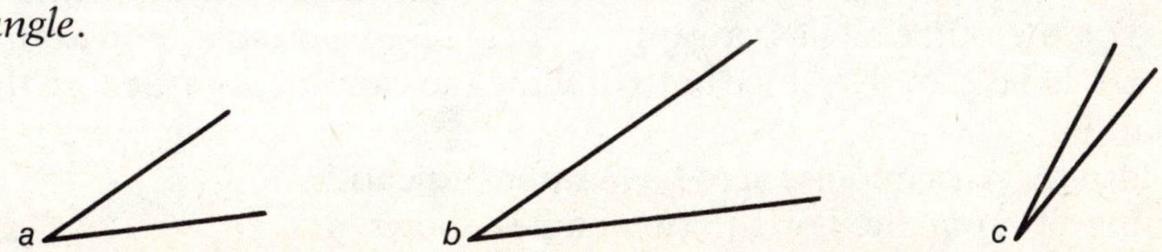

$\angle a$ and $\angle b$ are about the same size because the lines that form them are "open" about the same amount. The length of the lines has nothing to do with the size of the angles. In fact, $\angle c$ is smaller than the other two angles because the lines that form it are "closed" more.

The size or opening of an angle is measured in degrees (°). The name of an angle depends on the number of degrees, or the size of the opening. Following is a list of angle names. In the illustrations, a small curve indicates the angle opening.

An ***acute angle*** has less than 90°. Examples:

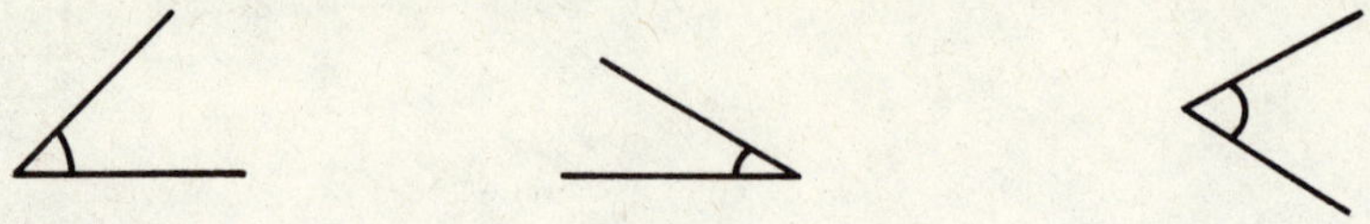

A ***right angle*** has exactly 90°. A right angle is often shown with a small square at the vertex. Examples:

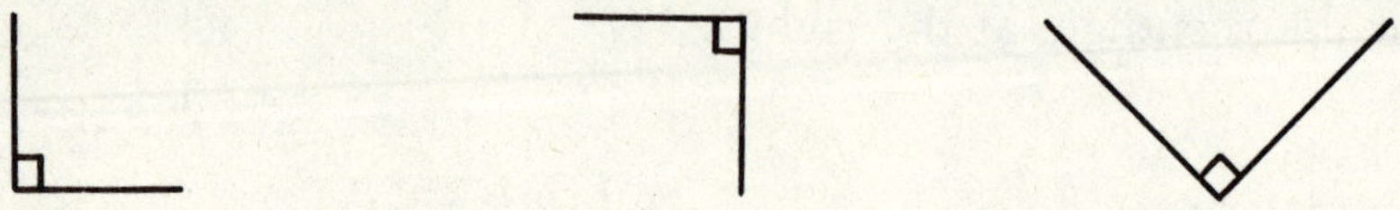

An ***obtuse angle*** has more than 90° but less than 180°. Examples:

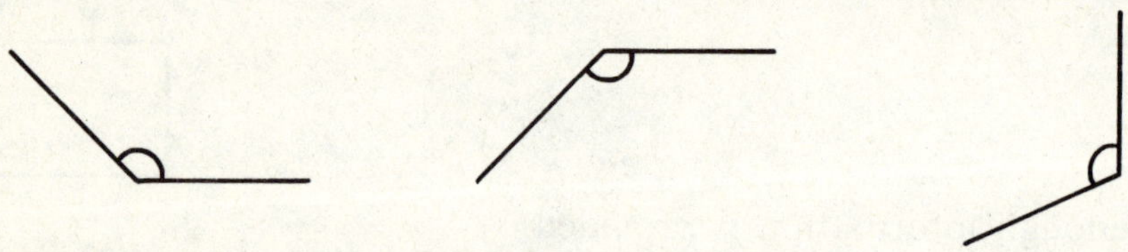

A ***straight angle*** has exactly 180°. Examples:

A ***reflex angle*** has more than 180° but less than 360°. Examples:

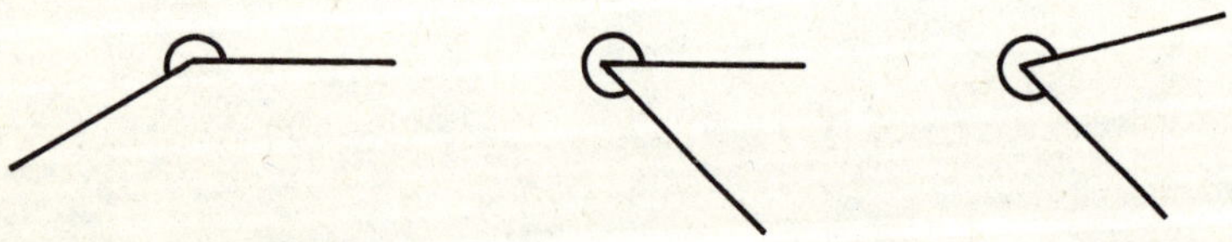

The angular measurement of a complete circle is 360°.

Notice that both the right angle and the straight angle contain exact numbers of degrees. For the other angles, a range of degrees is possible. For example, an acute angle could be 2° or 89°, since both of these measurements are more than 0° and less than 90°.

Notice also that it is important to have some indication of the opening of an angle. For the two angles shown at the right, the lines that form the angles look the same. However, the angle at the left is an acute angle, and the angle at the right is reflex.

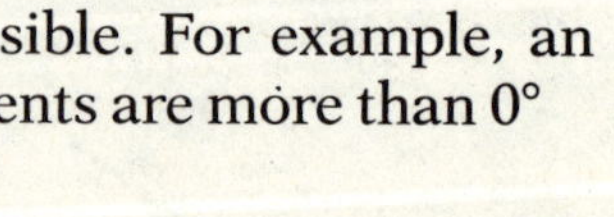

Memorize the names of the angles before you try the next exercise.

Geometry Exercise 10

Directions: For items 1 through 12, identify each angle by name (acute, right, obtuse, straight, or reflex.).

1. 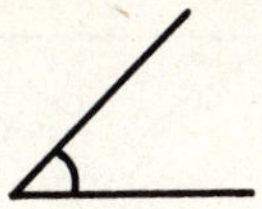

2. 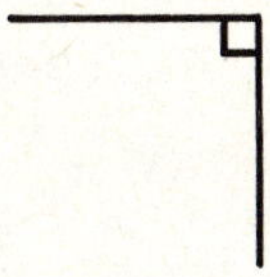

3. 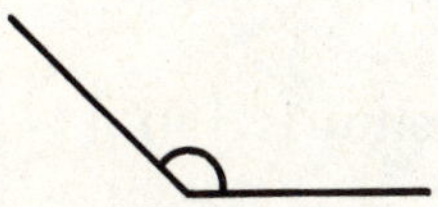

4. 

5. 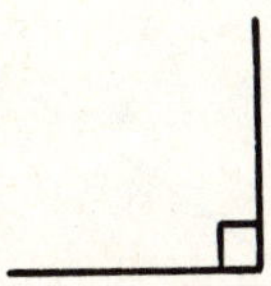

6. 

7. 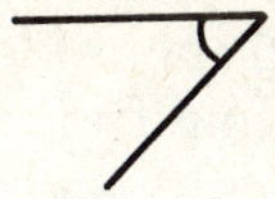

8. 

9. 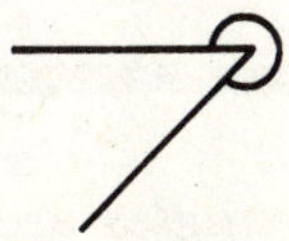

10. 

11. 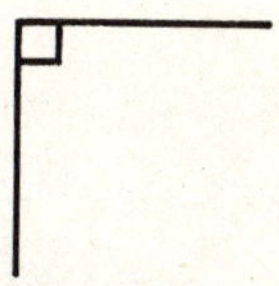

12. 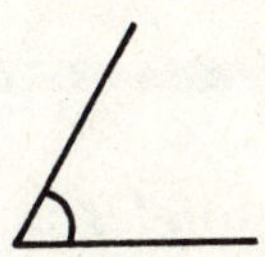

Directions: For items 13 through 21, tell what kind of angle contains each of the following numbers of degrees.

13. 40°

14. 100°

15. 180°

16. 220°

17. 90°

18. 83°

19. 175°

20. 300°

ANSWERS START ON PAGE 370.

Angle Relationships

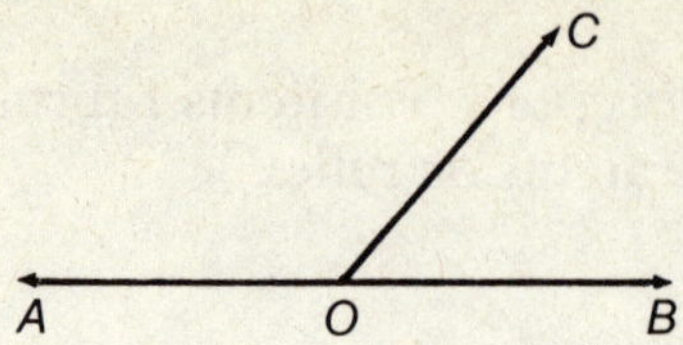

Angles are sometimes referred to with letters. For example, $\angle AOB$ refers to the straight angle shown at the right. The angle can also be called $\angle BOA$. In both cases, the vertex O is in the middle.

Another line extending from point O forms two new angles. $\angle BOC$ is acute, and $\angle AOC$ is obtuse.

Suppose $\angle BOC$ measures 50°. To find the measurement of $\angle AOC$ take 50° from 180°, since the straight angle $\angle AOB$ measures 180°.

$$\angle AOC = 180° - 50 = 130°$$

$\angle AOC$ and $\angle BOC$ are called supplementary angles.

Supplementary angles are two or more angles whose sum is 180°.
$\angle A + \angle B = 180°$.

In the picture at the right, $\angle DEF$ is a right angle. Suppose $\angle GEF = 35°$. To find $\angle DEG$, take 35° from 90°.

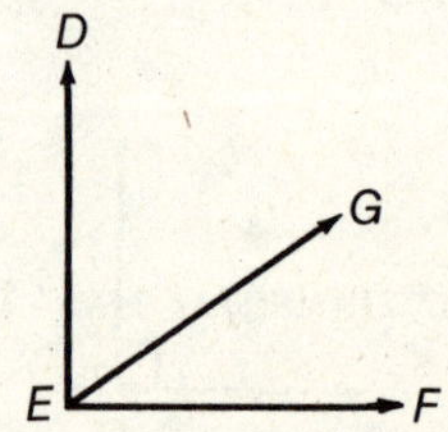

$$\angle DEG = 90° - 35° = 55°$$

$\angle DEG$ and $\angle GEF$ are called complementary angles.

Complementary angles are two or more angles whose sum is 90°.
$\angle A + \angle B = 90°$.

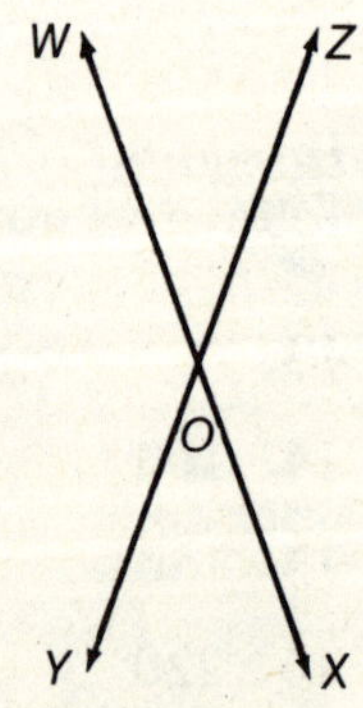

In the picture at the right, two lines intersect (cross) at point O. Suppose $\angle WOZ$ measures 70°. $\angle WOY$ must measure $180° - 70° = 110°$ because $\angle WOY$ and $\angle WOZ$ are supplementary.

In the picture at the right, $\angle WOY$ and $\angle ZOX$, which both measure 110°, are called **vertical angles**; $\angle WOZ$ and $\angle YOX$, which both measure 70°, are also called vertical angles.

Vertical angles are the two angles opposite each other when two straight lines intersect. Vertical angles are equal.

EXAMPLE 1: Find the supplement of an angle measuring 65°.

SOLUTION. Supplementary angles add up to 180°. Take 65° from 180°. The supplement of a 65° angle is 115°.
$180° - 65° = 115°$

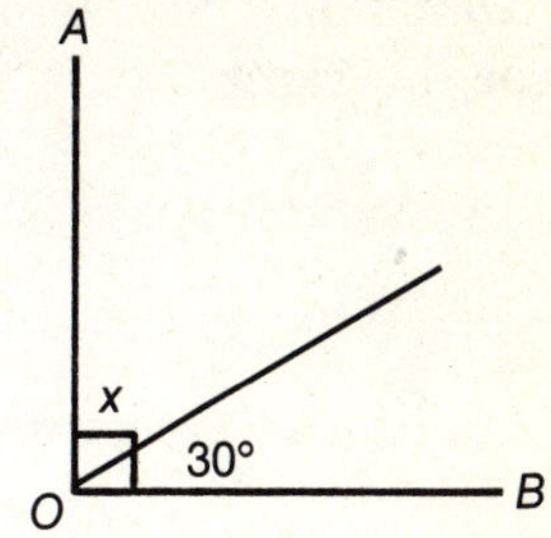

EXAMPLE 2: $\angle AOB$ in the figure at the right measures 90°. Find the measurement of angle x.

SOLUTION. Take 30° from 90°. $\angle x$ measures 60°.
$90° - 30° = 60°$

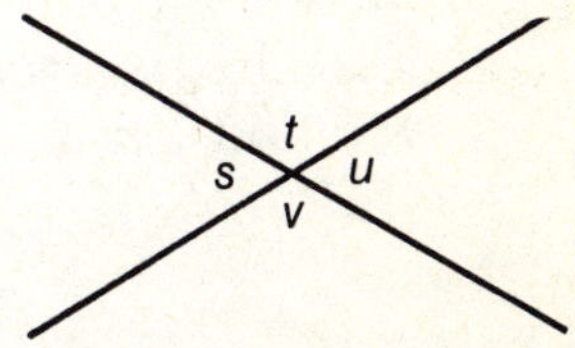

EXAMPLE 3: Which angle is vertical to $\angle u$ in the picture at the right?

SOLUTION. $\angle s$ is vertical to $\angle u$, because $\angle s$ is opposite $\angle u$.

You learned earlier that parallel lines are lines that run side by side and never cross. When two parallel lines are intersected by a third line, called a ***transversal***, a set of four angles is created at each point of intersection. In the figure at the right, lines m and n are parallel. Line t is the transversal. For this situation, the following statements are true:

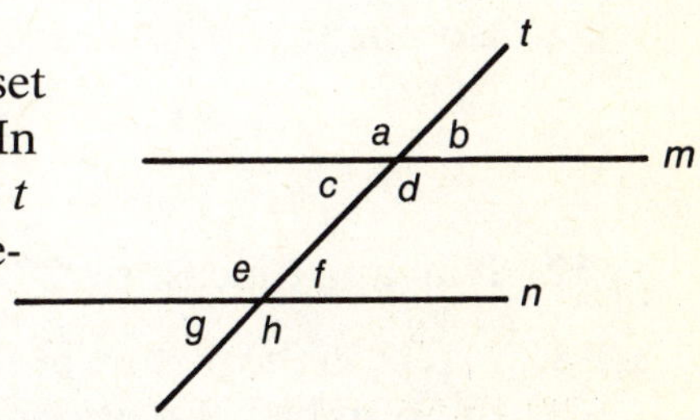

- The four acute angles (b, c, f, and g) are equal.
- The four obtuse angles (a, d, e, and h) are equal.
- Each acute angle is supplementary to each obtuse angle.

EXAMPLE 4: In the picture at the right, lines a and b are parallel. How many degrees are in $\angle x$?

SOLUTION. Each acute angle is equal to 50°. $\angle x$ is a supplementary obtuse angle. Subtract 50° from 180°.
$\angle x = 180° - 50° = 130°$

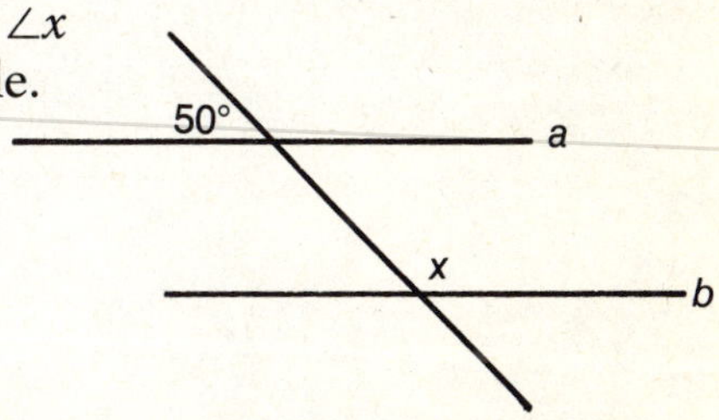

Geometry Exercise 11

Directions: Solve each problem.

1. How many degrees are in the complement of a 48° angle?
2. How many degrees are in the supplement of a 48° angle?

3. Find the complement of an angle measuring 25°.

4. Find the supplement of an angle measuring 63°.

Directions: Problems 5 through 8 refer to the illustration at the right.

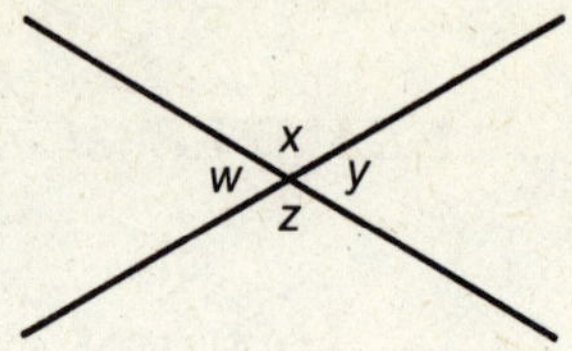

5. Which angle is vertical to $\angle y$?

6. What is the sum of $\angle w$, $\angle x$, $\angle y$, and $\angle z$?

7. If $\angle w$ measures 75°, how many degrees are in $\angle x$?

8. If $\angle x$ measures 119°, how many degrees are in $\angle w$?

9. In order to hang her mother's painting safely, Wendy stretched wires in the angles shown at right. If $\angle b$ is 80° and $\angle a$ and $\angle c$ are equal, how many degrees is $\angle a$?

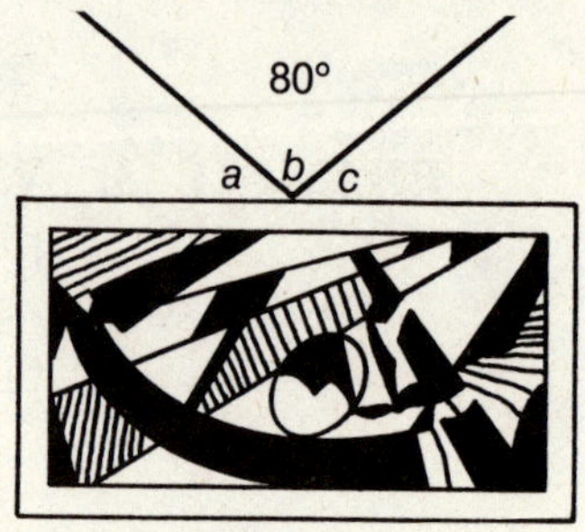

10. The supplement of an acute angle is always what kind of angle?

11. How many degrees are in $\angle m$ in the picture at the right?

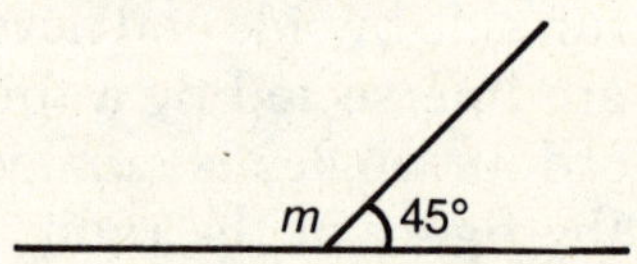

12. If an angle measures 37°, an angle vertical to it must measure how many degrees?

13. $\angle XOZ$ in the picture at the right measures 110.5°. Find the number of degrees in $\angle YOZ$.

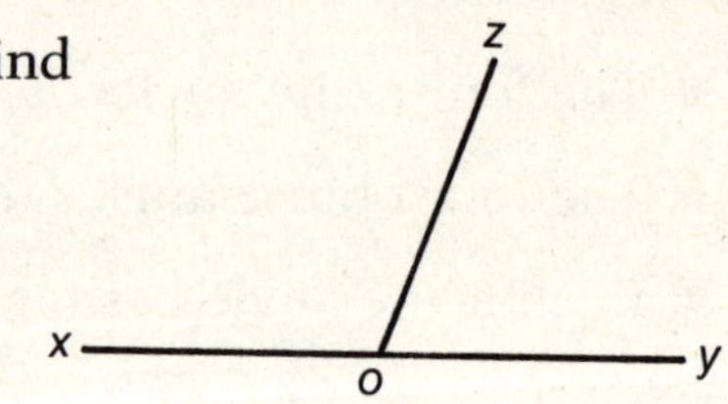

14. A surveyor stretched a wire line across a street as shown at right. If $\angle a$ is 30°, how many degrees is $\angle z$?

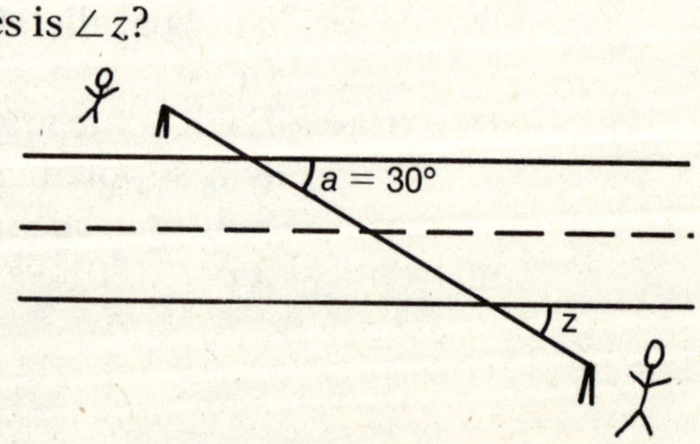

15. How many degrees are in $\angle b$ at the right?

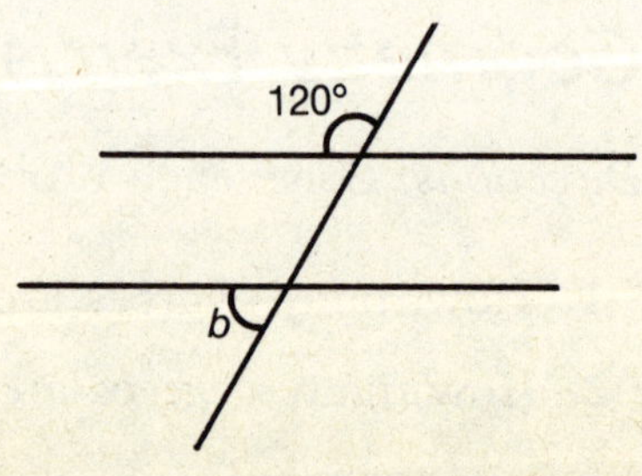

ANSWERS AND SOLUTIONS START ON PAGE 370.

Triangles

You have already learned that a triangle is a three-sided figure. Each of the three points where the sides meet is called a vertex of a triangle. The three angles inside a triangle add up to 180°. Following are the most important types of triangles. Learn these names before you go on to the next exercise.

An ***equilateral triangle*** has three equal sides. The name *equilateral* means equal sides. This triangle also has three equal angles. Each angle is $\frac{1}{3}$ of 180°, or 60°. An equilateral triangle can also be called *equiangular,* which means "equal angles." Examples of equilateral triangles:

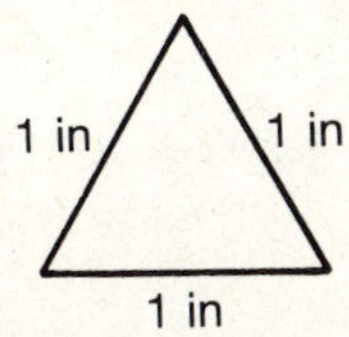

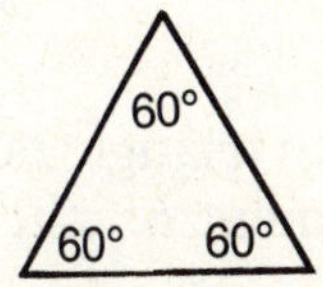

An ***isosceles triangle*** has two equal sides. The third side may be longer or shorter than the other two sides. An isosceles triangle also has two equal angles. The two equal angles are called the base angles. The angle with a different measurement is called the vertex angle. Examples of isosceles triangles:

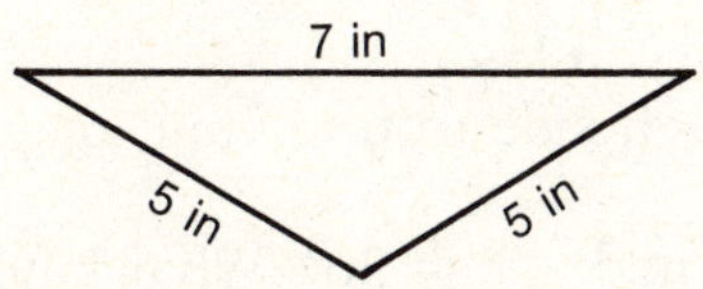

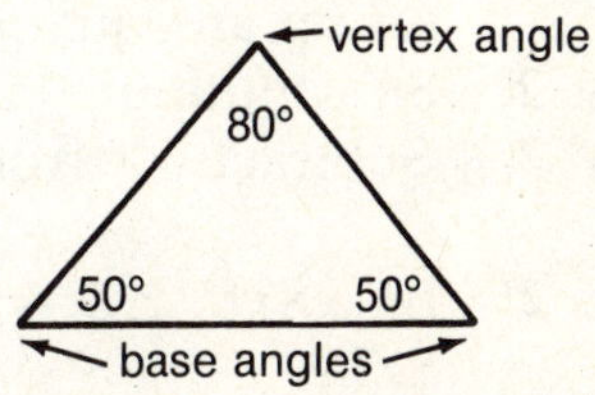

A ***scalene triangle*** has three sides with different measurements. It also has three angles of different sizes. Examples of scalene triangles:

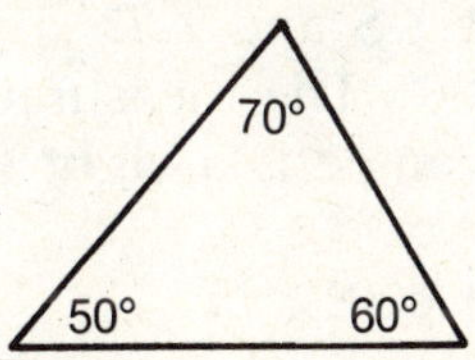

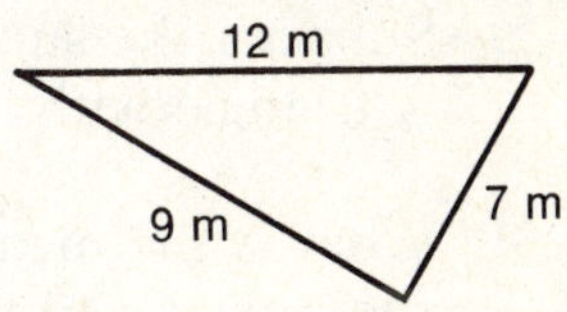

A ***right triangle*** has one right angle. The long side across from the right angle is called the ***hypotenuse***. The other two sides are called the **legs**. A right triangle can be either scalene or isosceles. When a scalene or an isosceles triangle contains one right angle, the triangle is called a right triangle. Examples of right triangles:

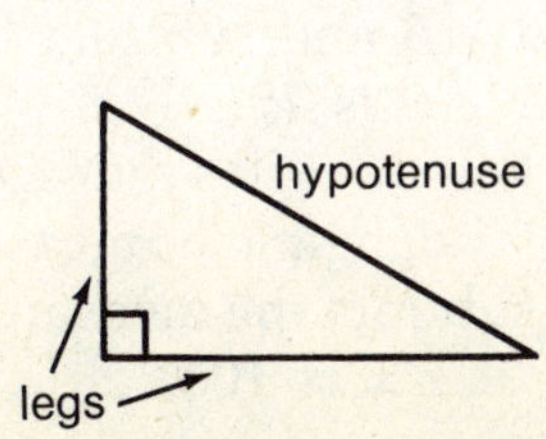

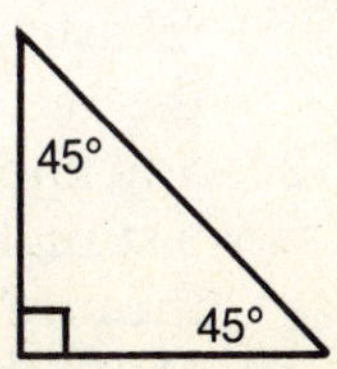

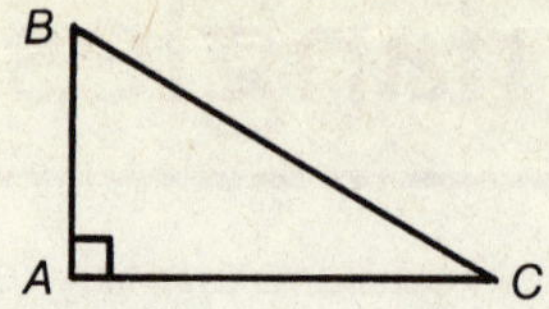

The symbol △ stands for the word *triangle*. A side of a triangle is sometimes referred to by the two letters that form the end points of each side. For △*ABC*, side *BC* is the hypotenuse, and sides *AB* and *AC* are the legs of this right triangle.

An angle of a triangle can be referred to by the letter that forms the vertex of the angle. For △*ABC*, ∠*A* is the right angle. Sometimes three letters are used to refer to an angle of a triangle. The letter of the vertex must be in the middle. For the triangle above, ∠*A* is the same as ∠*BAC* or ∠*CAB*.

Some questions on the GED Test will require you to find the measurement of a given angle. In some problems, you will be given a drawing of the angle. In other problems, you will simply use your knowledge of triangle facts to find an angle's measurement.

EXAMPLE 1: In triangle *PQR*, ∠*P* = 35° and ∠*Q* = 45°. Find the measurement of ∠*R*.

$$\angle P + \angle Q + \angle R = 180^\circ$$
$$35^\circ + 45^\circ + \angle R = 180^\circ$$
$$80^\circ + \angle R = 180^\circ$$
$$\angle R = 180^\circ - 80^\circ$$
$$\angle R = 100^\circ$$

STEP 1. You know that all three angles of a triangle add up to 180°, so set up an equation.
STEP 2. Substitute the values you know.
STEP 3. Subtract 80° from each side to find your answer: **100°**.

EXAMPLE 2: In △*STU*, ∠*S* = 60° and ∠*T* = 30°. What kind of triangle is *STU*?

(1) $\begin{array}{r} 60^\circ \\ +\ 30^\circ \\ \hline 90^\circ \end{array}$ **(2)** $\begin{array}{r} 180^\circ \\ -\ 90^\circ \\ \hline 90^\circ \end{array}$

STEP 1. Add the measurements of ∠*S* and ∠*T*.
STEP 2. Subtract the sum from 180°. The measurement of ∠*U* is 90°. A triangle with one right angle is a **right triangle**.

EXAMPLE 3: What is the measurement of ∠*x* in the drawing to the right?

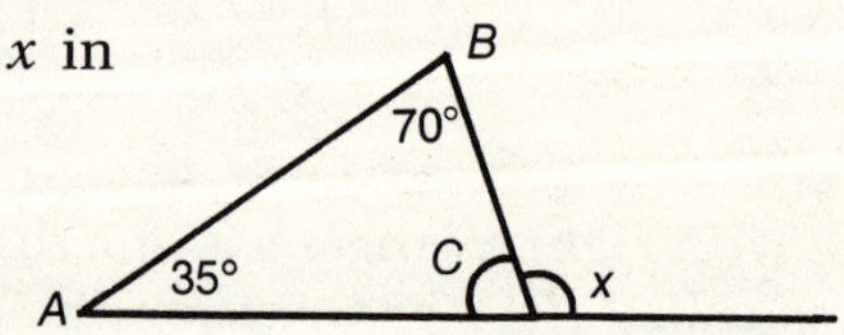

(1) $\begin{array}{r} 35 \\ +\ 70 \\ \hline 105 \end{array}$ $\begin{array}{r} 180 \\ -\ 105 \\ \hline 75 \end{array}$ **(2)** $\begin{array}{r} 180 \\ -\ 75 \\ \hline 105 \end{array}$

STEP 1. You know that ∠*A*, ∠*B*, and ∠*C* must add up to 180°. Add the measurement of ∠*A* and ∠*B* and subtract from 180 to find the measurement of ∠*C*. ∠*C* is 75°.
STEP 2. You know from your work with angles that ∠*C* and ∠*x* make a straight angle and that straight angles measure 180°. To find the measurement of ∠*x*, subtract the measurement of ∠*C* from 180°. The measurement of **∠*x* is 105°**.

Memorize the triangle definitions in this section before you try the next exercise.

Geometry Exercise 12

Directions: Solve each problem.

1. Identify each of the following triangles as right, isosceles, scalene, or equilateral.

a.
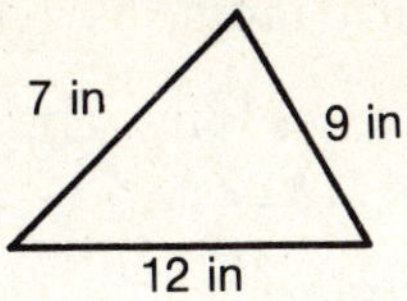

b.
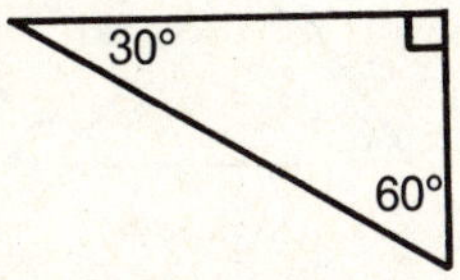

c.
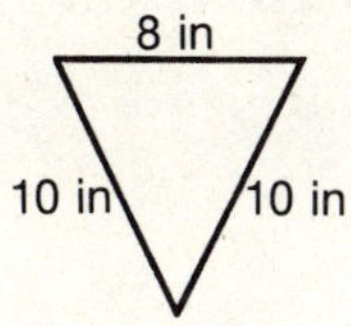

d.
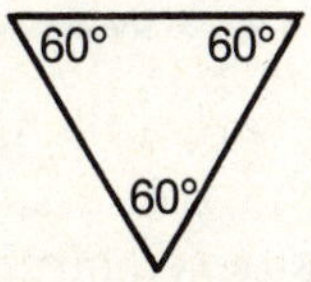

e.
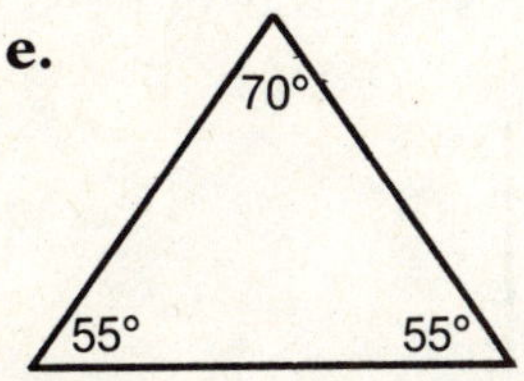

f.
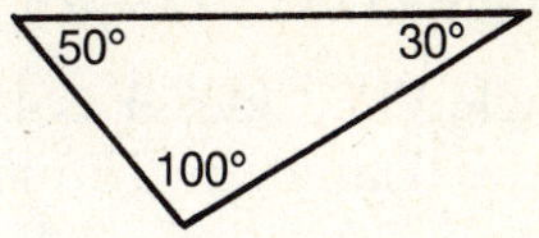

g.
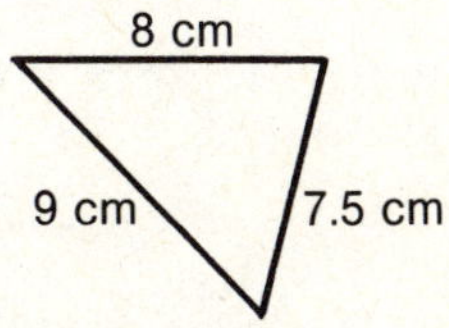

h.
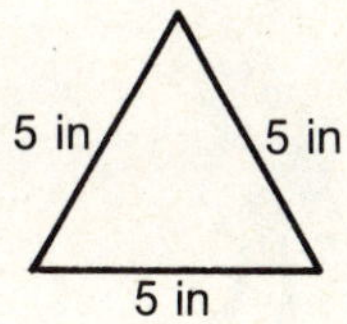

i.
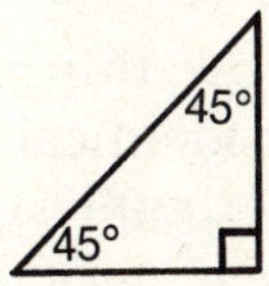

Directions: Questions 2 through 5 refer to the picture at the right.

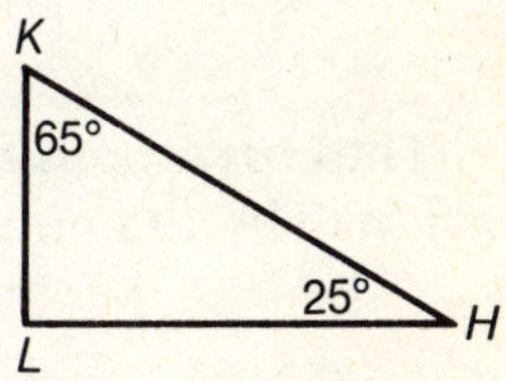

2. Find the measurement of $\angle L$.
3. What kind of triangle is $\triangle KLH$?
4. Side KH is called the ______________________.
5. In the triangle at the right, $\angle W = 65°$ and $\angle x = 50°$. Find the number of degrees in $\angle Y$.

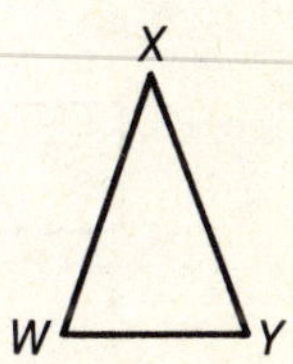

6. What kind of triangle is $\triangle WXY$ in problem 5?
7. In $\triangle ABC$, $\angle A = 30°$ and $\angle B = 60°$. How many degrees are there in $\angle C$?
8. What kind of triangle is $\triangle ABC$ in problem 7?
9. The vertex angle of an isosceles triangle measures 82°. How many degrees are there in each base angle?
10. Each base angle of an isosceles triangle measures 63°. How many degrees are there in the vertex angle?

11. In the triangle at the right, side DE measures 8 inches, and side DF measures 5 inches. The perimeter of the triangle is 21 inches. What kind of triangle is $\triangle DEF$?

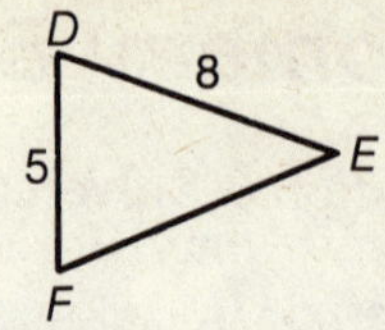

12. In triangle PQR, $\angle P = 25°$ and $\angle Q = 35°$. What kind of triangle is $\triangle PQR$?

13. In triangle XYZ, side XY is 4 inches long, and side YZ is 5 inches long. The perimeter of the triangle is 16 inches. What kind of triangle is $\triangle XYZ$?

14. What is the sum of the two acute angles in a right triangle?

15. Find how many degrees are in angle x in the picture at the right.

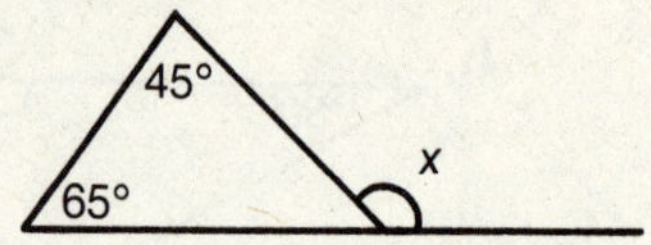

ANSWERS AND SOLUTIONS START ON PAGE 370.

Similarity

When you say that two things are *similar*, you mean that they are alike in some ways, but they are not identical. In geometry, similar figures have the same shape but different sizes. For example, the two triangles below are considered similar, even though they are different sizes. The reason is that the angles are the same size.

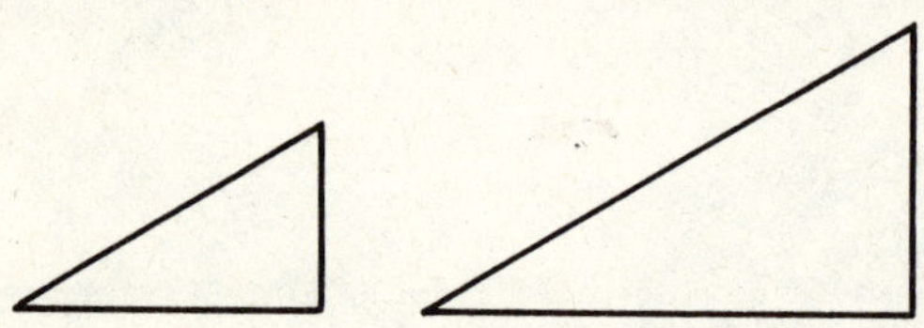

The ***corresponding*** (matching) sides of similar figures can be written as a proportion. The drawing below shows two similar rectangles.

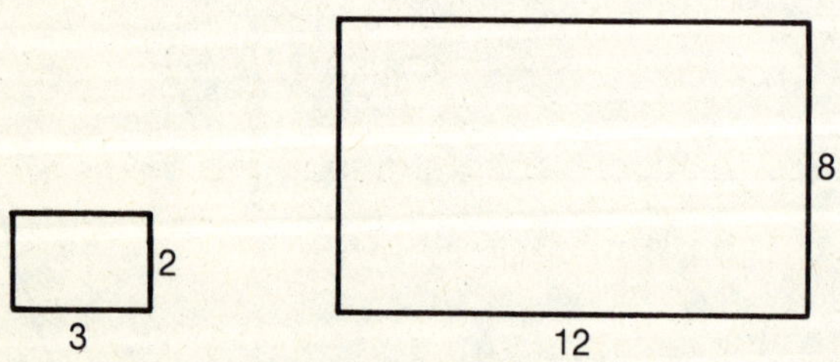

$$\frac{\text{length}}{\text{width}} \quad \frac{3}{2} = \frac{12}{8}$$

For the small rectangle above, the ratio of the length to the width is 3:2. The ratio of the length to the width (the corresponding sides) of the larger rectangle is 12:8, which can be reduced to 3:2. These two rectangles are similar because the corresponding sides are proportional. In other words, the ratio of the length to the width of each rectangle is 3:2.

Similar Triangles

Learning about similar triangles will help you solve some of the problems on the GED Test. Two triangles are similar if one of the following two statements is true:

Rules for Determining Similar Triangles

1. If any two angles of one triangle are equal to two angles of another triangle, the two triangles are similar.
2. If the sides of one triangle are proportional to the sides of another triangle, the two triangles are similar.

EXAMPLE 1: In triangle *DEF*, $\angle D = 50°$ and $\angle E = 75°$. In triangle *GHI*, $\angle G = 55°$ and $\angle I = 50°$. Are these triangles similar? If they are not similar, tell why.

(1) $\begin{array}{r} 50° \\ +\ 75° \\ \hline 125° \end{array} \quad \begin{array}{r} 180° \\ -\ 125° \\ \hline 55° \end{array} = \angle F$ **(2)** $\begin{array}{r} 55° \\ +\ 50° \\ \hline 105° \end{array} \quad \begin{array}{r} 180° \\ -\ 105° \\ \hline 75° \end{array} = \angle H$

STEP 1. Find $\angle F$. Subtract the sum of $\angle D$ and $\angle E$ from 180°.

STEP 2. Find $\angle H$. Subtract the sum of $\angle G$ and $\angle I$ from 180°. Each triangle has angles measuring 50°, 55°, and 75°. Since the triangles have equal angles, they are **similar**.

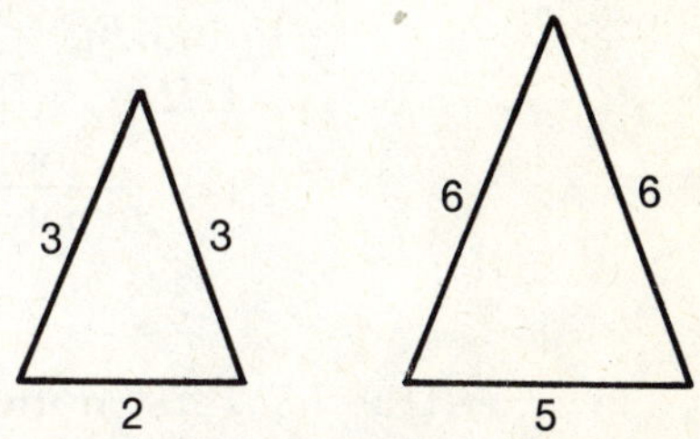

EXAMPLE 2: Are the two triangles pictured at the right similar? If they are not similar, tell why.

STEP 1. Set up a proportion with the corresponding sides.

$$\frac{3}{2} \stackrel{?}{=} \frac{6}{5}$$

STEP 2. Cross multiply.

$$3 \times 5 = 15$$
$$6 \times 2 = 12$$

STEP 3. Since the cross products are not equal, the two triangles are *not* similar.

Using Similar Triangles to Solve Word Problems

Once you know that two triangles are similar, you can use the following fact to solve many problems:

> Corresponding sides of similar triangles are proportional.

To be able to use this fact, you must be able to pick out which sides of two different triangles are *corresponding* sides. To do this easily, just remember that *corresponding sides are always opposite equal angles*. What are the corresponding sides in the figures below?

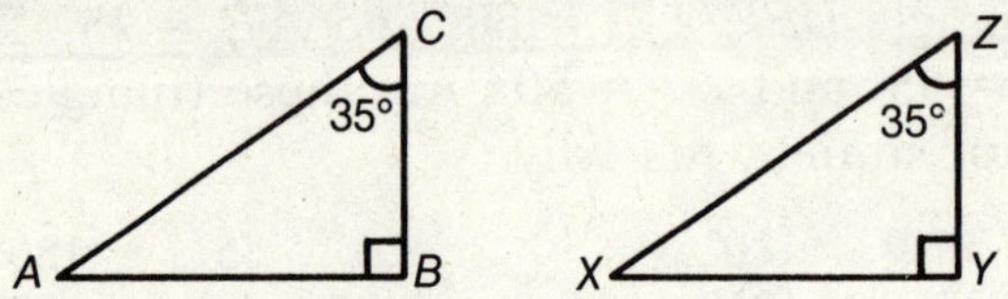

Side AB "corresponds" to side XY because both sides are across from angles measuring 35°. Sides BC and YZ are corresponding sides because they are opposite equal angles measuring 55°. (You know that these angles measure 55° because the third angle is a right angle, measuring 90°.) Finally, you know that sides AC and XZ are corresponding because the angles opposite them measure 90°.

EXAMPLE 3: In triangle ABC at the right, $AB = 9$ ft and $BC = 18$ ft. In triangle GHI, $GH = 5$ ft. Find the length of HI.

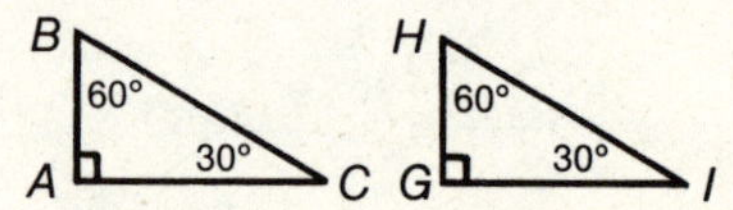

(1) $\dfrac{\text{short leg}}{\text{hypotenuse}} \quad \dfrac{9}{18} = \dfrac{5}{x}$ **(2)** $18 \times 5 = 90$ **(3)** $90 \div 9 = 10$ ft

STEP 1. First notice that these triangles are similar. Each has angles of 30°, 60°, and 90°. Set up a proportion with the corresponding sides of each triangle. Let x stand for the hypotenuse of triangle GHI.

STEP 2. Find the cross-product of 18 and 5.

STEP 3. Divide by the remaining number in the proportion, 9. The length of side HI is **10 feet.**

EXAMPLE 4: A vertical yardstick casts a two-foot shadow. At the same time, a building casts a 48-foot shadow. How tall is the building?

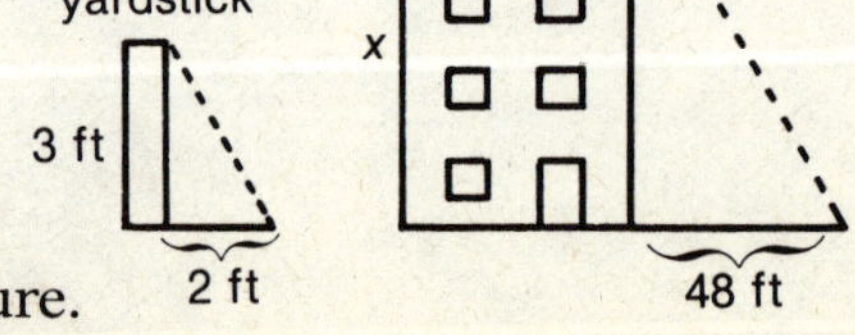

STEP 1. Draw a picture that shows the height and shadow of each figure.

STEP 2. Notice that you have two similar right triangles. The long leg of each triangle is the height of the object. The short leg is the length of the shadow. Make a proportion with the height and shadow of each object. Since the other measurements are given in feet, write the height of the yardstick as 3 feet. Let x stand for the height of the building.

$$\frac{\text{height}}{\text{shadow}} \quad \frac{3}{2} = \frac{x}{48}$$

STEP 3. Find the cross product of 3 and 48.

$$\begin{array}{r} 48 \\ \times\, 3 \\ \hline 144 \end{array}$$

STEP 4. Divide by the remaining number in the proportion, 2. The building is **72 feet** tall.

$$2\overline{)144}\ \ \text{72 ft}$$

Geometry Exercise 13

Directions: Solve each problem.

1. Are the rectangles pictured at the right similar? If they are not, tell why.

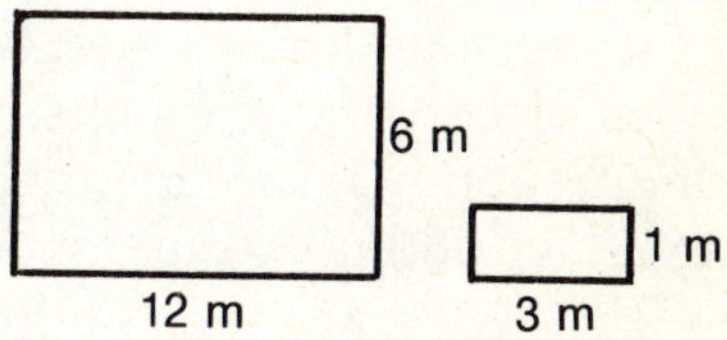

2. Are the rectangles pictured at the right similar? If they are not, tell why.

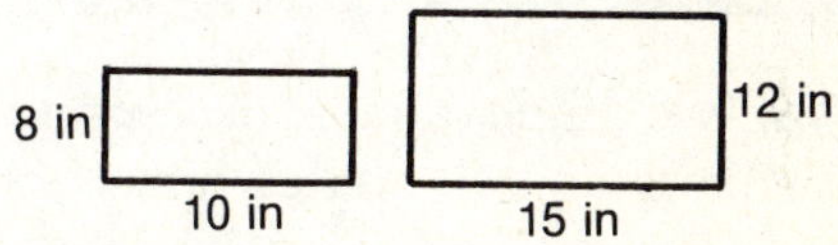

3. In triangle *MNO*, $\angle M = 45°$ and $\angle N = 85°$. In triangle *PQR*, $\angle P = 50°$ and $\angle Q = 45°$. Are these two triangles similar? If they are not, tell why.
4. In triangle *ABC*, $\angle A = 60°$ and $\angle B = 50°$. In triangle *DEF*, $\angle D = 50°$ and $\angle E = 80°$. Are these two triangles similar? If they are not, tell why.
5. Are the triangles pictured at the right similar? If they are not, tell why.

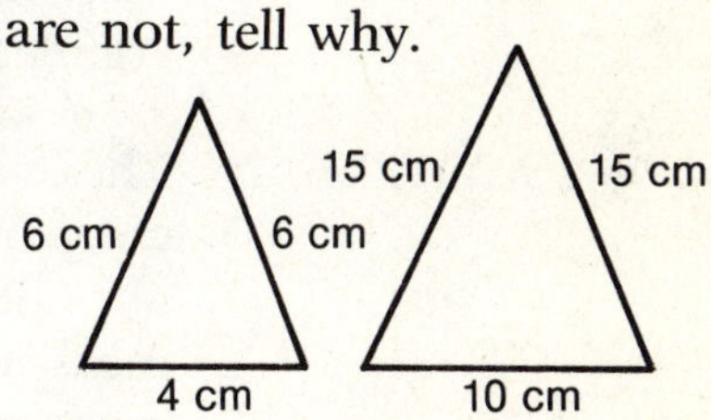

6. In the triangles pictured at the right, $\angle J$ and $\angle M$ are right angles. Are these triangles similar? If they are not, tell why.

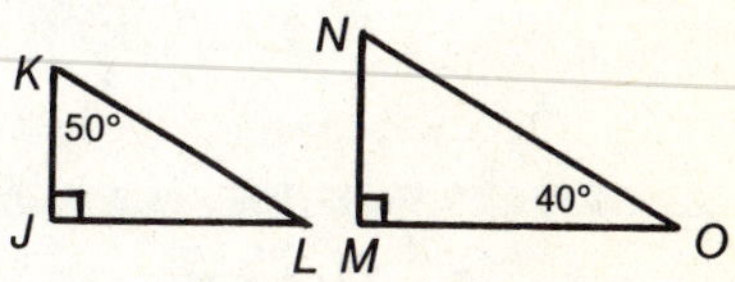

7. In triangle *JKL* above, $JK = 8$ inches and $JL = 12$ inches. In triangle *MNO*, $MN = 14$ inches. Find the length of side *MO*.
8. A six-foot-tall vertical post casts a five-foot shadow. At the same time, a tree casts a 65-foot shadow. How tall is the tree?
9. In the picture at the right, $\angle S$ and $\angle W$ are both right angles. Are the two triangles similar?
10. In the picture with problem 9, *VW* is the distance across a river. Find the measurement of *VW*.

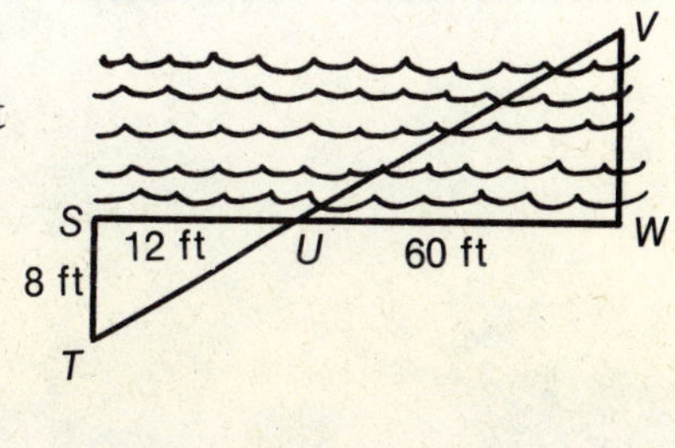

11. In the figure at the right, $\angle CDE$ and $\angle CAB$ are both right angles. Are triangles *CAB* and *CDE* similar?

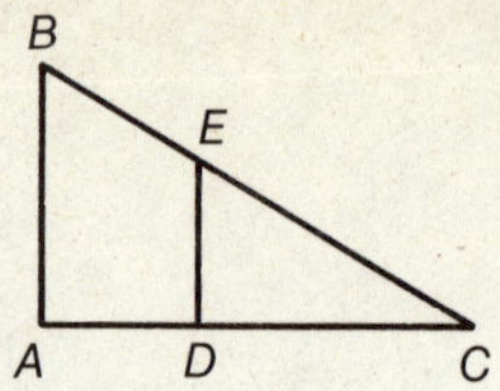

12. In problem 11, $CD = 15$ inches, $AD = 10$ inches, and $DE = 6$ inches. Find *AB*.

ANSWERS AND SOLUTIONS START ON PAGE 370.

Congruence

Congruent figures have the same shape and the same size. The corresponding angles of congruent figures are equal, and the corresponding sides are also equal. Congruent figures are identical. Remember: Corresponding sides are opposite equal angles. In addition, corresponding angles are opposite equal sides.

Following are the conditions that make two triangles congruent.

> If two angles and a corresponding side of one triangle are the same measurements as two angles and corresponding side of another triangle, the triangles are congruent. This is called the **angle, side, angle** requirement.
>
> Remember "ASA"

EXAMPLE 1: The two triangles at the right are congruent because two angles and a corresponding side are equal in each triangle.

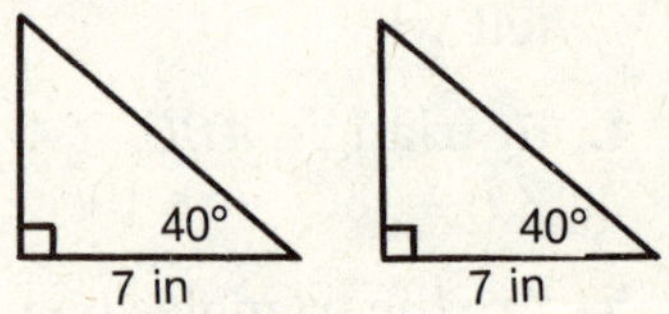

EXAMPLE 2: The two triangles at the right are *not* congruent because the equal sides are not corresponding. One side measuring 5 feet is opposite an angle measuring 45°. The other side measuring 5 feet is opposite an angle of 90°.

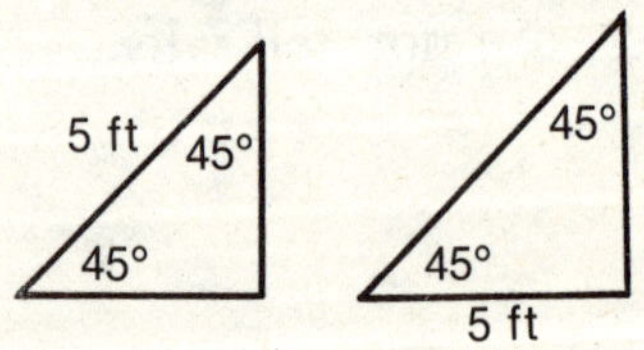

> If two sides and a corresponding angle of one triangle are the same measurements as two sides and corresponding angle of another triangle, the triangles are congruent. This is called the **side, angle, side** requirement.
>
> Remember "SAS"

EXAMPLE 3: The two triangles at the right are congruent because two sides and a corresponding angle are the same in both triangles.

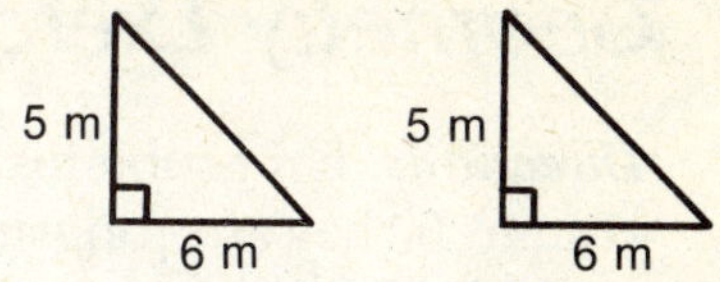

EXAMPLE 4: The two triangles at the right are *not* congruent because the equal sides are not corresponding.

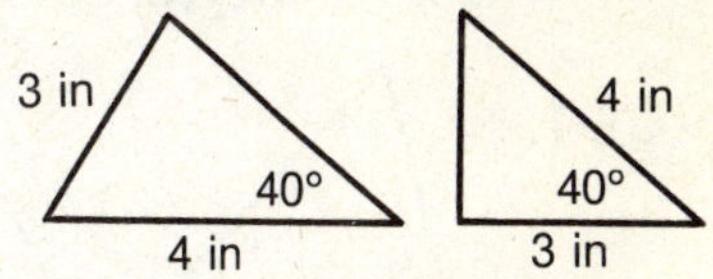

If the three sides of one triangle are the same measurements as the three sides of another triangle the two triangles are congruent. This is called the **side, side, side** requirement.

Remember "SSS"

EXAMPLE 5: The two triangles at the right are congruent because the three sides are the same in both triangles.

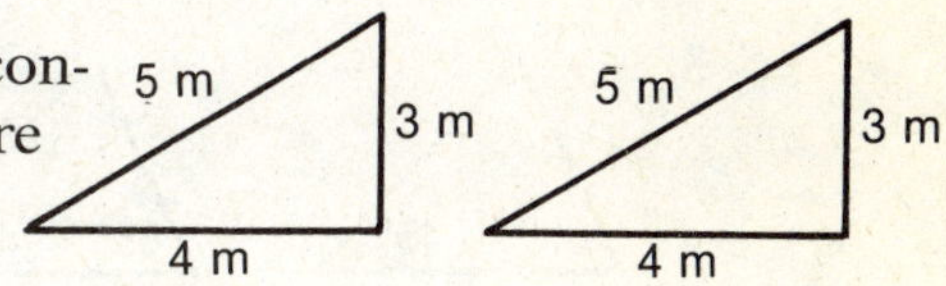

EXAMPLE 6: Are the triangles at the right congruent? If they are not, tell why.

SOLUTION. Find the missing angles: $\angle A$ is 90°, and $\angle E$ is 30°. Since the angles and corresponding sides are equal, the triangles are congruent.

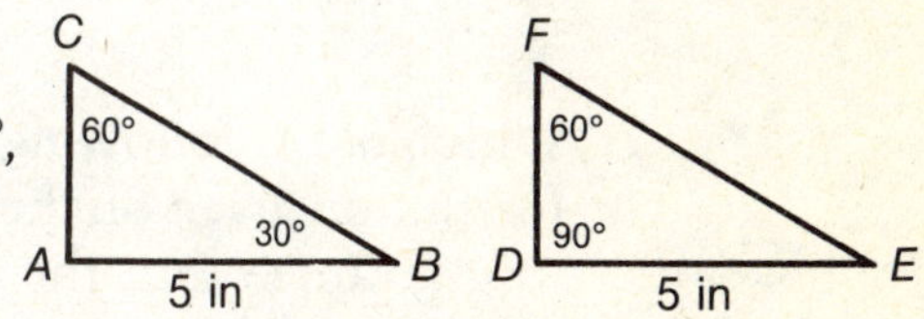

EXAMPLE 7: For the triangles at the right, $GH = JK$ and $GI = JL$. Of the following conditions, which together with the given information, is enough to guarantee that $\triangle GHI$ is congruent to $\triangle JKL$?

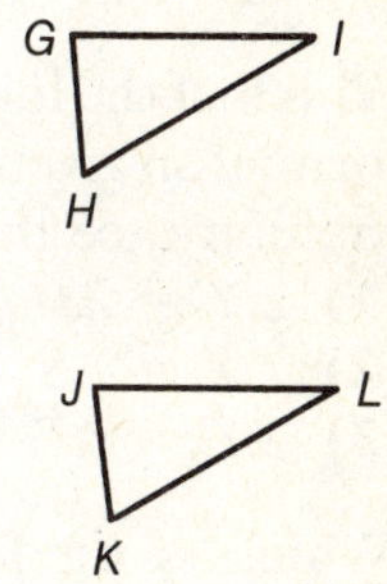

A. $HI = KL$
B. $\angle G = \angle K$
C. $\angle G = \angle J$

(1) A only
(2) B only
(3) C only
(4) A or B
(5) A or C

SOLUTION. Condition A is enough to fulfill the side, side, side requirement. Condition C is enough to fulfill the side, angle, side requirement. The best choice is answer **(5)**. Notice that condition B involves angles that are not corresponding.

Geometry Exercise 14

Directions: For problems 1 through 6, decide whether each pair of triangles is congruent. If they are congruent, tell which requirement the triangles fulfill (ASA, SAS, or SSS). If they are not congruent, tell why.

1.

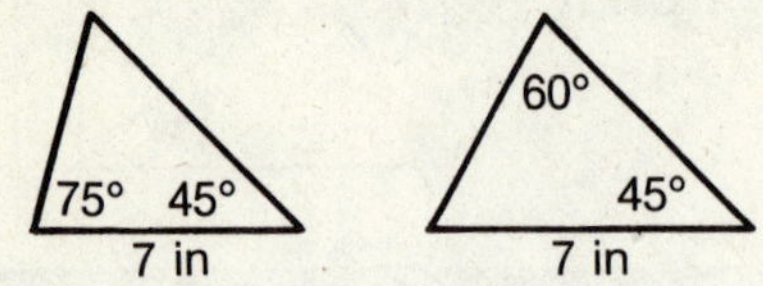

2.

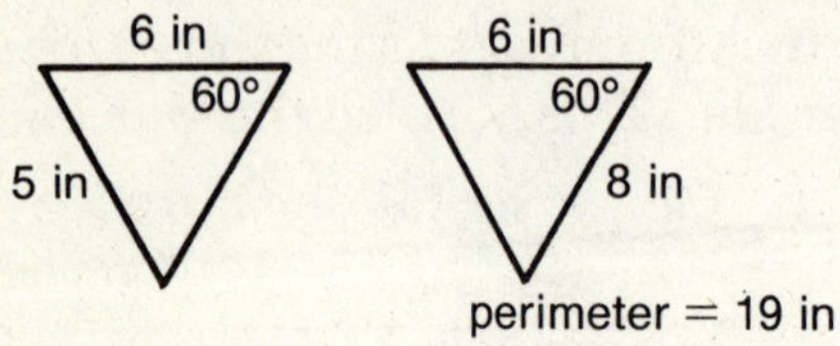

3.

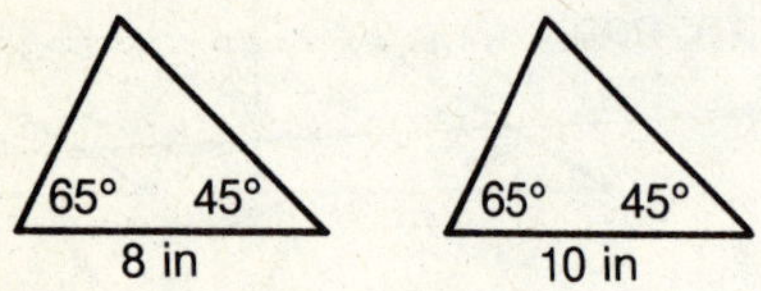

4.

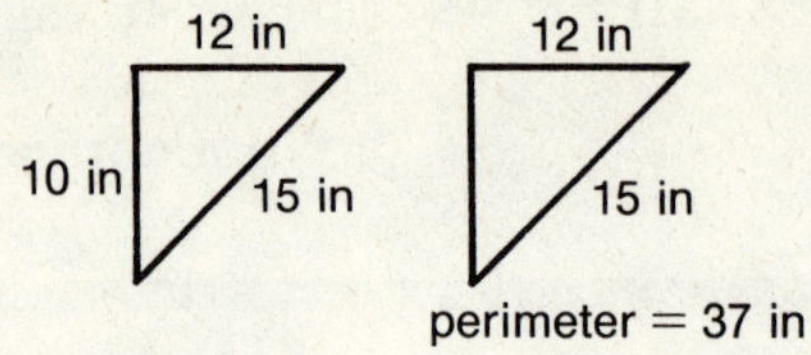

5.

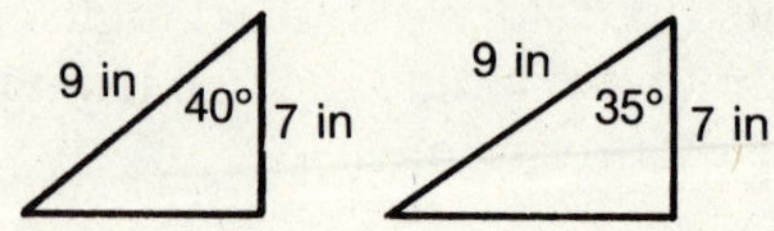

6.

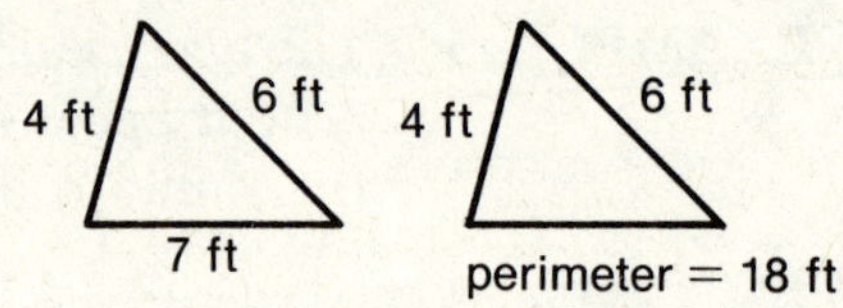

7. Which of the following conditions, along with the information given on the triangles at the right, is enough to guarantee that the triangles are congruent?
(1) $AC = DF$
(2) $\angle B = \angle E$
(3) $AB = EF$

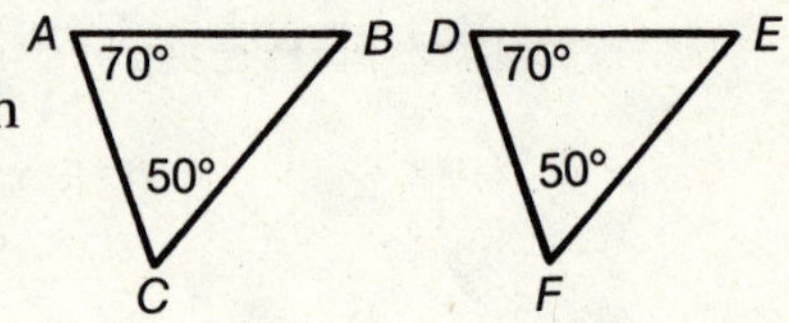

8. Which of the following conditions, along with the information given on the triangles at the right, is enough to guarantee that the triangles are congruent?
(1) $\angle Y = \angle U$
(2) $\angle X = \angle S$
(3) $\angle Z = \angle T$

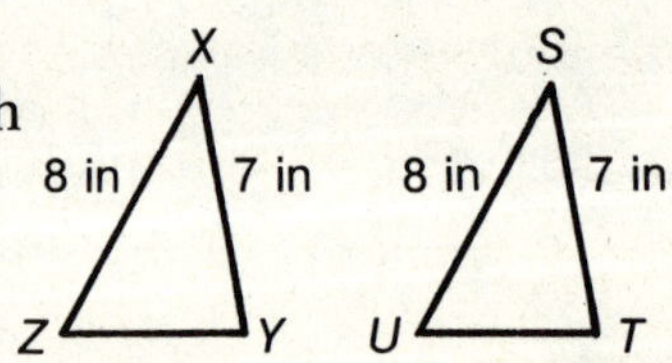

9. Which of the following conditions, along with the information given on the triangles at the right, is enough to guarantee that the triangles are congruent?
(1) $GI = JL$
(2) $GI = KL$
(3) $HI = JL$

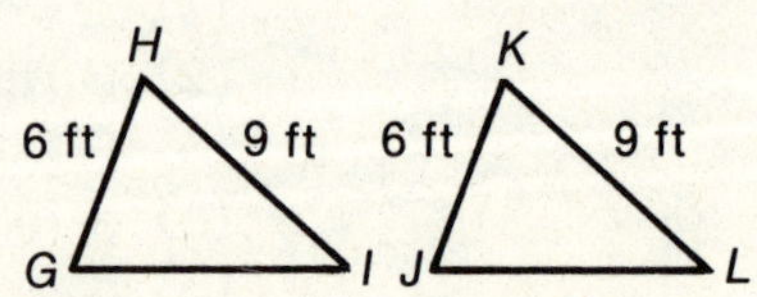

10. For the two triangles at the right, $MO = PR$ and $NO = QR$. Which of the following conditions, along with the information you already have, is enough to guarantee that $\triangle MNO$ is congruent to $\triangle PQR$?

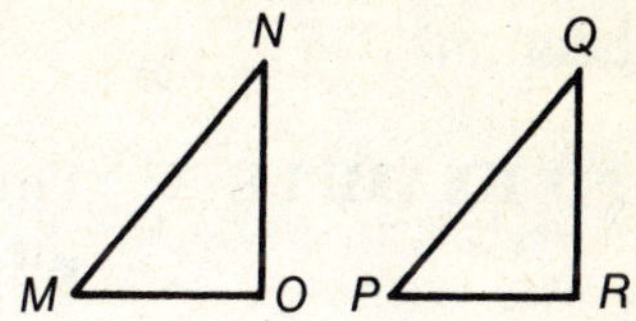

A. $\angle N = \angle R$
B. $MN = PQ$
C. $\angle O = \angle R$

(1) A only
(2) B only
(3) C only
(4) A or B
(5) B or C

ANSWERS AND SOLUTIONS START ON PAGE 371.

Pythagorean Relationship

A Greek mathematician named Pythagoras discovered a special relationship among the sides of a right triangle. We call his discovery the ***Pythagorean theorem*** or Pythagorean relationship.

With this theorem, you can find the length of any side of a right triangle if you know the lengths of the other two sides. In words, the theorem says:

> The square of the hypotenuse is equal to the sum of the squares of the other two sides.

Following is the formula for the Pythagorean relationship.

> $c^2 = a^2 + b^2$; where c = hypotenuse, and a and b are the legs of a right triangle.

EXAMPLE 1: Find the length of the hypotenuse in the triangle at the right.

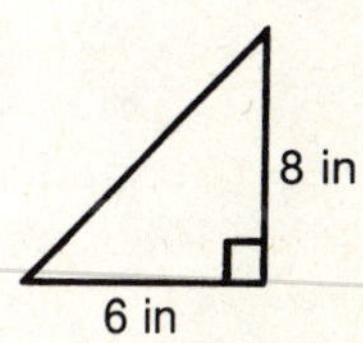

(1) $c^2 = a^2 + b^2$
$c^2 = 6^2 + 8^2$
$c^2 = 36 + 64$
$c^2 = 100$

(2) $c = \sqrt{100}$
$c =$ **10 in**

STEP 1. Replace a with 6 and b with 8 in the formula $c^2 = a^2 + b^2$.
STEP 2. The formula gives the value of the hypotenuse squared. To find the length of the hypotenuse, find the square root of 100.

To work with the Pythagorean relationship, you may need to review the section on squares and square roots starting on page 234.

In some problems, you may be given the length of the hypotenuse and the length of one of the legs. To find the length of the other leg, you can still use the Pythagorean theorem.

EXAMPLE 2: Find the length of the missing leg in the triangle at the right.

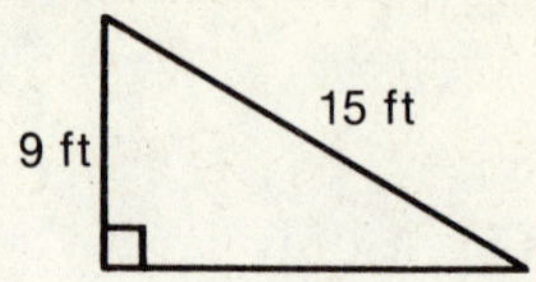

(1) $c^2 = a^2 + b^2$
$15^2 = a^2 + 9^2$

(2) $225 = a^2 + 81$

(3) $225 = a^2 + 81$
$-81 \qquad -81$
$144 = a^2$

(4) $\sqrt{144} = a$
$\mathbf{12 = a}$

STEP 1. Write down the Pythagorean theorem, and substitute in the values you know.
STEP 2. Find the values of the squares.
STEP 3. To get the unknown, a, alone on one side, subtract 81 from both sides.
STEP 4. To find a, find the square root of both sides of the equation.

TIP: In some problems, you will have to recognize that a figure is a right triangle. The picture or problem may say nothing about a right triangle, the hypotenuse, or legs. Drawing a picture may help you to see that the problem involves a right-triangle relationship.

EXAMPLE 3: A boat sails 20 miles east of port and then 15 miles south to an island. How far is the boat from the port if you measure in a straight line?

(1)
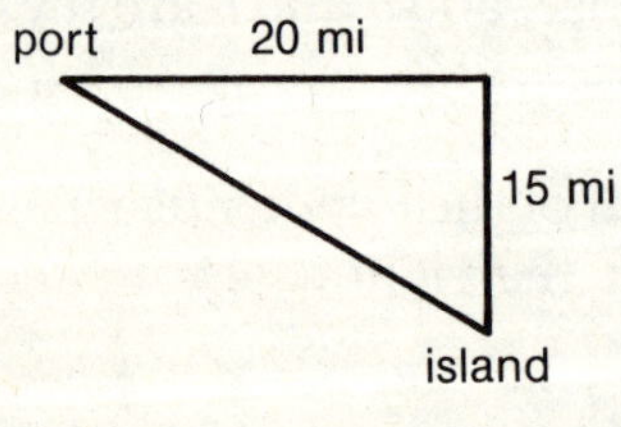

(2) $c^2 = a^2 + b^2$
$c^2 = 20^2 + 15^2$
$c^2 = 400 + 225$
$c^2 = 625$

(3) $c = \sqrt{625}$
$c = 25$ mi

STEP 1. Make a drawing to see how to solve the problem. East is normally to the right on a map, and south is toward the bottom. Notice that the actual distance from the port is the hypotenuse of a right triangle.
STEP 2. Replace a with 20 and b with 15 in the formula $c^2 = a^2 + b^2$.
STEP 3. Find the square root of 625. The boat is **25 miles** from the port.

Problem Solving for the GED

Using the Formulas Page

To make a flower vase, Tina chose a 20-cubic-inch ceramic mold. If the mold stood five inches tall and was 2 inches long, how many inches was the width?

(1) 1
(2) 2
(3) 5
(4) 10
(5) 20

In problems on the GED math test, you will not be told when to use the formulas page provided with your test booklet. Instead, you will need to infer from the problem that a formula would help you solve the problem. Then you will need to choose the right formula from the list and use it correctly.

Here are some hints to help you do this.

1. Read the problem carefully. If a geometric shape is being described, this is a clue that a formula may come in handy.

   In the problem above, you can tell by words such as *tall* and *long* and *width* that a shape is being described.

2. Decide what shape you are dealing with and what formula may apply. Remember that *perimeter* measures distance around, *area* measures a flat surface, and *volume* measures solid figures.

   Here is an additional clue:

   perimeter = plain units (inches)
   area = squared units (inches^2)
   volume = cubed units (inches^3)

   In this problem, you are given the clue *20 cubic inches*. This tells you that you will need a formula for *volume*. The shape being described is a *rectangular container*.

3. Find the correct formula and apply it. In this problem, we find that the formula for the volume of a rectangular container is $V = lwh$.

   $V = lwh$
   $20 = 2 \times w \times 5$
   $20 = 10 \times w$
   $2 = w$
   answer (2) **2**

Most of the information on the formulas page deals with geometric shapes. You have learned other formulas (such as the interest formula) which are also given on the formulas page. Remember to consult the formulas page when you need help with these problems as well.

Formulas

Description	Formula
AREA (A) of a:	
square	$A = s^2$; where s = side
rectangle	$A = lw$; where l = length, w = width
parallelogram	$A = bh$; where b = base, h = height
triangle	$A = \frac{1}{2}bh$; where b = base, h = height
circle	$A = \pi r^2$; where $\pi = 3.14$, r = radius
PERIMETER (P) of a:	
square	$P = 4s$; where s = side
rectangle	$P = 2l + 2w$; where l = length, w = width
triangle	$P = a + b + c$; where a, b, and c are the sides
circumference (C) of a circle	$C = \pi d$; where $\pi = 3.14$, d = diameter
VOLUME (V) of a:	
cube	$V = s^3$; where s = side
rectangular container	$V = lwh$; where l = length, w = width, h = height
cylinder	$V = \pi r^2 h$; where $\pi = 3.14$, r = radius, h = height
Pythagorean relationship	$c^2 = a^2 + b^2$; where c = hypotenuse, a and b are legs of a right triangle
distance (d) between two points in a plane	$d = \sqrt{(x_2 - x_1)^2 + (y_2 - y_1)^2}$; where (x_1, y_1) and (x_2, y_2) are two points in a plane
slope of a line (m)	$m = \frac{y_2 - y_1}{x_2 - x_1}$; where (x_1, y_1) and $(x_2\ y_2)$, are two points in a plane
mean	$mean = \frac{x_1 + x_2 + \ldots + x_n}{n}$; where the x's are the values for which a mean is desired, and n = number of values in the series
median	$median$ = the point in an ordered set of numbers at which half of the numbers are above and half of the numbers are below this value
simple interest (i)	$i = prt$; where p = principal, r = rate, t = time
distance (d) as function of rate and time	$d = rt$; where r = rate, t = time
total cost (c)	$c = nr$; where n = number of units, r = cost per unit

★ GED PRACTICE ★

Using the Formulas Page

The figure at right represents the area that Molly needs to cover with plywood. How many square inches of wood does she need? Use $\frac{22}{7}$ for π.

(1) 22
(2) 44
(3) 154
(4) 308
(5) 400

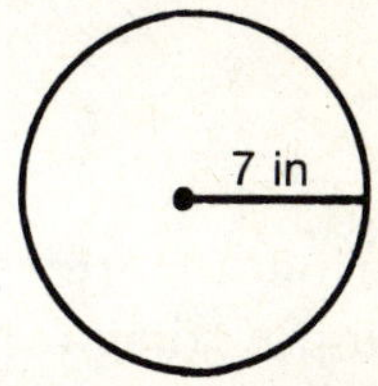

ANSWER AND SOLUTION ARE ON PAGE 371.

Geometry Exercise 15

Directions: Solve each problem. Use the formulas page when necessary.

1. Find the hypotenuse of the triangle at the right.

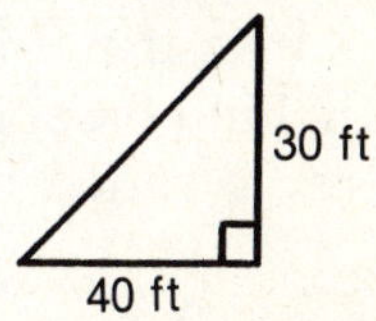

2. One leg of a right triangle measures 10 inches. The other measures 24 inches. Find the length of the hypotenuse.

3. In triangle *XYZ* at the right, side $XY = 12$ inches and side $YZ = 5$ inches. Find the length of *XZ*.

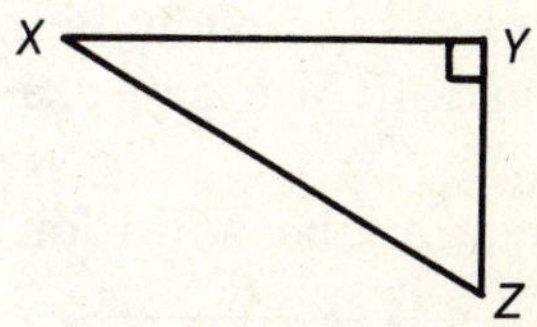

4. What is the length of the hypotenuse of a right triangle whose legs measure 12 yards and 16 yards?

5. Joanna drove 60 miles west from Bonesville to Mactown and then 45 miles south to Steven City. She returned to Bonesville on Route 3. How far is it from Steven City to Bonesville on Route 3?

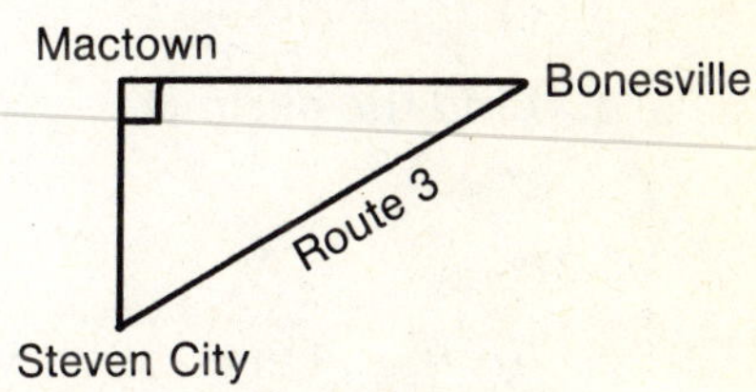

6. One leg of a triangle measures 18 yards. The hypotenuse measures 30 yards. Find the length of the other leg.

7. In triangle *ABC* at the right, $AB = 16$ feet and $AC = 34$ feet. Find the length of *BC*.

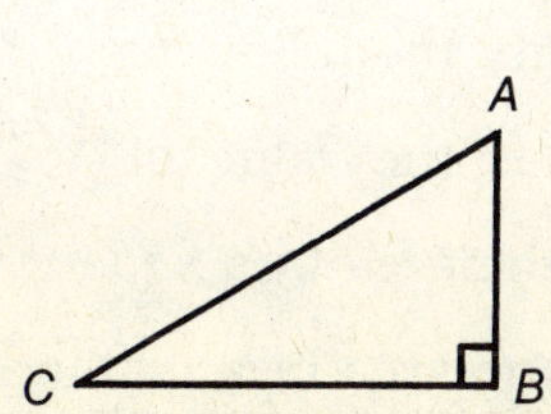

8. A 17-foot ladder just touches the bottom of a window. The bottom of the ladder is 8 feet from the bottom of the building. How far is the distance *d* from the ground to the bottom of the window?

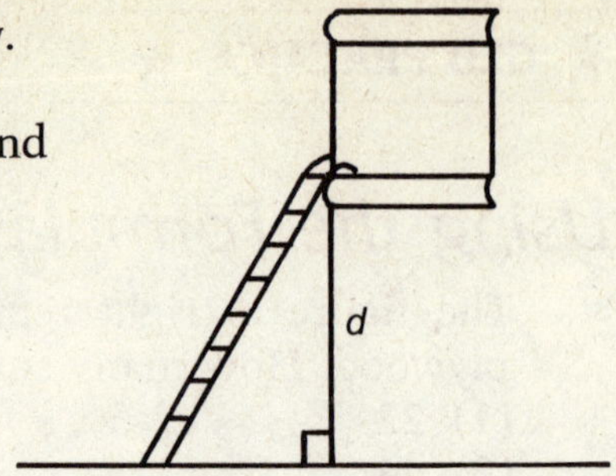

9. During a storm, Dick put tape on a picture window — taping diagonally from corner to corner. If his window measures 2.4 m by 1.8 m, how much tape did he use for each diagonal?

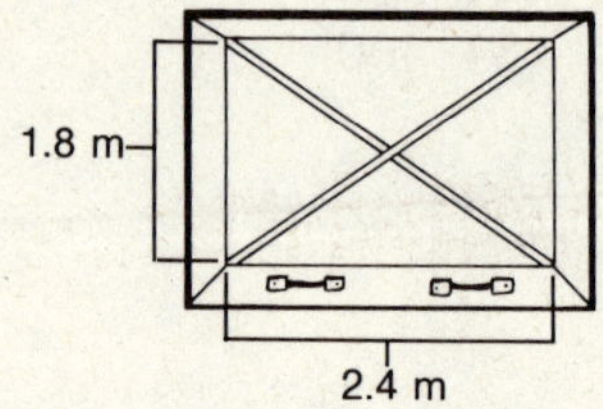

10. Louie drove 48 miles directly north and then 36 miles directly west. Find the shortest distance from the point where he ended up to his starting point.

11. On a bike trip, Manuel rode 15 miles directly west and then 36 miles directly south. Find the shortest distance from the point where he started to the point where he ended his trip.

ANSWERS AND SOLUTIONS START ON PAGE 371.

Geometry Exercise 16: Review

Directions: This review is in two parts. For questions 1 through 13, simply answer each question. You can use the formulas on page 270 for this exercise.

1. A rectangular solid for which every side has the same measurement is called a ________________.

2. The base of a square measures 3.2 meters. Find the perimeter of the square.

3. The base of a rectangle is 12 inches. The width is $4\frac{1}{4}$ inches. Find the perimeter of the rectangle.

4. Find the perimeter of the triangle pictured at the right.

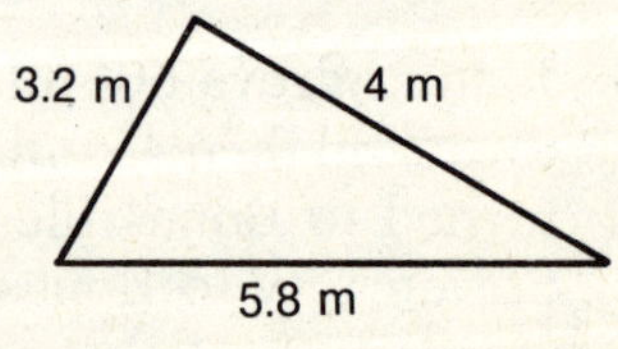

5. To the nearest foot, what is the circumference of a circle with a diameter of 30 feet?

6. What is the value of 20^3?

7. Simplify the following: $(.03)^2 + (.4)^2$

8. What is the value of $\sqrt{7{,}056}$?

9. Find the area of a square with a base that measures $5\frac{1}{2}$ inches.

10. Find the area of a rectangle with a base of 12.5 cm and a height of 7.4 cm.

11. Find the area of the triangle pictured at the right.

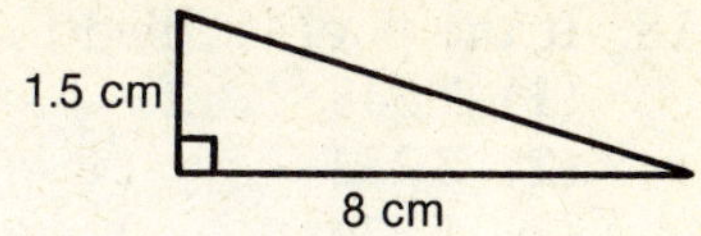

12. What is the area of a circle with a radius of 7 inches? Use $\pi = \frac{22}{7}$.

13. Find the volume of a rectangular container with a length of 15 inches, a height of 10 inches, and a width of $3\frac{1}{2}$ inches.

★ GED PRACTICE ★

Directions: Choose the correct answer for each of the following questions.

14. Sharon wants to put fencing around the rectangular garden in her yard. The garden measures 20 feet by 8 feet. She wants to leave a four-foot opening for a walkway into the garden. Which of the following expressions gives the number of feet of fencing she needs?

(1) 20 + 8
(2) (2 × 20) + (2 × 8)
(3) (2 × 20) + (2 × 8) − 4
(4) (20 × 8) − 4
(5) 2(20 − 8)

15. One side of the square pictured at the right is 12 inches. The length of the rectangle is 15 inches. Find the rectangle's height in inches if it has the same area as the square.

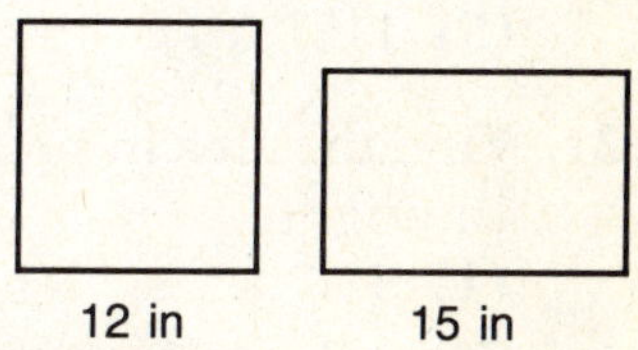

(1) 30
(2) 27
(3) 19.2
(4) 9.6
(5) 6

Directions: Questions 16 through 18 refer to the following situation.

The picture at the right shows the plan of a new pool at the Uptown Community Center. The depth of the pool will be six feet throughout. The edge of the pool will be surrounded by glazed tiles that are six inches wide and six inches long.

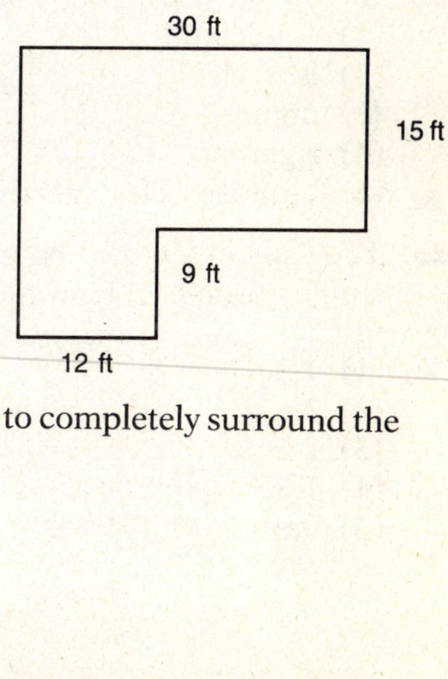

16. What is the smallest number of glazed tiles needed to completely surround the edge of the pool?

(1) 96
(2) 108
(3) 216
(4) 432
(5) Not enough information is provided.

17. Find the area in square yards of the bottom of the pool.

(1) 31
(2) 62
(3) 124
(4) 186
(5) 532

18. If the pool is filled to its limit, how many cubic feet of water will it hold?
(1) 3,348
(2) 2,232
(3) 1,674
(4) 1,116
(5) Not enough information is provided.

19. In the picture at the right $\angle BOC = 64°$. What kind of angle is $\angle AOB$?
(1) acute
(2) right
(3) obtuse
(4) straight
(5) reflex

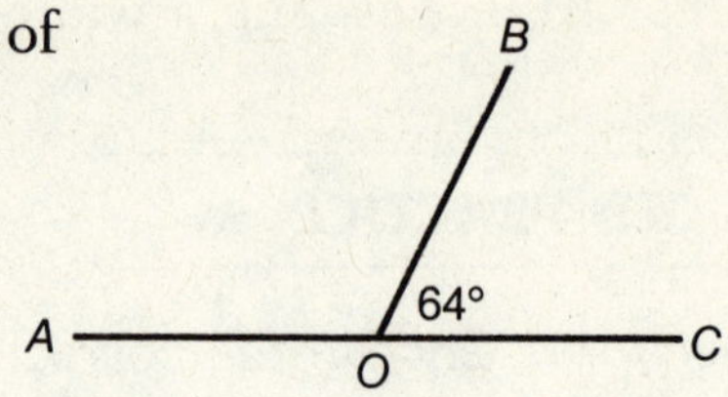

20. In the picture at the right, $\angle a = 65°$. How many degrees are there in $\angle d$?
(1) 15°
(2) 25°
(3) 35°
(4) 105°
(5) 115°

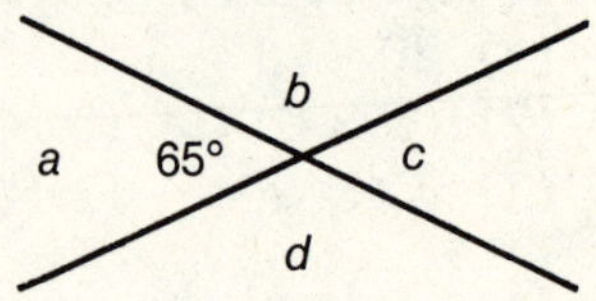

21. Find the area in square inches of the figure shown at the right.
(1) 130
(2) 120
(3) 100
(4) 98
(5) Not enough information is provided.

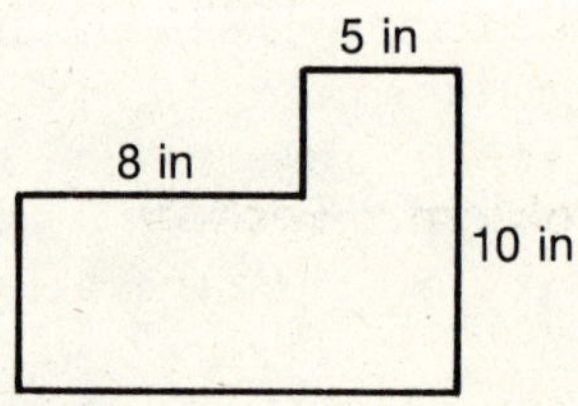

22. In triangle XYZ, $\angle X = 33°$ and $\angle Y = 57°$. What kind of triangle is $\triangle XYZ$?
(1) equilateral
(2) isosceles
(3) obtuse
(4) right
(5) both isosceles and right

23. The sails of the two boats pictured at the right are similar triangles. Find the height in feet of the small sail.
(1) 7.2
(2) 3.6
(3) 2.5
(4) 1.8
(5) 0.9

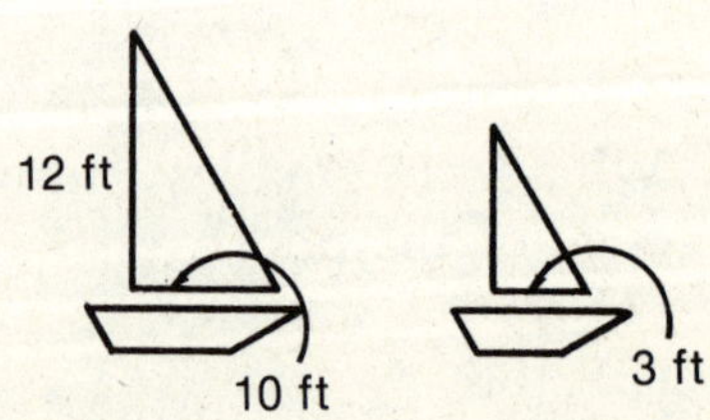

24. For the triangles at the right, $\angle G = \angle J$ and $\angle I = \angle L$. Which of the conditions listed below, together with the given information, is enough to guarantee that $\triangle GHI$ is congruent to $\triangle JKL$?

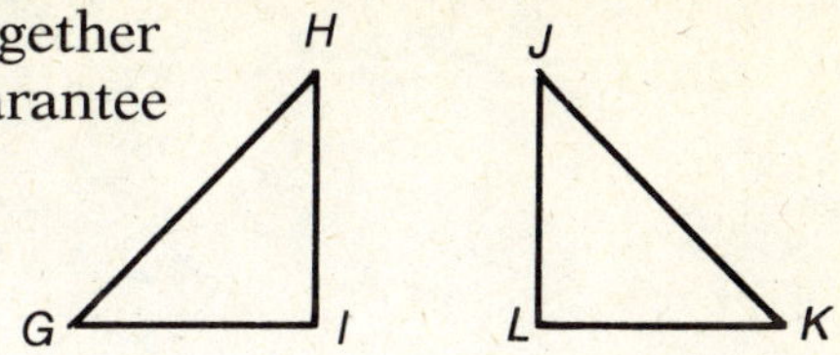

A. $GI = JL$
B. $GH = JK$
C. $\angle H = \angle K$

(1) A only
(2) B only
(3) C only
(4) A or B
(5) B or C

25. In the rectangle at the right, side *BC* measures 36 inches, and diagonal line *AC* measures 39 inches. Find the length in inches of side *AB*.

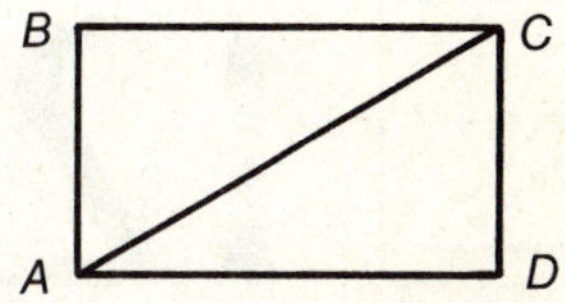

(1) 15
(2) 18
(3) 30
(4) 33
(5) 42

ANSWERS AND SOLUTIONS START ON PAGE 372.

Geometry Review Evaluation

Problem	Section	Starting Page
1	Common Geometric Figures	224
2–5	Perimeter and Circumference	227
6–7	Powers	234
8	Roots	236
9–12, 21	Area	238
13	Volume	244
14	Perimeter and Circumference/Set-up Questions	231/42
15	Area Word Problems	242
19	Angles	251
20	Angle Relationships	254
16–18	Perimeter, Area, and Volume Word Problems/ Item Sets	247/70
22	Triangles	257
23	Similarity	260
24	Congruence	264
25	Pythagorean Theorem	267

Passing score: 20 right out of 25 problems.

Your score: ______ right out of 25 problems.

If you had less than a passing score, review the sections for those problems you missed. Then redo these questions before going on to the next chapter.

If you had a passing score, correct any problem you got wrong. Then go on to the next chapter.

12
SPECIAL TOPICS IN ALGEBRA AND GEOMETRY

In this chapter, you will learn about several different topics in algebra and geometry. Any of the topics below could appear on the GED Math Test.

- The Number Line
- Signed Numbers
- Inequalities
- Monomials
- Factoring
- Using Algebra to Solve Geometry Problems
- Rectangular Coordinates
- Linear Equations
- Slope and Intercepts
- Quadratic Equations
- Simultaneous Equations

The Number Line

In algebra, you can use a set of numbers you have not used in arithmetic. In this section, you will learn about these numbers and how to use them.

The numbers we use in arithmetic can be represented on a line like this:

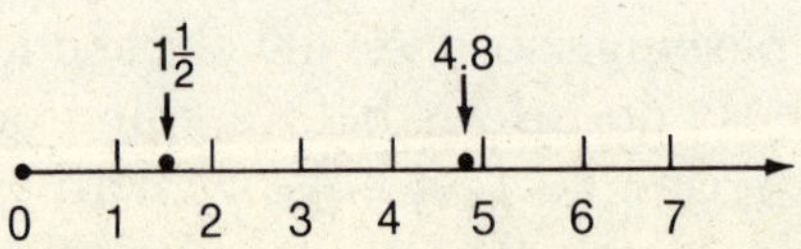

Fractions, decimals, and mixed numbers fit in the spaces between the whole numbers. Notice the points representing $1\frac{1}{2}$ and 4.8. Except for zero, all the numbers on this line are ***positive numbers***, which are numbers greater than zero. The arrow at the right end means that the numbers go on and on.

In algebra we extend this system to include numbers that are less than zero. These are called ***negative numbers***. Temperatures on a thermometer are the most familiar use of positive and negative numbers. When it is 70° outside, we mean that it is 70° above zero. In algebra we call this "positive 70." When it is −20° outside, we mean that it is 20° below zero. In algebra we call this "negative 20."

We can extend the line shown above to include both positive and negative numbers:

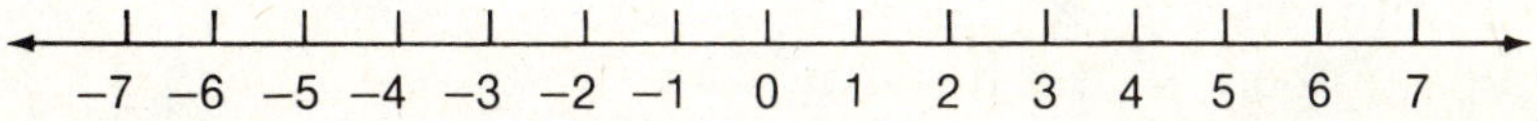

This is called the ***number line***. The arrows at both ends mean that both the positive and negative values go on and on.

The positive numbers represent numbers above zero. The negative numbers represent numbers below zero. Zero (0) has no sign. It is neither positive nor negative.

To move to the right along the number line means to find an increasing amount such as a gain in weight or a rise in temperature. To move to the left along the number line means to find a decreasing amount such as a loss of weight or a drop in temperature.

> Positive numbers do not have to be written with a plus sign (+). The number 8 is understood to be +8. Negative numbers, however, must always have a minus (−) sign in front of them.

Special Topics Exercise 1

Directions: Use the number line below to tell which letter corresponds to each of the following numbers. Problem 1 is already done as an example. The letter *J* corresponds to +6.

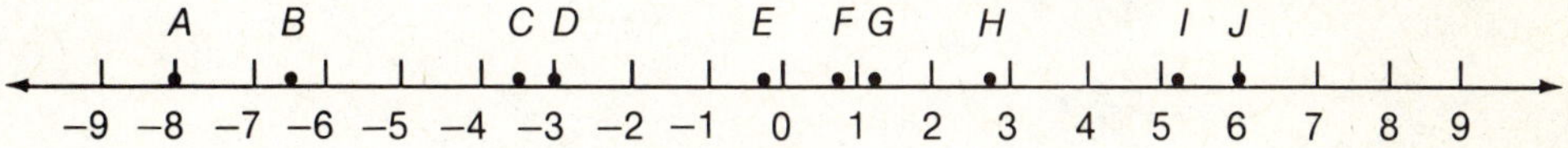

1. +6

2. −3

3. $-6\frac{1}{2}$

4. −8

5. $+\frac{2}{3}$

6. $\frac{5}{4}$

7. $+\frac{16}{3}$

8. 2.75

9. −3.5

10. −0.2

11. Mark each of the following numbers on the number line below. The first number, $-1\frac{2}{3}$, has been marked as an example.

$-1\frac{2}{3}$, $3\frac{1}{2}$, +6, −4.25, $\frac{4}{3}$, −5.2, −7.1, +6.4

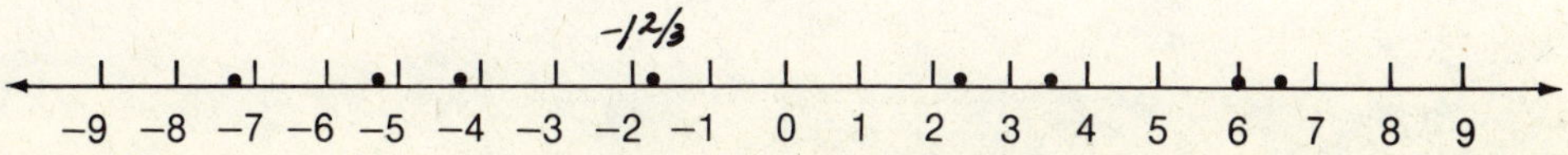

ANSWERS AND SOLUTIONS START ON PAGE 373.

Addition of Signed Numbers

The entire set of positive and negative numbers is called ***signed numbers***. In the next four sections, you will learn to add, subtract, multiply, and divide signed numbers.

Rules for Adding Two Signed Numbers

1. If the two numbers have the same signs, find the sum of the numbers, and give the answer that sign:
   - Positive numbers give a positive sum.
   - Negative numbers give a negative sum.
2. If the two numbers have different signs, find the difference between the two numbers (subtract the smaller from the larger), and *give the answer the sign of the larger number.*

In the next examples, you will see problems involving addition of signed numbers. A negative sign looks very much like a minus sign. To avoid confusion, signed numbers are sometimes written inside parentheses.

EXAMPLE 1: Find $(-16) + (-8)$.

$$\begin{array}{r} -16 \\ \underline{+\ -8} \\ -24 \end{array}$$

STEP 1. Since the numbers have the same signs, find their sum.
STEP 2. Since they are both negative numbers, the answer is negative.

The last example could represent a drop in temperature. If the temperature is $-16°$ and then drops another $8°$ $(-8°)$, the temperature will be $-24°$.

EXAMPLE 2: Add $+23$ and -54.

$$\begin{array}{r} -54 \\ \underline{+23} \\ -31 \end{array}$$

STEP 1. Since the numbers have different signs, find the difference. Subtract the smaller number (23) from the larger (54).
STEP 2. Give the answer the sign of the larger number, 54. The answer is negative.

This example could represent the following situation: Lois has paid back \$23 (+23) of the \$54 (−54) she borrowed from her sister. The amount she still owes is \$31. Her debt is written as −\$31.

It is also possible to add more than two signed numbers.

Rules for Adding More than Two Signed Numbers

1. Find the sum of the positive numbers, and make that sum positive.
2. Find the sum of the negative numbers, and make that sum negative.
3. Find the difference between the two sums, and give the answer the sign of the larger sum.

In the next examples, watch for the different ways to write addition of signed numbers.

EXAMPLE 3: $(-9) + (10) + (-8) + (+4) = ?$

(1) $\begin{array}{r} +10 \\ \underline{+4} \\ +14 \end{array}$ **(2)** $\begin{array}{r} -9 \\ \underline{-8} \\ -17 \end{array}$ **(3)** $\begin{array}{r} -17 \\ \underline{+14} \\ -3 \end{array}$

STEP 1. Find the sum of the positive numbers 10 and 4, and make the sum positive.
STEP 2. Find the sum of the negative numbers −9 and −8, and make the sum negative.
STEP 3. Find the difference between 17 and 14, and give the answer a negative sign because 17 is larger.

Notice in the last example that the numbers inside the parentheses are either positive or negative. The plus signs between the parentheses indicate addition.

EXAMPLE 4: Add: $-5 + 4 + 7 + (-6)$.

(1) $\begin{array}{r} +4 \\ \underline{+7} \\ +11 \end{array}$ **(2)** $\begin{array}{r} -5 \\ \underline{-6} \\ -11 \end{array}$ **(3)** $\begin{array}{r} +11 \\ \underline{-11} \\ 0 \end{array}$

STEP 1. Find the sum of the positive numbers, and make the sum positive.
STEP 2. Find the sum of the negative numbers, and make the sum negative.
STEP 3. Find the difference between the two sums. Remember that 0 has no sign.

Special Topics Exercise 2

Directions: Solve each problem.

1. $(-9) + (-3)$
2. $6 + (-8) = ?$
3. Find the sum of $+12$ and $-20\frac{1}{2}$.
4. Find the sum of $+12\frac{1}{2}$ and -20.
5. $(+5) + (-8) + (+9) =$

6. Find the sum of +8, −6, and −5.

7. −18 + (−2) + 6 = ?

8. Add −6, 8, −4, and +11.

9. 40 + (−7) + (−5) + (−3) + (−10) + (−2) = ?

10. Find the sum of +3, −10, and +7.

11. (−10) + (5) + (−3) + (−9) = ?

12. (−9) + (−4) + (8) + (7) + (−2) = ?

13. At 5 A.M. the temperature was −10°. By 8 A.M. the temperature had dropped another 4 degrees. What was the 8 A.M. temperature?

14. Jack weighed 180 pounds in January. In February he gained 15 more pounds. Then he dieted during March and April and lost 37 pounds. How much did he weigh at the end of April?

15. Mercedes had $235 in her checking account. She deposited $55 and wrote checks for $30, $150, and $120. Did she have enough money in her account to cover the checks?

ANSWERS AND SOLUTIONS START ON PAGE 373.

Subtraction of Signed Numbers

With signed numbers, it is possible to do subtraction problems that you could not do before. For example, you can take 9 from 4. Start at +4 on the number line, and move 9 spaces to the left.

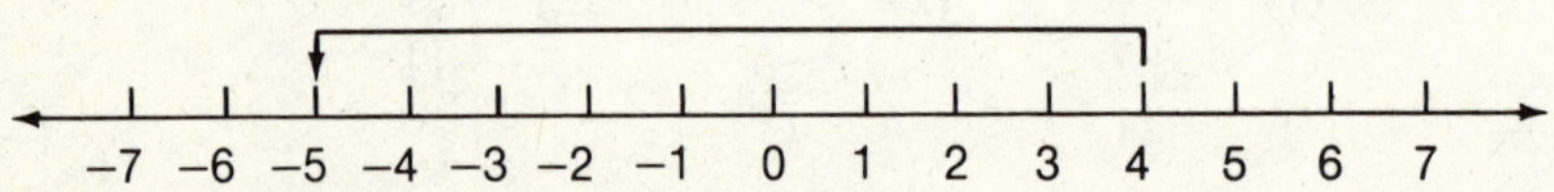

You should be at −5. In algebra, this problem is written as (+4) − (+9) = (−5). The minus sign between parentheses indicates subtraction. When you subtract signed numbers, you are finding the distance between them on the number line.

The only difference between adding and subtracting signed numbers is that a subtraction sign changes the sign of the number following it. After you change the sign, follow the rules for adding signed numbers. For example, (+4) − (+9) becomes + 4 − 9. Find the difference between 4 and 9 and give the answer a negative sign because 9 is larger than 4.

Rules for Subtracting Signed Numbers

1. Change the sign of the number being subtracted and drop the subtraction sign.
2. Follow the rules for adding signed numbers.
   a. If their signs are the same, add the numbers together and give the sum that sign.
   b. If their signs are different, find the difference between them and give the sum the sign of the larger number.

EXAMPLE 1: What is $(-8) - (+3)$?

$$(-8) - (+3) = -8 - 3 = -11$$

STEP 1. The minus sign outside the parentheses means that $+3$ is being subtracted. Change $+3$ to -3, and drop the subtraction sign.

STEP 2. Since both signs are now the same, find the sum of the numbers, and make the sum negative.

EXAMPLE 2: Find $(-10) - (-2)$.

$$(-10) - (-2) = -10 + 2 = -8$$

STEP 1. The minus signs outside the parentheses means that -2 is being subtracted. Change -2 to $+2$, and drop the subtraction sign and the parentheses.

STEP 2. Since the signs are different, find the difference between the two numbers and give the answer the sign of the larger number.

EXAMPLE 3: Simplify $(+6) - (-4) + (-2) - (+5)$.

STEP 1. The two minus signs outside parentheses mean that -4 and $+5$ are being subtracted. Change -4 to $+4$ and $+5$ to -5. Then change the minus signs to plus signs.

$$(+6) - (-4) + (-2) - (+5) = +6 + 4 - 2 - 5$$

STEP 2. Find the sum of the positive numbers, and make the sum positive.

$$\begin{array}{r} +6 \\ \underline{+4} \\ +10 \end{array}$$

STEP 3. Find the sum of the negative numbers, and make the sum negative.

$$\begin{array}{r} -2 \\ \underline{-5} \\ -7 \end{array}$$

STEP 4. Find the difference between 7 and 10, and give the answer the sign of the larger number. The answer is a positive 3 $(+3)$.

$$\begin{array}{r} +10 \\ \underline{-7} \\ +3 \end{array}$$

Special Topics Exercise 3

Directions: Solve each problem.

1. $(+6) - (+4) =$

2. $(-8) - (+3) =$

3. $(-9) - (-8) =$

4. $(+10) - (-9) =$

5. $(+8) - (7) =$

6. $(-9) - (-9) =$

7. $(-10) - 12 =$

8. $(+6) - (-3) + (-2) =$

9. $(-9) - (+4) - (+10) =$

10. $(-15) - (20) + (+6) =$

11. $(-8) + (-13) - (+6) =$

12. $(-3) + (-4) - (-5) - (-6) =$

13. The coastline closest to Rainbow Valley is 22 feet above sea level, and Rainbow Valley is 43 feet below sea level. How many feet lower than the coastline is Rainbow Valley?

14. At dawn it was 7° above zero. By dusk the temperature had dropped to −17°. By how much did the temperature drop from dawn to dusk?

ANSWERS AND SOLUTIONS START ON PAGE 374.

Multiplication of Signed Numbers

The rules for multiplying signed numbers are easy. Before learning the rules, however, study the following applications of multiplying signed numbers.

> **TIP:** Remember that if there is no sign between parentheses, multiply.

If you gain two pounds a week for five weeks, you will weigh ten pounds *more* than you weigh now. In algebra, this is $(+2)(+5) = +10$.

If you lose two pounds a week for five weeks, you will weigh ten pounds *less* than you weigh now. In algebra, this is $(-2)(+5) = -10$.

If you have been gaining two pounds a week for five weeks, you weighed ten pounds less five weeks ago. In algebra, this is $(+2)(-5) = -10$.

If you have been losing two pounds a week for five weeks, you weighed ten pounds more five weeks ago. In algebra, this is $(-2)(-5) = +10$.

The pattern to these examples is simple: When the signs of two numbers being multiplied are alike, the answer is positive. When the signs of the two numbers being multiplied are not alike, the answer is negative.

Rules for Multiplying Two Signed Numbers

1. Multiply the two numbers.
2. a. If the signs of the two numbers are alike, make the product positive.
   b. If the signs of the two numbers are different, make the product negative.

EXAMPLE 1: What is the product of (-8) and (-7)?

$(-8)(-7) = +56$

SOLUTION. Since the signs are alike, the answer is positive.

EXAMPLE 2: What is $(12)(-3)$?

$(12)(-3) = -36$

SOLUTION. Since the signs are different, the answer is negative.

EXAMPLE 3: Find $-3 \cdot 10$.

$-3 \cdot 10 = -30$

SOLUTION. Remember that the dot between the numbers is another symbol for multiplication. Since the signs are different, the answer is negative.

EXAMPLE 4: What is $(-6)(+2)(-3)(+4)$?

```
(−6)(+2) (−3)(+4)        (−6)(+2) (−3)(+4)
   \  /     \  /            \  /
  (−12)    (−12)           (−12)  (−3)(+4)
      \     /                  \  /
       144                     (36)      (+4)
                                   \    /
                                    144
```

SOLUTION. Remember that the rules are for multiplying only two numbers at a time. -6 times $+2$ is negative because the signs are different. The product, -12, times -3 is positive because the signs are alike. The product, $+36$, times $+4$ is also positive because the signs are alike.

The following rule is a shortcut for multiplying more than two signed numbers.

Rules for Multiplying More than Two Signed Numbers

1. Multiply all the numbers together.
2. a. If the problem has an even number of negative signs, the final product is positive.
   b. If the problem has an odd number of negative signs the final product is negative.

EXAMPLE 5: Find $(-2)(-1)(+3)(-5)$.

$$(-2)(-1)(+3)(-5) = -30$$

SOLUTION. This problem has three negative signs. Since three is an odd number, the answer is negative. The final product is -30.

Special Topics Exercise 4

Directions: Solve each problem.

1. $(-2)(+9) =$

2. $(-6)(-6) =$

3. $(+5)(-9) =$

4. $(+8)(3) =$

5. $(-\frac{3}{4})(12) =$

6. $(+\frac{2}{3})(-\frac{3}{4}) =$

7. $-8 \cdot 6 \cdot -2 =$

8. $-7 \cdot -103 =$

9. $+5 \cdot +4 \cdot -2 =$

10. $-8 \cdot -\frac{5}{8} =$

11. $(4)(-2)(-1)(-6)(4) =$

12. Find the product of -14 and $+2$.

13. What is the product of -2 and 6?

14. The Sunshine Day Care Center has been getting three new children every month. How many fewer children were in the school six months ago?

ANSWERS AND SOLUTIONS START ON PAGE 374.

Division of Signed Numbers

The rules for dividing signed numbers are similar to the rules for multiplying signed numbers.

Rules for Dividing Signed Numbers

1. Divide or reduce the numbers.
2. a. If the signs are alike, make the quotient positive.
   b. If the signs are different, make the quotient negative.

EXAMPLE 1: What is $\frac{+30}{-6}$?

$\frac{+30}{-6} = -5$

SOLUTION. Since the signs are different, the quotient is negative.

EXAMPLE 2: What is $\frac{-28}{-12}$?

$\frac{-28}{-12} = \frac{-28 \div 4}{-12 \div 4} = \mathbf{\frac{7}{3}}$ or $\mathbf{2\frac{1}{3}}$

SOLUTION. Reduce the fraction. Since the signs are alike, the quotient is positive. Answers in the form of improper fractions are common in algebra. Both $2\frac{1}{3}$ and $\frac{7}{3}$ could appear as answer choices on the GED Test, so you will need to check to see which is appropriate.

EXAMPLE 3: Simplify $\frac{+9}{-15}$.

$\frac{+9}{-15} = -\frac{3}{5}$

SOLUTION. In this example, you cannot divide. Simply reduce the fraction. Since the signs are different, the answer is negative.

Special Topics Exercise 5

Directions: Solve each problem.

1. $\frac{-40}{-20} =$
2. $\frac{-12}{+6} =$
3. Simplify $\frac{72}{-9}$.
4. $\frac{+16}{-24} =$
5. $\frac{-15}{+5} =$
6. $\frac{30}{-36} =$
7. Find the value of $\frac{-8}{-1}$.
8. $\frac{-13}{26} =$
9. $\frac{144}{-24} =$
10. $\frac{-108}{-9} =$
11. $\frac{-48}{-60} =$
12. $\frac{+65}{-5} =$

13. $\frac{-36}{-24} =$

14. $\frac{-63}{+35} =$

15. $\frac{75}{-100} =$

16. $\frac{-15}{-150} =$

ANSWERS AND SOLUTIONS START ON PAGE 374.

Signed Numbers Review

The next exercise will give you a chance to review the four basic operations with signed numbers. In some problems you will have to perform more than one operation. Review the order of operations you learned earlier.

1. Do all work in parentheses.
2. Find powers and roots.
3. Do all multiplication and division.
4. Do all addition and subtraction.

EXAMPLE: $(-4)(-2) - (5)(-3) =$

(1) $(-4)(-2) - (5)(-3) =$

$+8 - (-15) =$

(2) $+8 + 15 = +23$

STEP 1. First find the two products.

STEP 2. Since the second product, -15, is being subtracted, change -15 to $+15$. Then follow the rules for addition of signed numbers. Since the numbers have the same sign, find the sum and give the answer that sign.

Special Topics Exercise 6

Directions: Solve each problem.

1. $(15) - (-9) =$

2. $(-6)(+20) =$

3. $-4 + 3 - 7 =$

4. $\frac{-96}{-8} =$

5. $(-8) - (12) =$

6. $\frac{1,000}{-10} =$

7. $(-29) + (-14) =$

8. $(-20) - (-21) =$

9. $(20) + (-12) - (+15) =$

10. $(-3)(\frac{2}{3})(-10) =$

11. $\frac{-14}{21} =$

12. $(-1)(\frac{7}{8}) =$

13. $(-3)(5) + \frac{6}{-2} =$

14. $(\frac{-8}{4}) + (\frac{15}{-3}) =$

15. $(-6) + (\frac{30}{5}) =$

16. $(15)(\frac{-2}{3}) - 8 =$

17. $-4 + (-8)(-6) =$

18. $(\frac{-36}{9}) - (-10) =$

19. $(12)(-2) + (-1)(-15) =$

20. $(\frac{10}{-15}) - (\frac{18}{27}) =$

ANSWERS AND SOLUTIONS START ON PAGE 374.

Inequalities

You learned in the chapter on the Basics of Algebra that an equation is a statement that two amounts are equal. An ***inequality*** is a statement that two amounts are *not* equal. Following are four symbols used to write inequalities.

Symbol	Meaning	Example	
$<$	less than	$3 < 4$	"3 is less than 4."
$>$	more than	$5 > 2$	"5 is more than 2."
$\leq$	less than or equal to	$m \leq 6$	"m is less than or equal to 6." This means that m can be 6 or any number less than 6, including negative numbers.
$\geq$	more than or equal to	$x \geq 8$	"x is more than or equal to 8." This means that x can be 8 or any number larger than 8.

Solving inequalities is like solving equations. You can perform inverse operations on both sides of inequalities to find the value of the unknown.

EXAMPLE 1: Solve for m in $6m \leq 42$.

$$\frac{6m}{6} \leq \frac{42}{6}$$

$$m \leq 7$$

SOLUTION. Divide both sides by 6. The solution is $m \leq 7$. This inequality is true for 7 and every number less than 7.

EXAMPLE 2: For the inequality $n + 2 > 5$, could n be 2?

$$\begin{array}{rcr} n + 2 & > & 5 \\ +2 & & -2 \\ \hline n & > & 3 \end{array}$$

SOLUTION. Subtract 2 from both sides. The solution is $n > 3$. This inequality is true for every number larger than 3. The number 2 is *not* a solution. n could not be 2.

Special Topics Exercise 7

Directions: Solve each inequality.

1. $c - 3 < 21$
2. $p + 1 > 7$
3. $9w \leq 27$
4. $8x \geq 20$
5. $\frac{s}{6} < 4$
6. $2a > -42$
7. For the inequality $m - 6 > 1$, could m be equal to 7?
8. For the inequality $\frac{1}{2}r \leq 4$, could r equal $\frac{1}{2}$?
9. For the inequality $d + 7 \leq 2$, could d be -6?
10. For $2f < 12$, could f be 4?

ANSWERS AND SOLUTIONS START ON PAGE 375.

Solving Longer Inequalities

You can solve long inequalities just as you solve long equations. Use inverse operations to simplify the inequalities, and remember to take care of addition or subtraction first.

EXAMPLE: Solve $9x - 7 > 11$

(1)
$$\begin{array}{rcl} 9x - 7 & > & 11 \\ +7 & & +7 \\ \hline 9x & > & 18 \end{array}$$

(2)
$$\frac{9x}{9} > \frac{18}{9}$$
$$x > 2$$

STEP 1. Add 7 to both sides of the inequality.
STEP 2. Divide both sides by 9. The solution is $x > 2$.

Special Topics Exercise 8

Directions: Solve each inequality.

1. $5m - 4 \leq 26$
2. $3n + 2 > 14$
3. $4p - 3 < 15$
4. $\frac{x}{3} + 5 \geq 7$
5. $7c - 3 \leq 5c + 15$
6. $8y + 1 < y + 22$
7. $3(s - 2) \geq 2s + 10$
8. $5(a + 3) < 2(a - 6)$

ANSWERS AND SOLUTIONS START ON PAGE 375.

Multiplying and Dividing Monomials

A ***monomial*** is an algebraic expression with only one term. A term is a letter, a number, or a group of letters and numbers not separated by plus or minus signs. Three examples of monomials are $-5x$, $2a^2$, and $64y$.

So far in this book you have added and subtracted monomials. You already know that $2x + 3x = 5x$ and that $14a - a = 13a$.

When you multiply or divide monomials, the power of the same letters changes. Look at the following rules and examples.

Rule for Multiplying Monomials

For the numbers, follow the rules for multiplying signed numbers. For like letters, *add* the exponents.

EXAMPLE 1: Find $5x \cdot 3x$.

$$5x \cdot 3x = 15x^{1+1} = 15x^2$$

SOLUTION. Multiply the numbers and add the exponents of the letters.

TIP: Any letter written without an exponent is understood to be to the first power.

EXAMPLE 2: Find $(-3a^2c)(4a^3c^4)$.

$$(-3a^2c)(4a^3c^4) = -12a^{2+3}c^{1+4} = -12a^5c^5$$

SOLUTION. Multiply -3 and 4. Add the exponents of a $(2 + 3 = 5)$. Add the exponents of c. $(1 + 4 = 5)$.

Rule for Dividing Monomials

Follow the rules for dividing signed numbers, and *subtract* the exponents of like letters.

EXAMPLE 3: Find $\dfrac{27y^4z^2}{3y^2z}$

$$9y^{4-2}z^{2-1} = 9y^2z$$

SOLUTION. Divide 27 by 3. Subtract the exponent of the bottom y from the exponent of the top y $(4 - 2 = 2)$. Then subtract the exponent of the bottom z from the exponent of the top z $(2 - 1 = 1)$.

Special Topics Exercise 9

Directions: Solve each problem.

1. $4x \cdot 3x =$
2. $(-4a^2)(-6a^3) =$
3. $m^3 \cdot m^4 =$
4. $a^2 \cdot a =$
5. $(5b^2)(b^4) =$
6. $10x \cdot \frac{1}{2}x =$
7. $4x^2y/xy =$
8. $6x^4y^4/6x^2y =$
9. $\dfrac{18b^3x^2}{2b^2x^2}$
10. $\dfrac{12t^3v^4}{6t^2v^3}$

ANSWERS AND SOLUTIONS START ON PAGE 376.

Factoring

Factors are numbers or algebraic terms that, when multiplied together, give a product. For example, the numbers 2 and 5 are factors of 10 because $2 \times 5 = 10$.

The fraction $\frac{1}{2}$ is also a factor of 10. However, we usually write only *whole* prime number factors other than the number 1. A ***prime number*** is a number that can be divided by only itself and the number one. For example, 5 is a prime number, because no whole number other than 5 or 1 divides into it evenly. But 6 is not a prime number, because both 2 and 3 divide into it evenly.

EXAMPLE 1: Find the prime factors of 8.

$8 = 4 \times 2$

$8 = 2 \times 2 \times 2$

STEP 1. Find two whole numbers that multiply together to give 8. $4 \times 2 = 8$.

STEP 2. Check to see whether there are prime factors of either 4 or 2. $4 = 2 \times 2$. Therefore, the prime factors of 8 are $\mathbf{2 \times 2 \times 2}$.

EXAMPLE 2: Find the prime factors of 12.

$12 = 6 \times 2$

$12 = 3 \times 2 \times 2$

STEP 1. Find two whole numbers that multiply together to give 12. $6 \times 2 = 12$.

STEP 2. Check to see whether there are prime factors of either 6 or 2. $6 = 3 \times 2$. The prime factors of 12 are $\mathbf{3 \times 2 \times 2}$.

Special Topics Exercise 10

Directions: Write each of the following whole numbers as a product of prime factors.

1. 9

2. 10

3. 18

4. 20

5. 24

6. 30

7. 40

8. 50

ANSWERS AND SOLUTIONS START ON PAGE 376.

Factoring with Variables

The expression $10x - 6$ has two terms. A ***term*** is a letter, a number, or a group of letters and numbers not separated by plus or minus signs. In many expressions with two terms, you can factor out a whole number.

To factor out a whole number, find a number that divides evenly into each term of an expression. Write the number that divides evenly into each term on the outside of a set of parentheses. The result of dividing each term by that number goes inside the parentheses.

EXAMPLE: Factor the expression $10x + 6$.

$10x + 6 = 2(5x + 3)$

SOLUTION. Look for a number that divides evenly into both $10x$ and 6. The number 2 divides evenly into both terms. Write 2 outside a set of parentheses. Write the result of dividing both $10x$ and 6 by 2 on the inside of the parentheses: $\mathbf{2(5x + 3)}$.

Special Topics Exercise 11

Directions: Factor a whole number out of each of the following expressions. Then write the factored expression.

1. $4n + 4$

2. $3p - 6$

3. $15a - 10$

4. $14c + 35$

5. $6b + 8$

6. $6f - 30$

7. $36y - 9$

8. $16k + 56$

ANSWERS AND SOLUTIONS START ON PAGE 376.

Factoring Out the Variable

The expression $x^2 - 4x$ has two terms. There is no whole number other than the number 1 that divides evenly into both terms. However, you can divide both terms by x. As you know from your work with monomials, when you divide x^2 by x, you subtract exponents to get x. When you divide x by x, you get 1.

To factor out a letter from an algebraic expression, divide each term in the expression by the letter that can be found in each term.

EXAMPLE 1: Factor the expression $x^2 - 4x$.

$x^2 - 4x = x(x - 4)$

SOLUTION. Divide each term by the letter that is in each term: x divided into x^2 is x. x divided into $-4x$ is -4.

EXAMPLE 2: Factor the expression $m^2 - m$.

$m^2 - m = m(m - 1)$

SOLUTION. Divide each term by the letter m: m divided into m^2 is m, and m divided into m is 1.

Special Topics Exercise 12

Directions: Factor out a letter from each of the following expressions.

1. $c^2 + 8c$
2. $y^2 - 5y$
3. $m^2 + 3m$
4. $a^2 - a$
5. $d^2 + 4d$
6. $n^2 - 8n$
7. $p^2 - 2p$
8. $s^2 + 9s$

ANSWERS AND SOLUTIONS START ON PAGE 376.

Simplifying Square Roots

There is another useful application of factoring that you may use on the GED. Remember that when you learned to find square roots earlier in this book, all the problems had exact solutions. For example, $\sqrt{81} = 9$ and $\sqrt{144} = 12$. Numbers like 81 and 144 are called ***perfect squares*** because their square roots are whole numbers.

Most numbers do not have whole number square roots. For many numbers, however, you can use factoring to simplify the square root.

Rule for Simplifying Square Roots

1. Try to find a factor for the number that is a perfect square.
2. Write the square root of the perfect square outside the $\sqrt{}$ sign, and leave the other factor inside the $\sqrt{}$ sign.

EXAMPLE 1: Simplify the square root of 200.

$$\sqrt{200} = \sqrt{100 \cdot 2} = 10\sqrt{2}$$

STEP 1. Think about the factors of 200. There are many: 2, 4, 5, 20, and so on. The combination of factors that includes the largest perfect square is 100 and 2. 200 is equal to $\sqrt{100}$ multiplied by $\sqrt{2}$.

STEP 2. Put the whole number square root of 100 outside the $\sqrt{\ }$ sign. Leave 2 inside the $\sqrt{\ }$ sign, because there is no whole number square root of 2.

Note: The answer $10\sqrt{2}$ may look strange. The square root of 2 is a mixed decimal that is approximately equal to 1.41. This means that the square root of 200 is approximately 10 × 1.41, or 14.1. If we multiply 14.1 by 14.1 we get:

```
   14.1
 × 14.1
 ------
   14 1
  564
 141
 ------
 198.81, which is approximately 200.
```

EXAMPLE 2: Simplify $\sqrt{24}$.

$$\sqrt{24} = \sqrt{4} \cdot \sqrt{6} = 2\sqrt{6}$$

STEP 1. Think of the factors of 24. The combination that includes a perfect square is 4 and 6, so $\sqrt{24}$ is equal to $\sqrt{4}$ multiplied by $\sqrt{6}$.

STEP 2. Write the whole number square root of 4 outside the $\sqrt{\ }$ sign. Leave 6 inside the $\sqrt{\ }$ sign.

Special Topics Exercise 13

Directions: Simplify each of the following square roots.

1. $\sqrt{27}$

2. $\sqrt{8}$

3. $\sqrt{20}$

4. $\sqrt{150}$

5. $\sqrt{48}$

6. $\sqrt{50}$

7. $\sqrt{12}$

8. $\sqrt{32}$

ANSWERS AND SOLUTIONS START ON PAGE 376.

Using Algebra to Solve Geometry Problems

Now that you know how to multiply monomials and simplify square roots, you are ready to solve some of the more complex geometry problems that may appear on the GED Test. Look at the next examples carefully to see how to substitute algebraic expressions into geometry formulas.

EXAMPLE 1: The area of a rectangle is 162 in^2. The length of the rectangle is twice as long as the width. Find the width of the rectangle.

(1) width $= x$
length $= 2x$

(2) $A = lw$
$162 = 2x \cdot x$
$162 = 2x^2$

(3) $\frac{162}{2} = \frac{2x^2}{2}$
$81 = x^2$
$\sqrt{81} = x$
$9 = x$

STEP 1. Let x stand for the width and $2x$ stand for the length.
STEP 2. Substitute 162 for A, $2x$ for l, and x for w in the formula for the area of a rectangle.
STEP 3. Then solve the equation. Notice that the value of x is the square root of 81. The solution is $\boldsymbol{x}$ = **9 inches.**

EXAMPLE 2: The perimeter of a rectangle is 74. The length of the rectangle is 5 more than the width. Find the length of the rectangle.

(1) width $= w$
length $= w + 5$

(2) $P = 2l + 2w$
$74 = 2(w + 5) + 2w$

(3) $74 = 2w + 10 + 2w$
$74 = 4w + 10$
$-10 \quad\quad -10$
$64 = 4w$
$\frac{64}{4} = \frac{4w}{4}$
$16 = w$

(4) $w + 5 =$
$16 + 5 = 21$

STEP 1. Let w stand for the width and $w + 5$ for the length.
STEP 2. Substitute 74 for P and $(w + 5)$ for l, in the formula for the perimeter of a rectangle.
STEP 3. Solve the equation. The solution for w is 16.
STEP 4. Substitute 16 for w in the expression $w + 5$ to find the length of the rectangle. The length is 21.

Special Topics Exercise 14

Directions: Solve each problem.

1. The area of a rectangle is 32 square feet. The width is half of the length. Find the length.
2. The area of a triangle is 16 square inches. The height of the triangle is half of the base. Find the height of the triangle.
3. A rectangle has an area of 96. The width is $\frac{2}{3}$ of the length. Find the length.
4. A rectangle has an area of 98 square meters. The length is twice the width. Find the measurement of the length.
5. A rectangle has a perimeter of 48 feet. The length is twice the width. Find the measurement of the length.
6. A rectangle has a perimeter of 35 meters. The width is $\frac{3}{4}$ of the length. Find both the width and the length.
7. A triangle has an area of 24. The base is three times the height. Find the base.
8. A rectangle has a perimeter of 60 inches. The ratio of the width to the length is 2:3. Find the measurement of the length. (*Hint*: Let $2x$ stand for the width.)
9. A rectangle has an area of 6,000 square feet. The ratio of the width to the length is 3:5. Find the measurement of both the width and the length.
10. The area of the triangle at the right is 216. The ratio of *AB* to *AC* is 3:4. Find the length of *AB*.

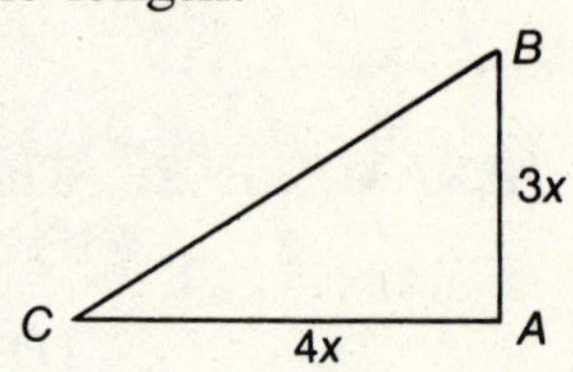

11. For the rectangle at the right, the ratio of *WX* to *WZ* is 5:6. The perimeter of the rectangle is 110 feet. Find the length of *WZ*.

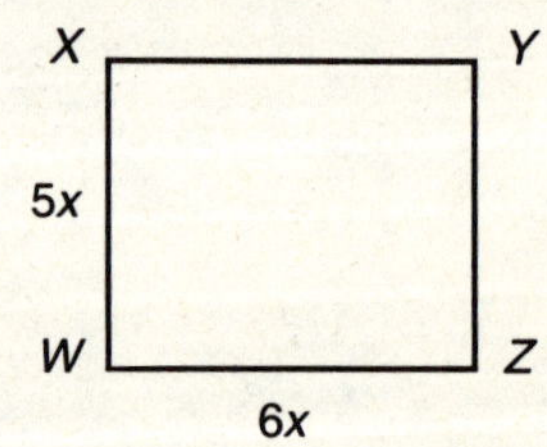

12. The perimeter of the rectangle at the right is 34 feet. *MP* is 3 feet longer than *MN*. Find the measurement of *MN*.

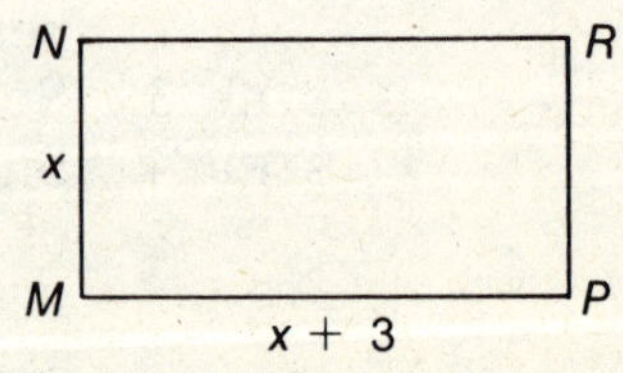

ANSWERS AND SOLUTIONS START ON PAGE 376.

Rectangular Coordinates

A graph called the ***rectangular coordinate system*** is a useful method for showing many algebraic relationships. The rectangular coordinate system is a flat surface or plane. The plane is divided by a horizontal line called the x-axis and a vertical line called the y-axis. Each of these two lines is simply a number line like the ones you saw earlier. The lines intersect at the number 0, which is called the ***origin***.

A point anywhere on the graph can be identified by a pair of numbers called the ***coordinates*** of the point. The coordinates are written inside parentheses in the order (x, y). The x-coordinate has a positive value for points to the right of the y-axis and a negative value for points to the left of the y-axis. The y-coordinate has a positive value for points above the x-axis and a negative value for points below the x-axis.

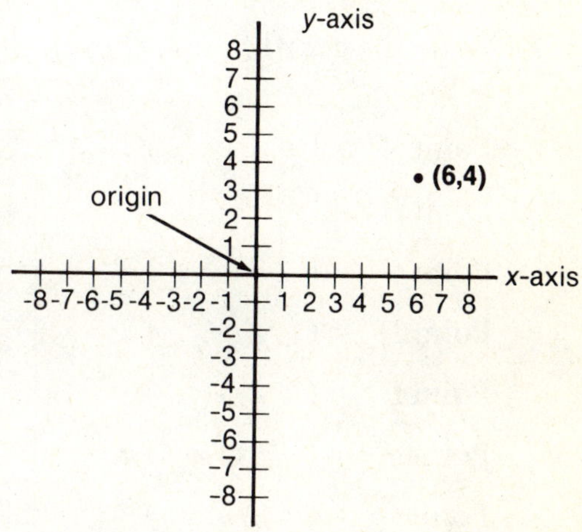

The illustration above shows a rectangular coordinate system with one point labeled (+6, +4). These coordinates mean that the point is 6 units to the right of the y-axis and 4 units above the x-axis.

Use the figure at the right for Examples 1 through 4.

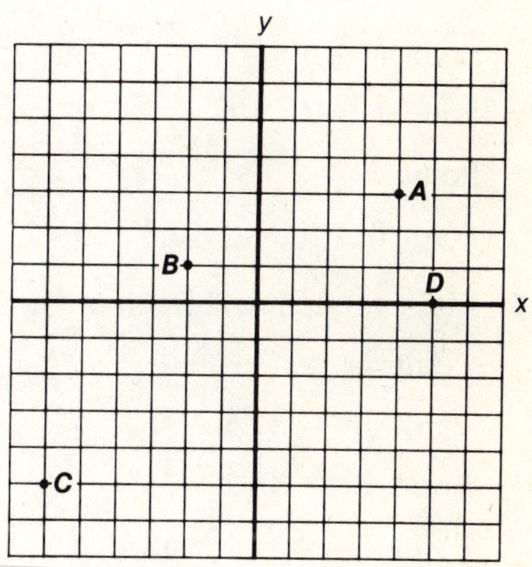

EXAMPLE 1: What are the coordinates of point A?

(+4, +3)

SOLUTION. First tell how far point A is to the right of the y-axis. It is 4 units to the right, or +4. This is the x-coordinate. Then tell how far point A is above the x-axis. It is 3 points above, or +3. This is the y-coordinate. The coordinates are (4, 3).

EXAMPLE 2: What are the coordinates of point B?

(−2, +1)

SOLUTION. Point B is 2 units to the left of the y-axis, or −2, and one unit above the x-axis, or +1. The coordinates are (−2, 1).

EXAMPLE 3: What are the coordinates of point C?

(−6, −5)

SOLUTION. Point C is 6 units to the left of the y-axis, or −6, and 5 units below the x-axis, or −5. The coordinates are (−6, −5).

EXAMPLE 4: What are the coordinates of point *D*?

(+5, 0)

SOLUTION. Point *D* is 5 units to the right of the *y*-axis, or +5. It is neither above nor below the *x*-axis. The coordinates are (5, 0).

Special Topics Exercise 15

Directions: Solve the following problems.

1. Write the coordinates for each point shown on the rectangular coordinate system below.

Point *A* = ( )

Point *B* = ( )

Point *C* = ( )

Point *D* = ( )

Point *E* = ( )

Point *F* = ( )

Point *G* = ( )

Point *H* = ( )

Point *I* = ( )

Point *J* = ( )

Point *K* = ( )

Point *L* = ( )

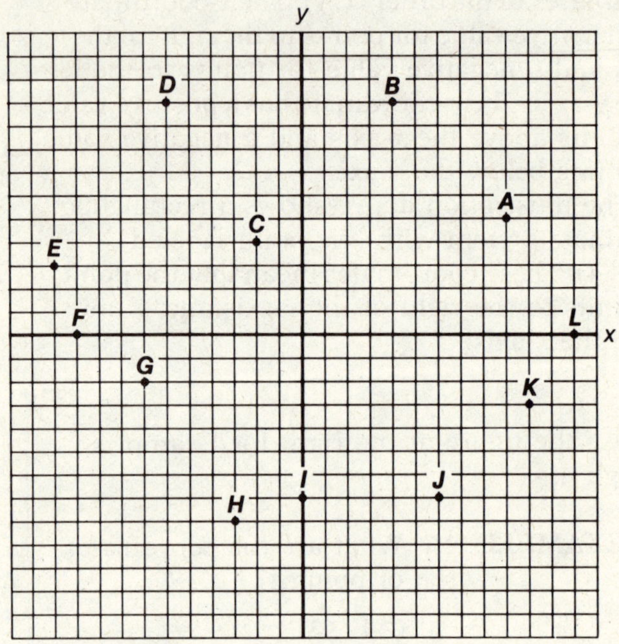

2. Mark the following points on the coordinate system below.

Point A = (+6, +4)

Point B = (−5, +3)

Point C = (−2, −7)

Point D = (+1, −6)

Point E = (−8, 0)

Point F = (+3, 0)

Point G = (0, −4)

Point H = (0, +7)

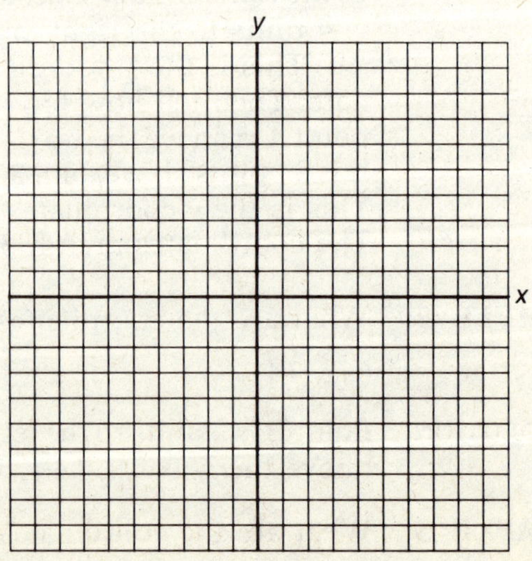

ANSWERS AND SOLUTIONS START ON PAGE 377.

Finding Distances on the Rectangular Coordinate System

Distance on the rectangular coordinate system is not measured in inches, feet, and meters as you used in geometry. Instead, it is measured in units of whole numbers as you will see in this section. On the GED you may be asked to find the distance between two points on the rectangular coordinate system, or **plane** as it is sometimes called.

In some cases, you can simply count the number of spaces between two points.

EXAMPLE 1: What is the distance from A to B on the graph at the right?

$4 + 1 = 5$

SOLUTION. A is 4 units above the x-axis, and B is 1 unit below. The distance between them is the total: $4 + 1 = 5$.

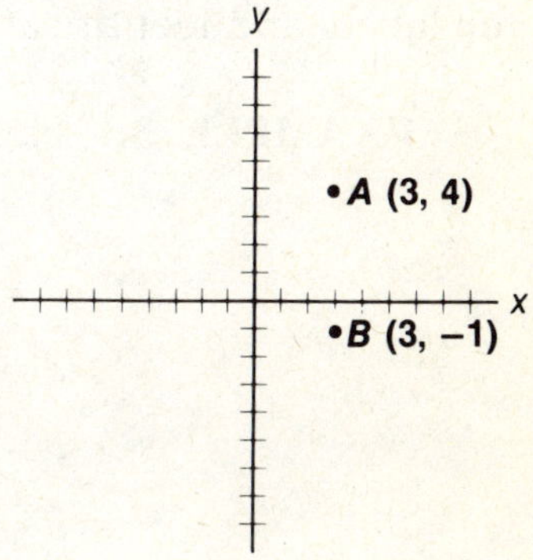

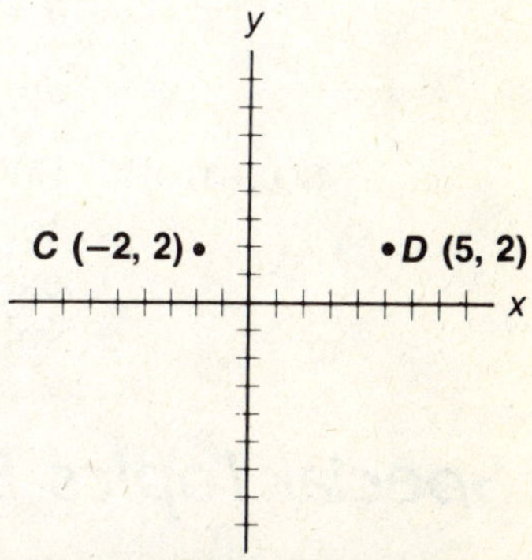

EXAMPLE 2: What is the distance from C to D on the graph at the right?

$2 + 5 = 7$

SOLUTION. C is 2 units to the left of the y-axis, and D is 5 units to the right. The distance between them is $2+5 = 7$.

> **TIP:** Distance is *always* positive regardless of the signs of the coordinates.

The distance between S and T in the figure at the right is more complicated. The distance is the hypotenuse of a right triangle, where the legs are shown by the two dotted lines. There is a special formula for finding the distance between any two points on the rectangular coordinate system. This formula looks complicated, but it is based on the Pythagorean relationship that you learned on page 267.

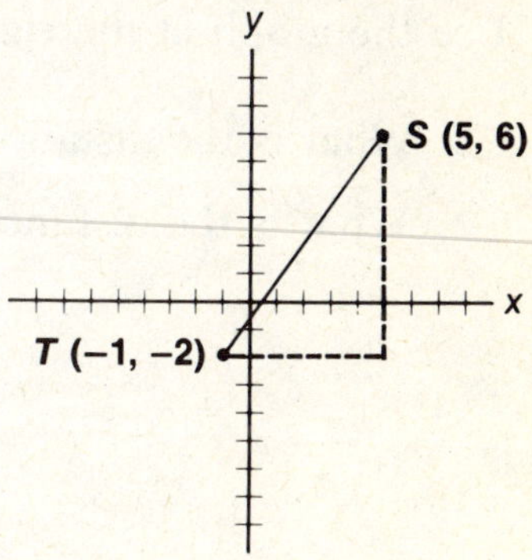

Distance between two points in the plane	$d = \sqrt{(x_2 - x_1)^2 + (y_2 - y_1)^2}$; where (x_1, y_1) and (x_2, y_2) are two points on the plane.

Point T on the last graph has the coordinates $(-1, -2)$. This is (x_1, y_1) in the formula. Point S has the coordinates $(5, 6)$. This is (x_2, y_2) in the formula.

It is important to realize that the small numbers *1* and *2* below a variable are not involved in any of your computations. They are called ***subscripts*** and are used for labels and identification only.

EXAMPLE 3: Find the distance from S to T in the preceding figure.

$$d = \sqrt{(x_2 - x_1)^2 + (y_2 - y_1)^2}$$
$$d = \sqrt{(5 - (-1))^2 + (6 - (-2))^2}$$
$$d = \sqrt{(5 + 1)^2 + (6 + 2)^2}$$
$$d = \sqrt{6^2 + 8^2}$$
$$d = \sqrt{36 + 64}$$
$$d = \sqrt{100}$$
$$d = 10$$

SOLUTION. Substitute 5 for x_2, -1 for x_1, 6 for y_2, and -2 for y_1 in the formula for the distance between two points. Notice how the extra parentheses are used to show that -1 and -2 are being subtracted. The distance between S and T is 10.

Special Topics Exercise 16

Directions: Solve each problem. Use the formula $d = \sqrt{(x_2 - x_1)^2 + (y_2 - y_1)^2}$ for the distance between two points on a plane, when necessary. Approximate if your answer is not a perfect square.

Use the graph at the right for questions 1 and 2.

1. What is the distance from point A to point C?
2. What is the distance from B to C?

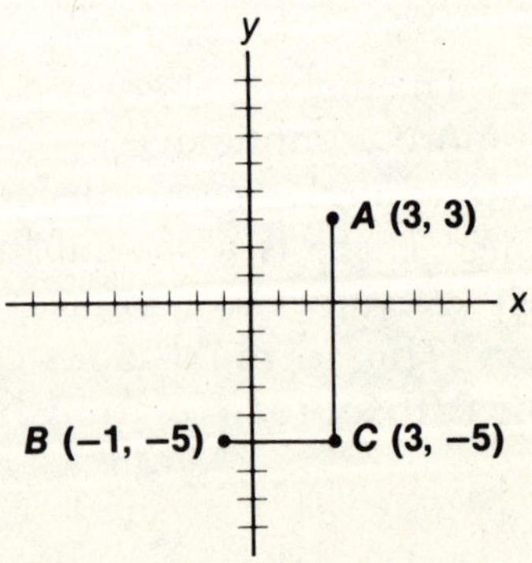

Directions: Use the graph at the right for questions 3 and 4.

3. Find the distance from E to F.

4. What is the distance from D to F?

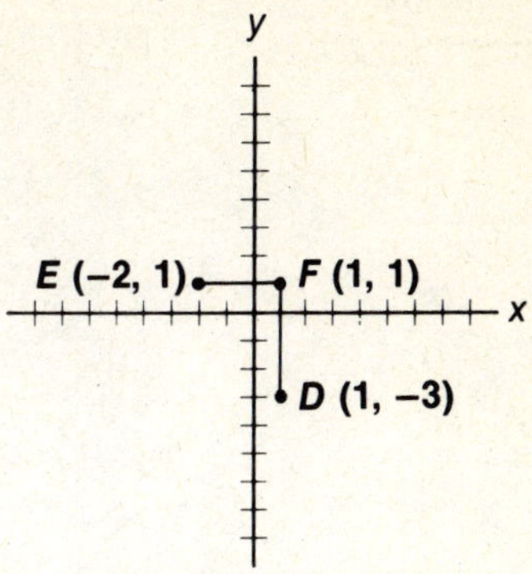

5. What is the distance from point G to point H in the graph at the right?

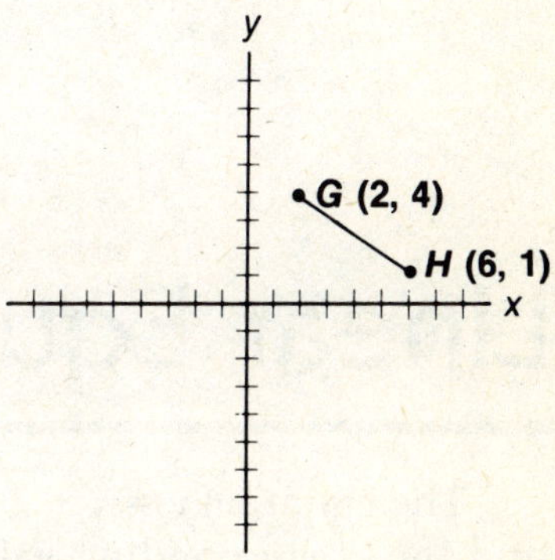

6. Find the distance from I to J in the graph at the right.

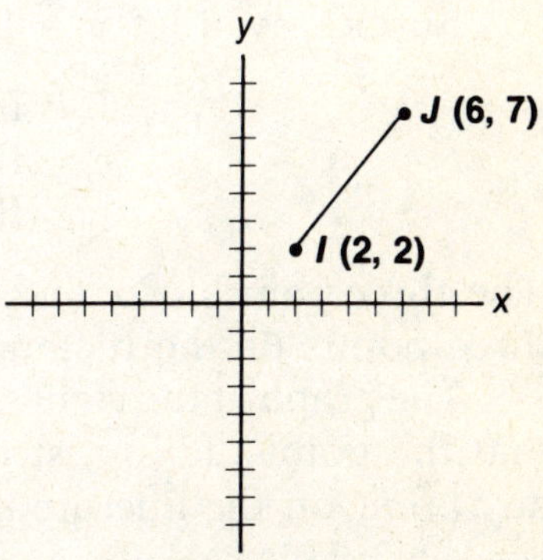

7. What is the distance from point K to point L on the graph at the right?

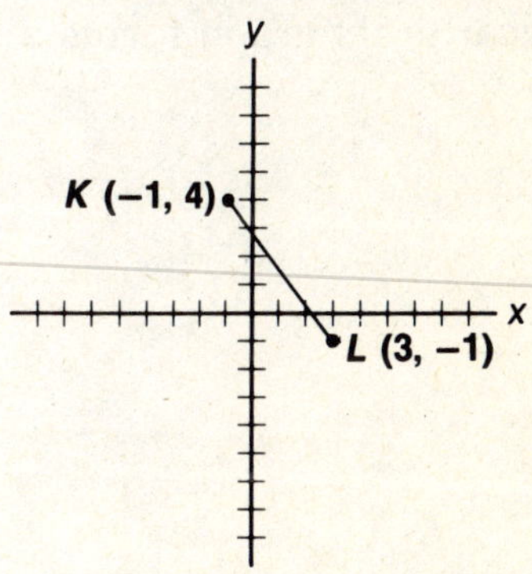

8. What is the distance from M to N on the graph at the right?

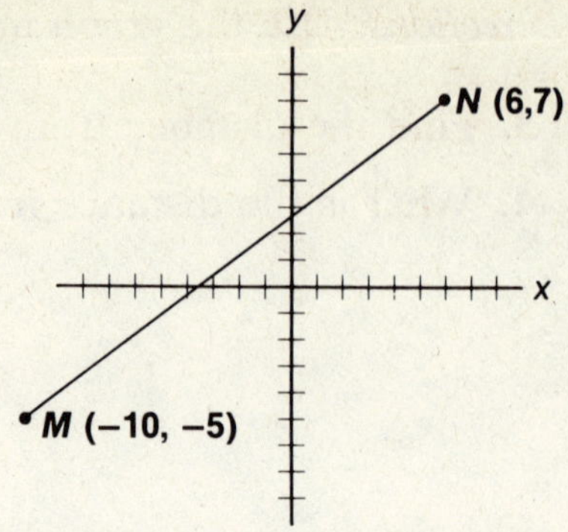

ANSWERS AND SOLUTIONS START ON PAGE 377.

Linear Equations

The equation $y = x + 2$ has many solutions. If you select values for x, you can find the corresponding values for y by substituting and solving the equation.

Suppose you let $x = 1$, 3, and -2 for the equation $y = x + 2$:

$$\text{If } x = 1, \text{ then } y = 1 + 2 = 3.$$
$$\text{If } x = 3, \text{ then } y = 3 + 2 = 5.$$
$$\text{If } x = -2, \text{ then } y = -2 + 2 = 0.$$

The three values of x and their corresponding values of y are the coordinates for three points on the rectangular coordinate plane: (1, 3) (3, 5), and (−2, 0).

The graph at the right shows these three points. Notice that the points lie in a straight line. The coordinates of any point on the line are solutions to the equation $y = x + 2$. This type of equation is called a ***linear equation*** because its graph forms a straight line.

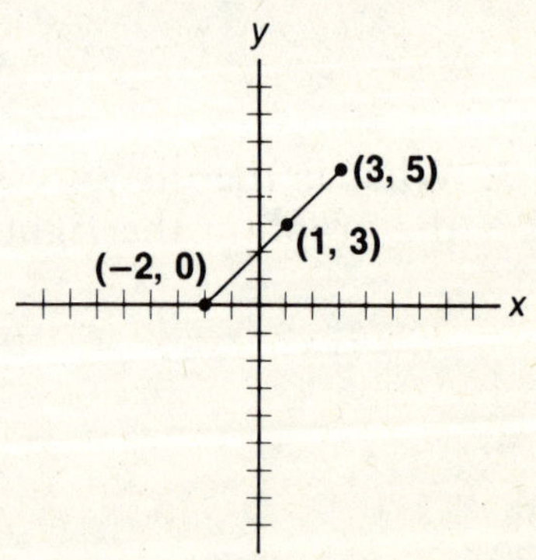

EXAMPLE 1: Make a graph of the equation $y = 2x - 1$. Let $x = 2$, 5, and -1.

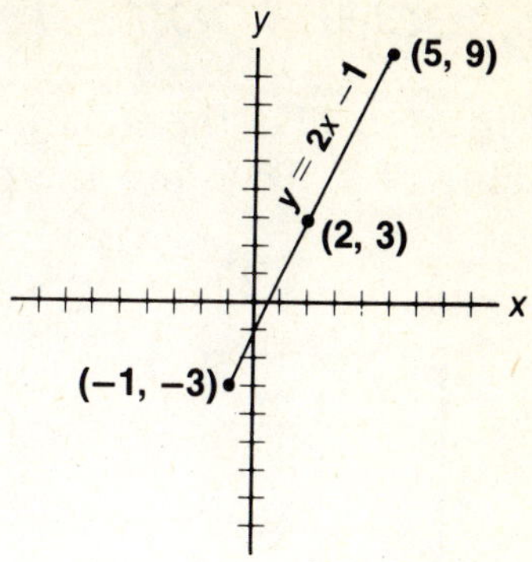

If $x = 2$, then $y = 2(2) - 1$
$= 4 - 1$
$= 3.$

If $x = 5$, then $y = 2(5) - 1$
$= 10 - 1$
$= 9.$

If $x = -1$, then $y = 2(-1) - 1$
$= -2 - 1$
$= -3.$

x	y
2	3
5	9
−1	−3

STEP 1. Substitute each value of x into the equation to find the corresponding value of y.
STEP 2. Make a table of the x and y values.
STEP 3. Put these points on the coordinate plane, and connect them with a straight line.

On the GED you will not have to make a graph of an equation, but you may need to know whether certain points lie on the line of an equation.

To find out whether a point is on the graph of an equation, substitute the x-coordinate of the point into the equation. If the solution to the equation is the same as the y-coordinate, then the point is on the graph of the equation.

EXAMPLE 2: Is the point (2, 3) on the graph of $y = 2x - 1$?

$x = 2x - 1$
$y = 2(2) - 1 = 4 - 1 = 3$

SOLUTION. Substitute 2 for x in $y = 2x - 1$. This gives y the value of 3. Therefore, point (2, 3) is on the graph of $y = 2x - 1$.

EXAMPLE 3: Is the point $(-2, -4)$ on the graph of the equation $y = 2x - 1$?

$y = 2x - 1$
$y = 2(-2) - 1 = -4 - 1 = -5$

SOLUTION. Substitute -2 for x in $y = 2x - 1$. This gives y the value of -5. Point $(-2, -4)$ is not on the graph of $y = 2x - 1$.

Special Topics Exercise 17

Directions: For the equations in problems 1 through 4, fill in the table for each given value of x. Then put the points on the graph and connect them.

1. $y = x + 3$

x	y
1	
4	
−3	

2. $y = 2x - 3$

x	y
5	
3	
0	

3. $y = \frac{x}{3} + 1$

x	y
6	
3	
−3	

4. $y = -x + 5$

x	y
8	
5	
1	

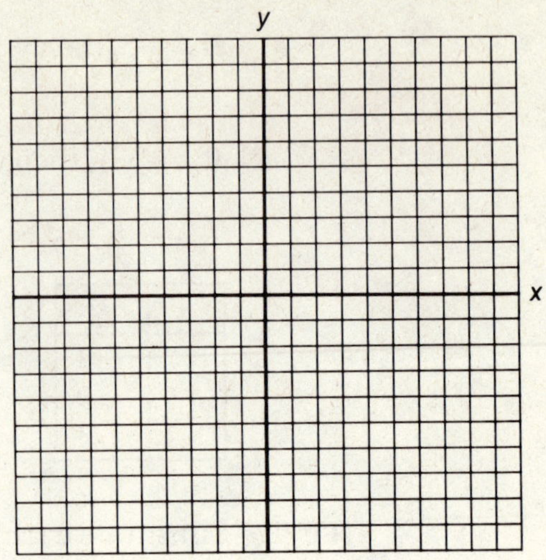

5. Is the point (2, 5) on the graph of the equation $y = 3x - 1$?

6. Is the point (3, 4) on the graph of the equation $y = 4x - 5$?

7. Is the point (6, 7) on the graph of the equation $y = \frac{x}{2} + 4$?

8. Is the point (3, 2) on the graph of the equation $y = \frac{2}{3}x + 1$?

9. Is the point (8, −5) on the graph of the equation $y = -x + 3$?

10. Is the point (6, −2) on the graph of the equation $y = -\frac{x}{2} + 1$?

ANSWERS AND SOLUTIONS START ON PAGE 378.

Slope and Intercepts

Slope is a measure of how "steep" a line is. Slope tells something about what a linear equation looks like.

- A line that goes up from left to right has a positive slope.
- A line that goes down from left to right has a negative slope.
- A horizontal line has a zero slope.
- A vertical line has an undefined slope. It has no meaning.

Look at the graphs of the two linear equations at the right. The line for the equation $y = x + 3$ has a positive slope. It rises from left to right. The line for the equation $y = -x - 1$ has a negative slope. It goes down from left to right.

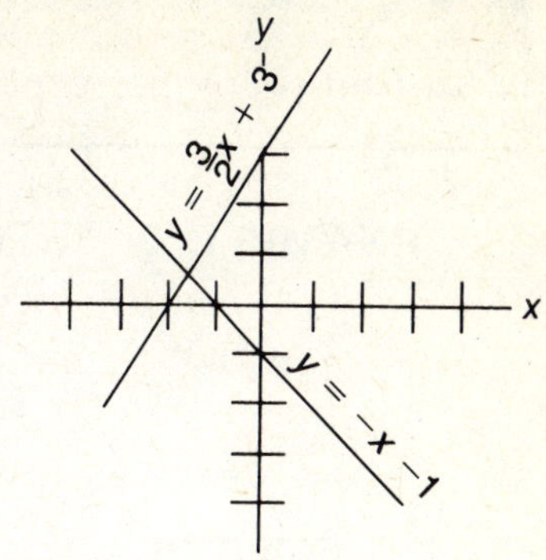

Look at the graphs of the two linear equations at the right. The line for the equation $y = -2$ has a zero slope. The line for the equation $x = 2$ has an undefined slope.

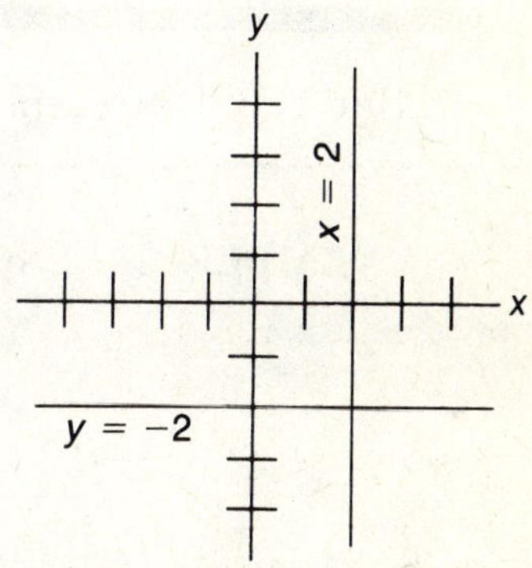

Positive and negative slopes have exact values. A high value for a slope means the line is very steep. To find the exact value of the slope of a line, substitute values into the following formula.

slope $= \dfrac{y_2 - y_1}{x_2 - x_1}$; where (x_1, y_1) and (x_2, y_2) are the coordinates of two points on the graph of a linear equation.

In words this formula means: Divide the difference between two y-values by the difference between two x-values.

EXAMPLE 1: What is the slope of a line that passes through the points (2, 5) and (6, 13)?

$$\text{slope} = \frac{13 - 5}{6 - 2} = \frac{8}{4} = 2$$

SOLUTION. Subtract the y-values, $13 - 5 = 8$, and subtract the x-values, $6 - 2 = 4$. Then divide 8 by 4. The slope is +2.

An ***intercept*** tells where a line crosses an axis. On the graph at the right, the line labeled m crosses the y-axis at +3. The coordinates of the y-intercept are (0, 3). The line crosses the x-axis at −2. The coordinates of the x-intercept are (−2, 0).

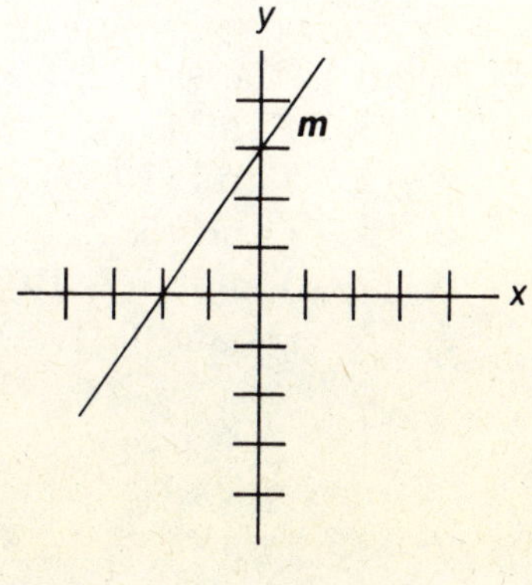

To find the y-intercept of a linear equation, substitute 0 for x, and solve for y.

EXAMPLE 2: What is the y-intercept of the equation $y = 4x - 5$?

$$y = 4x - 5$$
$$y = 4(0) - 5$$
$$= 0 - 5$$
$$= -5$$

SOLUTION. Replace x with 0 and solve for y in $y = 4x - 5$. The coordinates of the y-intercept are **(0, −5)**.

To find the x-intercept of a linear equation, substitute 0 for y and solve for x.

EXAMPLE 3: What is the x-intercept of the equation $y = 3x - 6$?

$$y = 3x - 6$$
$$0 = 3x - 6$$
$$+6 \quad\quad +6$$
$$6 = 3x$$
$$2 = x$$

SOLUTION. Replace y with 0 and solve for x in $y = 3x - 6$. The coordinates of the x-intercept are (2, 0).

Special Topics Exercise 18

Directions: Solve each problem.

1. Fill in each blank with the letter that corresponds to a line on the graph at the right.

a. The slope of line ______ is positive.

b. The slope of line ______ is negative.

c. The slope of line ______ is zero.

d. The slope of line ______ is undefined.

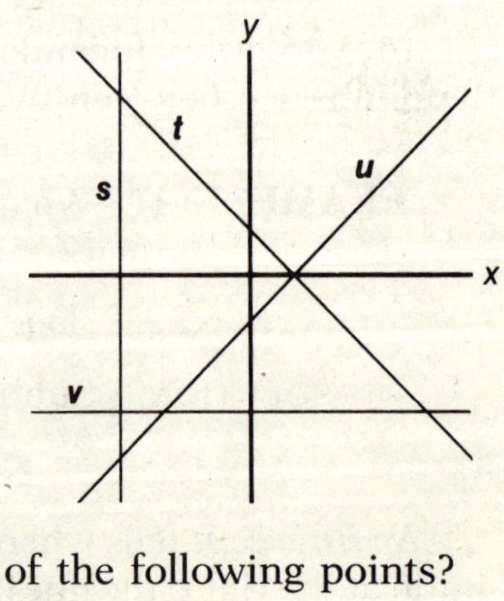

2. What is the slope of a line that passes through each of the following points?
a. (3, 2) and (5, 8)
b. (2, 5) and (10, 9)
c. (3, 6) and (5, 2)
d. (2, 4) and (−7, 10)
e. (3, −3) and (−1, 5)

3. Write the coordinates for the x-intercept and the y-intercept for each of the following equations.

	x-intercept	y-intercept
a. $y = 3x - 9$		
b. $y = x - 6$		
c. $y = \frac{x}{2} + 5$		
d. $y = \frac{3}{4}x + 12$		
e. $y = 2x - 1$		

ANSWERS AND SOLUTIONS START ON PAGE 379.

Quadratic Equations

One type of problem that you may see on the GED Mathematics Test is the quadratic equation. A ***quadratic equation*** is an equation where the unknown is raised to the second power.

You have already used equations that are raised to the second power. The Pythagorean relationship and the formula for the distance between two points on the rectangular coordinate plane are examples.

Quadratic equations have two solutions. Think about the example $x^2 = 9$. Both $+3$ and -3 are solutions to this equation. $(+3)(+3) = +9$ and $(-3)(-3) = 9$. With the Pythagorean relationship and the distance between two points, we were interested in the positive solutions only.

There are different ways to solve a quadratic equation, and there is not space in this book to discuss each one. To succeed on problems using quadratics, you can use the answer solutions given with the problem. Look at the following example to see how this is done.

EXAMPLE: Which of the following expresses the solutions to $x^2 - 8x + 12 = 0$?

(1) $x = 3$ and $x = 4$
(2) $x = 6$ and $x = 2$
(3) $x = -3$ and $x = 4$
(4) $x = 6$ and $x = -2$
(5) $x = -6$ and $x = -2$

STEP 1. Try the solutions in answer choice (1). First substitute 3 into the equation.

$$x^2 - 8x + 12 = 0$$
$$3^2 - 8(3) + 12 = 0$$
$$9 - 24 + 12 = 0$$
$$-24 + 21 = 0 \quad \textbf{Wrong}$$

Since $-24 + 21$ does not equal 0, answer choice (1) is wrong.

STEP 2. Try the solutions in answer choice (2). First substitute 6 into the equation.

$$x^2 - 8x + 12 = 0$$
$$6^2 - 8(6) + 12 = 0$$
$$36 - 48 + 12 = 0$$
$$-48 + 48 = 0 \quad \textbf{Correct}$$

Since $-48 + 48 = 0$, you know that the first solution, $x = 6$, in answer choice (2) is correct. Now substitute the other value, $x = 2$, into the equation.

$$x^2 - 8x + 12 = 0$$
$$2^2 - 8(2) + 12 = 0$$
$$4 - 16 + 12 = 0$$
$$-16 + 16 = 0 \quad \textbf{Correct}$$

Since $-16 + 16 = 0$, you know that $x = 2$ is also a solution to the equation. This means that answer choice (2) is correct.

For more practice, try substituting the other answer choices for the example into the equation.

TIP: Remember that you must check *both* values for each solution.

Special Topics Exercise 19

Directions: For these problems, choose the correct solution to each quadratic equation.

1. $x^2 - 14x + 48 = 0$
(1) $x = 4$ and $x = 12$
(2) $x = 3$ and $x = -16$
(3) $x = 6$ and $x = 8$
(4) $x = 2$ and $x = -24$

2. $x^2 - 3x - 10 = 0$
(1) $x = 5$ and $x = -2$
(2) $x = 10$ and $x = 1$
(3) $x = 2$ and $x = -5$
(4) $x = -10$ and $x = -1$

3. $x^2 - x - 12 = 0$
(1) $x = 2$ and $x = 6$
(2) $x = 12$ and $x = -1$
(3) $x = -4$ and $x = 3$
(4) $x = 4$ and $x = -3$

4. $x^2 - 9x + 8 = 0$
(1) $x = 8$ and $x = 1$
(2) $x = 4$ and $x = -2$
(3) $x = -4$ and $x = 2$
(4) $x = -8$ and $x = 1$

5. $x^2 - 15x + 44 = 0$
(1) $x = 2$ and $x = 22$
(2) $x = 4$ and $x = 11$
(3) $x = -4$ and $x = 11$
(4) $x = 44$ and $x = -1$

6. $x^2 + 11x + 18 = 0$
(1) $x = 6$ and $x = -3$
(2) $x = -1$ and $x = 18$
(3) $x = -9$ and $x = -2$
(4) $x = -6$ and $x = 3$

ANSWERS AND SOLUTIONS START ON PAGE 380.

Scientific Notation

As you know, our number system is based on multiples of ten. Though this system is usually easy to work with, it can be difficult and time consuming to write extremely large or extremely small numbers. For this reason, scientists and mathematicians often use a shortened method of writing numbers called ***scientific notation***. This method of expressing numbers uses the number 10 and an exponent to simplify large or small numbers. Remember, you learned about exponents, or "powers," in Chapter 11 of this book.

Before you practice scientific notation, study the following chart. This chart lists some powers of ten with both positive *and* negative exponents. Negative exponents indicate numbers smaller than one, or fractional powers of ten. For instance, $10^{-1} = \frac{1}{10} = .1$. As you can see, the negative sign indicates a reciprocal. (The reciprocal of 10 is $\frac{1}{10}$.)

$10^0 = 1$	
$10^1 = 10$	$10^{-1} = \frac{1}{10} = .1$
$10^2 = 100$	$10^{-2} = \frac{1}{100} = .01$
$10^3 = 1000$	$10^{-3} = \frac{1}{1000} = .001$
$10^4 = 10,000$	$10^{-4} = \frac{1}{10,000} = .0001$
$10^5 = 100,000$	$10^{-5} = \frac{1}{100,000} = .00001$

Rules for Writing Numbers in Scientific Notation

1. Ignore the zeros in the number and represent it as a whole number or mixed decimal between 1 and 10.
2. Next to this number, write a multiplication sign and 10 to the correct power, which will be either positive or negative. To find the correct power, count the number of places the decimal point must move to get from the original number to the number between one and ten. If the point must move to the left, use a positive exponent. If the point must move to the right, use a negative exponent.

EXAMPLE 1: Express the number 45,000,000 in scientific notation.

STEP 1. Take off the zeros and represent 45,000,000 as a whole number or mixed decimal between 1 and 10.

A decimal point should be inserted between 4 and 5 to get 4.5.

STEP 2. Find the power of ten that, if multiplied by 4.5, equals 45,000,000.

To find the correct power, count the number of places the decimal point must move to get from 45,000,000 to 4.5. It must move 7 places to the left.

So, 45,000,000 = $\mathbf{4.5 \times 10^7}$

7 places left

EXAMPLE 2: Express .0024 in scientific notation.

STEP 1. Represent the number as a number between 1 and 10.

The decimal point should be inserted between 2 and 4 to get 2.4.

STEP 2. Set up a multiplication equation between 2.4 and the correct power of ten.

.0024 = $\mathbf{2.4 \times 10^{-3}}$

3 places right

Exercise 20: Scientific Notation

Directions: Express each of the numbers below in scientific notation.

1. 700

2. .0446

3. 95,000,000

4. .00066

5. 17,000

The numbers below are already written in scientific notation. Find the actual value that each represents.

6. 2.74×10^3

7. 9.1×10^{-2}

8. 4.882×10^4

9. 1.98×10^{-5}

10. 5.2×10^0

ANSWERS AND SOLUTIONS START ON PAGE 380.

Posttest Answer Grid

1 ① ② ③ ④ ⑤
2 ① ② ③ ④ ⑤
3 ① ② ③ ④ ⑤
4 ① ② ③ ④ ⑤
5 ① ② ③ ④ ⑤
6 ① ② ③ ④ ⑤
7 ① ② ③ ④ ⑤
8 ① ② ③ ④ ⑤
9 ① ② ③ ④ ⑤
10 ① ② ③ ④ ⑤
11 ① ② ③ ④ ⑤
12 ① ② ③ ④ ⑤
13 ① ② ③ ④ ⑤
14 ① ② ③ ④ ⑤
15 ① ② ③ ④ ⑤
16 ① ② ③ ④ ⑤
17 ① ② ③ ④ ⑤
18 ① ② ③ ④ ⑤
19 ① ② ③ ④ ⑤
20 ① ② ③ ④ ⑤
21 ① ② ③ ④ ⑤
22 ① ② ③ ④ ⑤
23 ① ② ③ ④ ⑤
24 ① ② ③ ④ ⑤
25 ① ② ③ ④ ⑤
26 ① ② ③ ④ ⑤
27 ① ② ③ ④ ⑤
28 ① ② ③ ④ ⑤
29 ① ② ③ ④ ⑤
30 ① ② ③ ④ ⑤
31 ① ② ③ ④ ⑤
32 ① ② ③ ④ ⑤
33 ① ② ③ ④ ⑤
34 ① ② ③ ④ ⑤
35 ① ② ③ ④ ⑤
36 ① ② ③ ④ ⑤
37 ① ② ③ ④ ⑤
38 ① ② ③ ④ ⑤
39 ① ② ③ ④ ⑤
40 ① ② ③ ④ ⑤
41 ① ② ③ ④ ⑤
42 ① ② ③ ④ ⑤
43 ① ② ③ ④ ⑤
44 ① ② ③ ④ ⑤
45 ① ② ③ ④ ⑤
46 ① ② ③ ④ ⑤
47 ① ② ③ ④ ⑤
48 ① ② ③ ④ ⑤
49 ① ② ③ ④ ⑤
50 ① ② ③ ④ ⑤
51 ① ② ③ ④ ⑤
52 ① ② ③ ④ ⑤
53 ① ② ③ ④ ⑤
54 ① ② ③ ④ ⑤
55 ① ② ③ ④ ⑤
56 ① ② ③ ④ ⑤

MATHEMATICS POSTTEST

The test that follows will give you a chance to see if you are prepared for the GED Mathematics Test. Like the GED Math Test, this test contains 56 multiple-choice questions. Answer every question. Use the formulas below when necessary. When you are finished, check your answers. The evaluation chart at the end of the answers will help you determine which areas to review before you take the GED Math Test.

Posttest Formulas

AREA (A) of a:	
square	$A = s^2$; where s = side
rectangle	$A = lw$; where l = length, w = width
parallelogram	$A = bh$; where b = base, h = height
triangle	$A = \frac{1}{2}bh$; where b = base, h = height
circle	$A = \pi r^2$; where $\pi = 3.14$, r = radius
PERIMETER (P) of a:	
square	$P = 4s$; where s = side
rectangle	$P = 2l + 2w$; where l = length, w = width
triangle	$P = a + b + c$; where a, b and c are the sides
circumference (C) of a circle	$C = \pi d$; where $\pi = 3.14$, d = diameter
VOLUME (V) of a:	
cube	$V = s^3$; where s = side
rectangular container	$V = lwh$; where l = length, w = width, h = height
cylinder	$V = \pi r^2 h$; where $\pi = 3.14$, r = radius, h = height
Pythagorean relationship	$c^2 = a^2 + b^2$; where c = hypotenuse, a and b are legs of a right triangle
distance (d) between two points in the plane	$d = \sqrt{(x_2 - x_1)^2 + (y_2 - y_1)^2}$; where (x_1, y_1) and (x_2, y_2) are two points in the plane
slope of a line (m)	$m = \frac{y_2 - y_1}{x_2 - x_1}$; where (x_1, y_1) and (x_2, y_2) are two points in a plane
mean	mean = $\frac{x_1 + x_2 + \cdots + x_n}{n}$; where the x's are the values for which a mean is desired, and n = number of values in the series
median	median = the point in an ordered set of numbers at which half of the numbers are above and half of the numbers are below this value
simple interest (i)	$i = prt$; where p = principal, r = rate, t = time
distance (d) as function of rate and time	$d = rt$; where r = rate, t = time
total cost (c)	$c = nr$; where n = number of units, r = cost per unit

1. What is the perimeter in meters of the figure shown at the right?
(1) 14
(2) 28
(3) 29
(4) 47.04
(5) 48

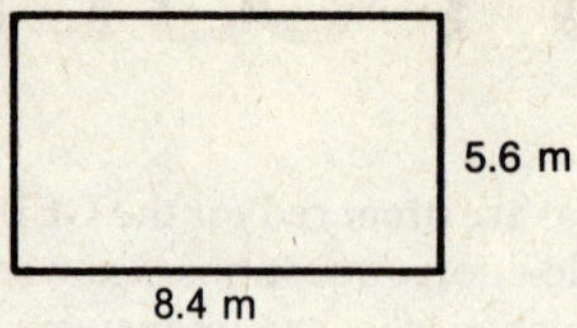

2. In the figure at the right, $\angle AOB$ is a right angle. Find the measurement of $\angle COD$.
(1) 25
(2) 35
(3) 45
(4) 55
(5) Not enough information is provided.

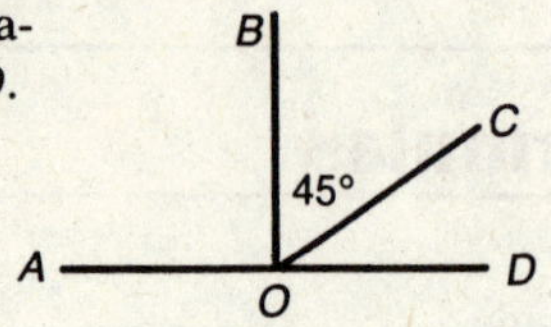

3. In the figure below, AB is parallel to CD, and $\angle 2 = 119°$. Find $\angle 8$.
(1) 71
(2) 61
(3) 29
(4) 19
(5) 9

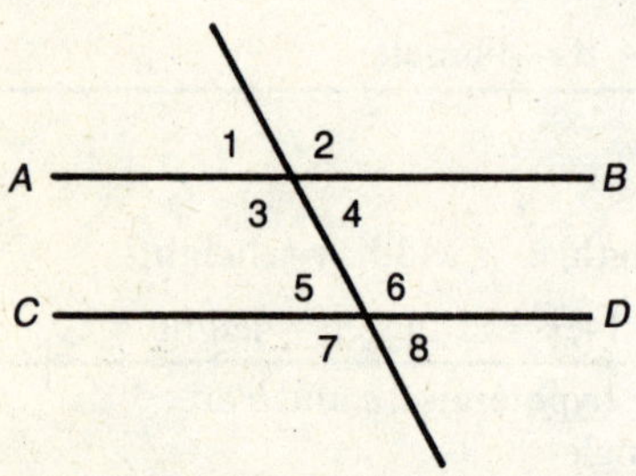

4. In the tool shop at Acme, Inc., 35 employees work 40 hours a week at an average wage of \$8 an hour. Which of the following tells the total weekly payroll for the employees in the tool shop at Acme?
(1) $(35 \times 40)/8$
(2) $35 \times 40 \times 8$
(3) $35 + 40 \times 8$
(4) $35(40 + 8)$
(5) $8(35 + 40)$

5. In $\triangle XYZ$, $\angle X = 65°$ and $\angle Y = 50°$. What kind of triangle is XYZ?
(1) right
(2) scalene
(3) isosceles
(4) equilateral
(5) Not enough information is provided.

6. Grace has to type a 2500-word report and a 500-word letter. She can type 65 words per minute. Which of the following expresses the number of minutes she will need to type both the report and the letter?
(1) $\frac{(2500 + 500)}{65}$
(2) $65 \times 2000 + 65 \times 500$
(3) $65(2500 + 500)$
(4) $\frac{2500 + 500}{65}$
(5) $\frac{2500}{65} + 500$

7. What is the area of the shaded part of the figure at the right?
(1) 1000 ft^2
(2) 886 ft^2
(3) 686 ft^2
(4) 569 ft^2
(5) 56.9 ft^2

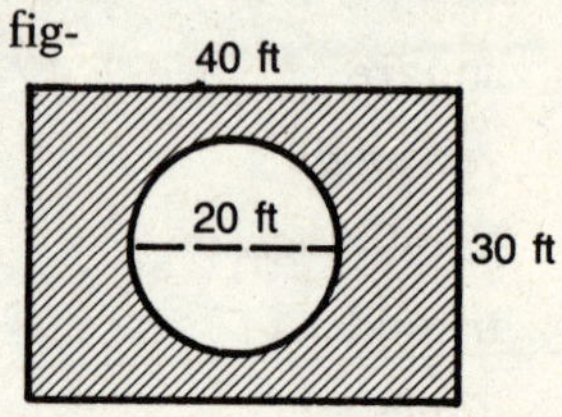

8. For a picnic, Anna bought 12 six-packs of soft drinks at \$2.40 a six-pack and 8 pounds of beef at \$3.50 a pound. Which of the following expresses the total cost of Anna's purchases?
(1) $(12 + 8)(2.40 + 3.50)$
(2) $(12 \times 8) + (2.40 \times 3.50)$
(3) $12(2.40 + 3.50)$
(4) $12 + 2.40 \times 8 + 2.50$
(5) $(12 \times 2.40) + (8 \times 3.50)$

9. In the picture at the right, $MN = 10$, $NO = 3$, and $QO = 48$. Find PQ.
(1) 480
(2) 360
(3) 320
(4) 240
(5) 160

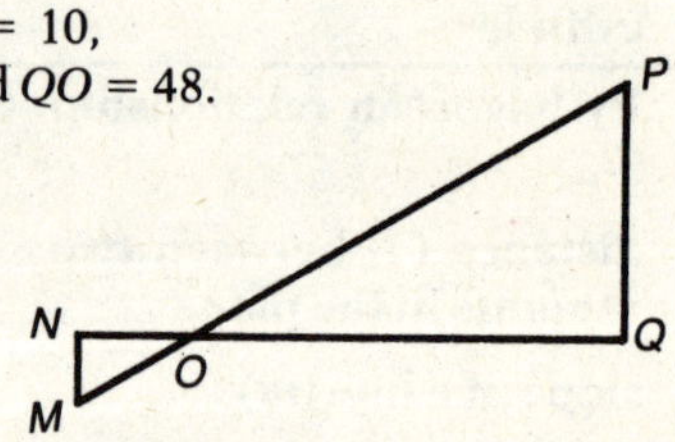

10. Solve for r in the equation $9r - 8 = 6r + 7$
(1) $\frac{1}{3}$
(2) 1
(3) 3
(4) 5
(5) 15

11. Find the volume of a cylindrical tank with a height of 15 feet and a radius of 5 feet. Round your answer off to the nearest cubic foot.
(1) 1178 ft^3
(2) 707 ft^3
(3) 471 ft^3
(4) 236 ft^3
(5) Not enough information is provided.

12. James, Steve, and Charlie worked together to build a garage. Steve worked twice as many hours as James, and Charlie worked 10 hours more than Steve. Altogether the three of them worked 100 hours. How many hours did Charlie work?
(1) 26
(2) 33
(3) 36
(4) 46
(5) 50

13. The ratio of male employees to female employees at a certain factory is 5:3. Altogether there are 96 employees. How many women work at the factory?
(1) 64
(2) 60
(3) 36
(4) 32
(5) 28

14. The height of a triangle is three times the base. The area of the triangle is 96. Find the height of the triangle.
(1) 4
(2) 8
(3) 16
(4) 24
(5) 48

15. Louie drove for 3 hours at an average speed of 18 mph and for 2 hours at 52 mph. Which of the following expresses the total distance he drove?
(1) $(3 \times 2) + (18 + 52)$
(2) $(3 \times 18) + (2 \times 52)$
(3) $\frac{18}{3} + \frac{52}{2}$
(4) $(3 + 18) \times (2 + 52)$
(5) $3(18 + 52) \times 2$

16. Geraldine bought 4 pounds of apples at $.89 a pound. Then she went back to the store to buy 5 more pounds of the apples at the same price. Which of the following expresses the total amount she paid for apples?
(1) $.89 \times 4 \times 5$
(2) $.89 + 4 + 5$
(3) $5 + (.89 \times 4)$
(4) $(4 + 5) + .89$
(5) $.89(4 + 5)$

17. To build a cabinet, Paul needs 8 pieces of lumber each $38\frac{1}{2}$ inches long. How many 10-foot boards should he buy?
(1) 1
(2) 2
(3) 3
(4) 4
(5) 5

18. For the two triangles pictured below, $DE = GH$ and $EF = HI$. Which of the following, along with the information given, is enough to make the triangles congruent?
A. $\angle D = \angle H$
B. $DF = GI$
C. $\angle E = \angle H$

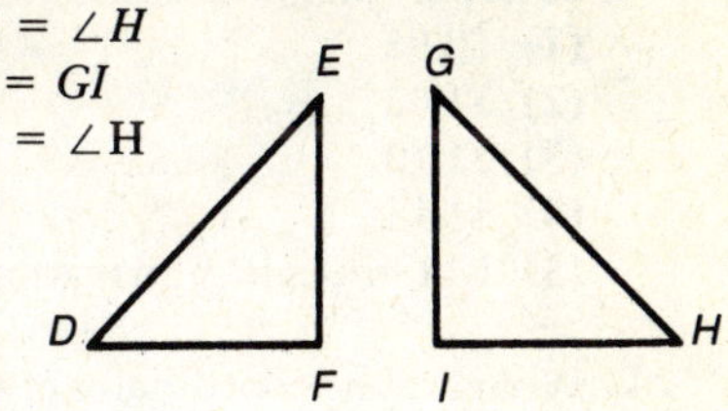

(1) A only
(2) B only
(3) C only
(4) A or B
(5) B or C

Directions: Use the following passage to answer questions 19–21.

Mr. and Mrs. Allen are planning to buy a house that is listed at $46,500. The day they sign papers for the house, they must pay a down payment of 10% of the price. That day they also have to pay $1200 to their lawyer, $2790 as a commission to the real estate agent, and $450 for other expenses involved in the purchase.

19. Find the total amount the Allens must pay the day they sign papers.
(1) $4440
(2) $6300
(3) $7890
(4) $8640
(5) $9090

20. The commission the real estate agent receives is what percent of the total list price of the house?
(1) 3%
(2) 5%
(3) 6%
(4) 10%
(5) Not enough information is provided.

21. The real estate agent must pay part of the commission to his employer. What amount does the agent receive for himself?
(1) $3035
(2) $2790
(3) $2092
(4) $1395
(5) Not enough information is provided.

22. The two figures below have equal areas. Find the side of the square.

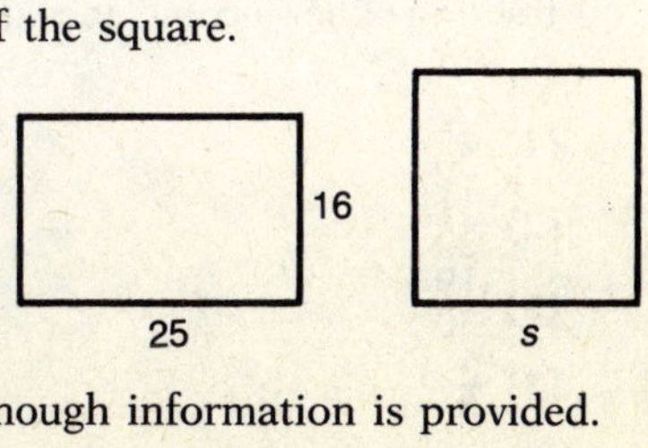

(1) 20
(2) 40
(3) 100
(4) 200
(5) Not enough information is provided.

23. Alvaro drove 236 miles on 18 gallons of gasoline and for another 135 miles on 10 gallons of gasoline. What was the price Alvaro paid per gallon of gasoline?
(1) \$0.95
(2) \$1.02
(3) \$1.10
(4) \$1.15
(5) Not enough information is provided.

24. What are the coordinates of the y-intercept of the equation $y = \frac{2}{3}x - 5$?
(1) $(0, \frac{2}{3})$
(2) $(\frac{2}{3}, 0)$
(3) $(0, -5)$
(4) $(-5, 0)$
(5) $(\frac{2}{3}, -5)$

25. At the Central County post office, the ratio of men to women employees is 4:3. 92 men work there. Which expression gives the total number of *people* who work at the post office?
(1) $(7 \times 92)/4$
(2) $(3 \times 92)/4$
(3) $(4 \times 92)/7$
(4) $(4 \times 7)/92$
(5) $(4 \times 92)/3$

26. Solve for w in $3w + 1 \leq 7(3 - w)$
(1) $w \leq 2$
(2) $w \leq 4$
(3) $w \leq 5$
(4) $w \leq 10$
(5) $w \leq 20$

27. Which of the following is the same as $x^2 - 8x$?
(1) $x \cdot -8$
(2) $(x - 8)(x + 8)$
(3) $x^2 + 8x$
(4) $x(1 - 8)$
(5) $x(x - 8)$

28. Which of the following is the same as $6a + 24b - 18$?
(1) $(6a + 4b)(4a - 3)$
(2) $6a(1 + 4b - 3)$
(3) $6(a + 4b - 3)$
(4) $(a + b)(6 + 4 - 18)$
(5) $6(1 + 4b - 3a)$

29. Marva spends $\frac{1}{3}$ of her income on rent, $\frac{1}{5}$ on food, and $\frac{3}{10}$ on loan payments. What total fraction of her income does she pay for these expenses?
(1) $\frac{8}{30}$
(2) $\frac{11}{20}$
(3) $\frac{19}{30}$
(4) $\frac{5}{6}$
(5) Not enough information is provided.

Directions: Use the following passage to answer questions 30–32.

Sam is building a cabinet. The materials he needs will cost \$480. To determine the price he will charge his customer, Sam adds in a profit of 20% of the cost of the materials and additional expenses of 10% of the cost of materials. Sam usually charges 40% of the total price at the beginning. The customer must pay the balance when the job is finished.

30. Find the total price of the cabinet.
(1) \$780
(2) \$672
(3) \$624
(4) \$504
(5) Not enough information is provided.

31. Find the down payment the customer must pay.
(1) \$374.40
(2) \$249.60
(3) \$230.40
(4) \$192.00
(5) Not enough information is provided.

32. Sam pays an assistant \$7.50 an hour. How much profit is left over for Sam after he pays his assistant?
(1) \$81.00
(2) \$88.50
(3) \$103.50
(4) \$117.30
(5) Not enough information is provided.

---

33. If Carmen cuts a 25-inch wire into pieces $\frac{1}{8}$ of an inch long, how many pieces of wire will we have?
(1) 3
(2) $3\frac{1}{8}$
(3) 20
(4) 200
(5) 250

34. A butcher sliced a roast weighing 4.56 pounds from a leg of lamb that weighs 7 pounds. What is the weight (in pounds) of the remaining leg of lamb?
(1) 1.08
(2) 2.36
(3) 2.44
(4) 3.44
(5) 11.56

35. Arrange the following lengths of 2″ × 4″ lumber in order from shortest to longest.

A: $2\frac{1}{2}$ feet

B: $33\frac{1}{3}$ inches

C: 3 feet $3\frac{1}{3}$ inches

D: 29 inches

E: 3 feet

(1) A, B, D, E, C
(2) B, D, A, A, C
(3) C, D, A, B, E
(4) D, A, B, E, C
(5) C, B, C, A, E

36. Alfredo drove 20 miles north and then 48 miles east. Find the shortest distance in miles Alfredo is from his starting point.
(1) 68
(2) 52
(3) 46
(4) 34
(5) 24

37. Which of the following expresses 67,500,000 in scientific notation?
(1) 6.75×10^{-7}
(2) 6.75×10^{6}
(3) 6.75×10^{7}
(4) 6.00×10^{-6}
(5) 6.57×10^{-7}

38. Which of the following points lies on the line of the graph of the equation $y = 3x - 2$?
(1) (2, 5)
(2) (1, −4)
(3) (4, 14)
(4) (3, 7)
(5) (0, 2)

Directions: Use the table below to answer questions 39 and 40.

Unemployment Rates

	1979	1982	1985
Men			
16–24 years	11.6%	20.3%	14.0%
25–54 years	3.6	9.1	5.4
55 years and older	2.8	5.8	4.0
Women			
16–24 years	12.5	17.0	13.1
25–54 years	5.1	8.5	5.8
55 years and older	3.1	5.0	3.7

Source: Bureau of Labor Statistics.

39. Based on the information in the table, the 1982 unemployment rate for 16- to 24-year-old males was about:
(1) four times the rate for 25- to 54-year-old males
(2) one-fourth of the rate for 25- to 54-year-old males
(3) the same as the rate for 25- to 54-year-old males
(4) about twice the rate for 25- to 54-year-old males
(5) about half the rate for 25- to 54-year-old males

40. For what year was the unemployment rate for 25- to 54-year-old women lower than the same age group for men?
(1) 1979
(2) 1982
(3) 1985
(4) both 1979 and 1982
(5) both 1982 and 1985

---

41. At a sale, Ruth bought four cans of pea soup, six cans of tomato soup, and ten cans of chicken soup. There were no labels on the soup cans.

The first can she opened turned out to be tomato soup. The second can she opened was chicken soup. What is the probability that the third can she opened will be pea soup?

(1) $\frac{3}{20}$

(2) $\frac{1}{6}$

(3) $\frac{1}{5}$

(4) $\frac{2}{9}$

(5) $\frac{1}{4}$

42. Jack sells used cars at his garage. The four cars he has for sale now are priced at \$2380, \$4950, \$8265, and \$5790, respectively. Find the median price of the cars.
(1) \$5320
(2) \$5346
(3) \$5370
(4) \$5400
(5) \$5790

43. The area of the triangle pictured at the right is 192. The ratio of side *AB* to side *AC* is 2:3. Find the length of side *AB*.
(1) 4
(2) 8
(3) 12
(4) 16
(5) 24

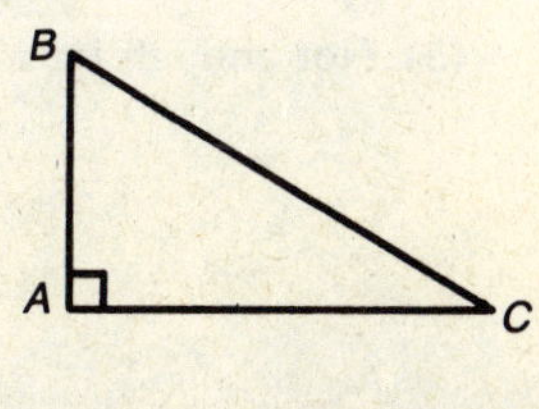

Directions: Use the diagram below to answer questions 44 and 45.

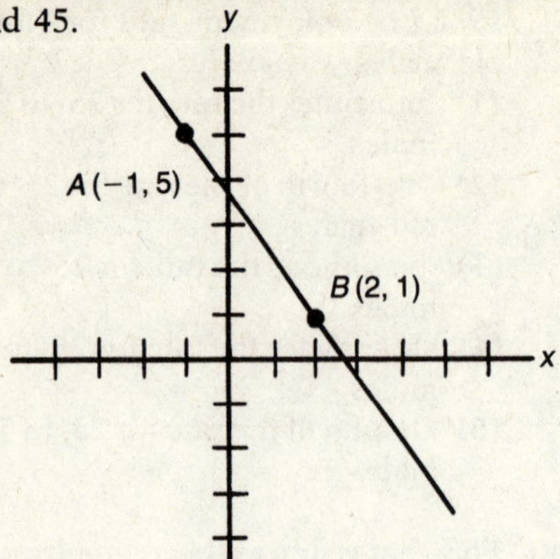

44. What is the slope of the line that passes through points *A* and *B*?

(1) $\frac{5}{2}$
(2) $-\frac{5}{2}$
(3) $-\frac{4}{3}$
(4) $\frac{3}{4}$
(5) $-\frac{2}{5}$

45. What is the distance between points *A* and *B*?
(1) 5
(2) 7
(3) 8
(4) 9
(5) 11

Directions: Use the table below to answer questions 46–48.

Corn Production for Four Farmers

	Farmer A	Farmer B	Farmer C	Farmer D
Yield (bushels/acre)	135	167	148	112
Cost ($/bushel)	$3.40	$2.20	$1.90	$1.40
Net ($/acre*)	−$135	$33.40	$74	$112

*Based on a price of $2.40/bushel.

46. Which farmer had the highest yield?
(1) A
(2) B
(3) C
(4) D
(5) Not enough information is provided.

47. If farmers C and D each have 200 acres of corn, how much more does farmer D net than farmer C?
(1) $4,600
(2) $7,600
(3) $8,120
(4) $20,320
(5) Not enough information is provided.

48. Based on the information in the table, which of the following statements is true?
(1) The lower the cost, the higher the profit.
(2) The higher the yield, the higher the profit.
(3) A higher yield results in lower profits.
(4) Higher profits come from higher costs.
(5) Higher profits come from lower yields.

49. Arrange the weights of the following packages in order from lightest to heaviest:
package A: .5 kilogram
package B: 485 grams
package C: 1.1 kilograms
package D: 0.65 kilogram
package E: 1200 grams

(1) B, A, D, C, E
(2) A, C, E, D, B
(3) A, B, D, C, E
(4) C, D, B, A, E
(5) C, B, A, E, D

50. Eighteen railroad ties, each 3.9 meters long, are laid end to end. How many meters do the ties cover in total?
(1) 21
(2) 21.9
(3) 39
(4) 70.2
(5) 702

Directions: Questions 51–53 refer to the following passage.

A new rock band, the Great Divide, plays twice a week at Jack's Bar. The price of admission to see a show is $7.50, and the band keeps 40% of the total ticket sales.

51. If 100 tickets were sold on Thursday night, how much money did the band keep?
(1) $100
(2) $300
(3) $400
(4) $450
(5) $3000

52. Kevin, the lead vocalist, wants to take \$240 of one week's earnings to do some local advertising. That week, the bar sold 400 tickets. What percentage of the band's profits will Kevin use for advertising?
(1) 10
(2) 15
(3) 20
(4) 25
(5) 30

53. Each week, Kevin pays band expenses, then divides up the band's profits evenly among the band members. During an especially successful week in March, the band made \$2350 after expenses. To find out how much money each band member earned, what more do you need to know?
(1) the band's expenses that week
(2) how many people attended the shows that week
(3) the hours worked by each band member
(4) the number of members in the band
(5) the amount that Kevin needed for advertising

---

54. A computer program can figure out how much lumber a manufacturer needs to buy by multiplying its amount of inventory by .75 and subtracting from that product the amount of lumber sold the previous month. If the manufacturer's lumber sales from January came to 22,500 square feet and its amount of inventory is 95,000 square feet, which of the following expressions gives the amount of lumber that the manufacturer needs to order in February?
(1) $.75 \times 22{,}500$
(2) $.75 \times 95{,}000$
(3) $(.75 \times 95{,}000) - 22{,}500$
(4) $(.75 \times 22{,}500) - 95{,}000$
(5) $\dfrac{95{,}000 - 22{,}500}{.75}$

55. The two figures below have the same area. What is the length of side *XY*?

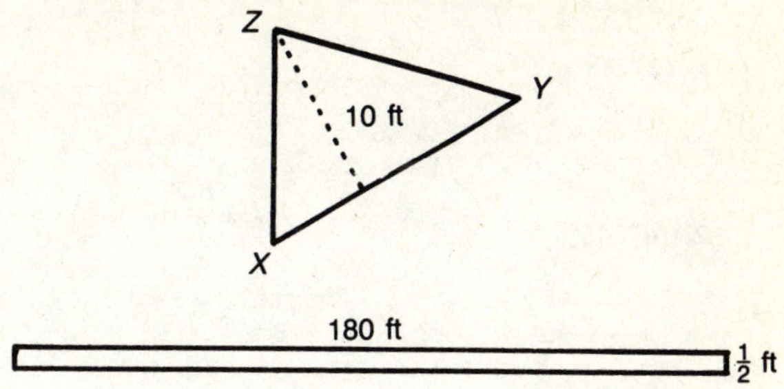

(1) 8 ft
(2) 9 ft
(3) 18 ft
(4) 27 ft
(5) Not enough information is provided.

56. Out of a total printing of 10,000 copies, 2600 newsletters were missing pages. Five hundred newsletters had smudged ink. What percent of the newsletters had smudged ink?
(1) 5
(2) 15
(3) 21
(4) 26
(5) 50

ANSWERS AND SOLUTIONS START ON PAGE 318.

POSTTEST ANSWERS AND SOLUTIONS

1. **(2) 28 m**
$p = 2l + 2w$
$p = 2(8.4) + 2(5.6)$
$p = 16.8 + 11.2 = 28$

2. **(4) 45°**
$90° - 45° = 45°$

3. **(2) 61°**
$180° - 119° = 61°$

4. **(2) 35 × 40 × 8**

5. **(3) isosceles**
$65° + 50° = 115°$ $180° - 115° = 65°$ Since two angles are equal, the triangle is isosceles.

6. **(1) (2500 + 500)/65**

7. **(2) 886 ft²**
Area of rectangle: $A = lw$
$A = 40 \times 30$
$A = 1200$
Area of circle: $A = \pi r^2$
$A = 3.14 \times 10^2$
$A = 3.14 \times 100$
$A = 314$
Shaded area: $1200 - 314 = 886$

8. **(5) (12 × 2.40) + (8 × 3.50)**

9. **(5) 160**
$\frac{3}{10} = \frac{48}{x}$ $48 \times 10 = 480$ $480 \div 3 = 160$

10. **(4) 5**
$$\begin{aligned} 9r - 8 &= 6r + 7 \\ -6r & \quad -6r \\ 3r - 8 &= +7 \\ +8 & \quad +8 \\ \frac{3r}{3} &= \frac{15}{3} \\ r &= 5 \end{aligned}$$

11. **(1) 1178 ft³**
$v = \pi r^2 h$
$v = 3.14 \times 5^2 \times 15$
$v = 3.14 \times 25 \times 15$
$v = 1177.5$ to the nearest ft³ $= 1178$

12. **(4) 46**
James's hours $= x$
Steve's hours $= 2x$
Charlie's hours $= 2x + 10$
$$\begin{aligned} x + 2x + 2x + 10 &= 100 \\ 5x + 10 &= 100 \\ -10 & \quad -10 \\ \frac{5x}{5} &= \frac{90}{5} \\ x &= 18 \end{aligned}$$
Charlie worked $2(18) + 10 = 46$ hr.

13. **(3) 36**
male 5 + female 3 = 8
$\frac{\text{female}}{\text{total}}$ $\frac{3}{8} = \frac{x}{96}$
$96 \times 3 = 288$ $288 \div 8 = 36$

14. **(4) 24**
base $= x$
height $= 3x$
$$\begin{aligned} A &= bh/2 \\ 96 &= (x \cdot 3x)/2 \\ \frac{2}{3} \cdot \frac{96}{1} &= \frac{3}{2}x^2 \cdot \frac{2}{3} \\ 64 &= x^2 \\ \sqrt{64} &= x \\ 8 &= x \end{aligned}$$
height $= 3(8) = 24$

15. **(2) (3 × 18) + (2 × 52)**

16. **(5) .89(4 + 5)**

17. **(3) 3**
First find how many inches are needed.
$38\frac{1}{2} \times 8 = 308$
Then find how many inches are in the 10-foot boards.
10 ft × 12 in = 120 in
Divide.
$308.00 \div 120 = 2.56$
(120 into 308.00: 240, 68 0, 60 0, 8 00, 7 20)
Paul will need 3 boards.

18. **(5) B or C**
Condition B fulfills the side, side, side requirement.
Condition C fulfills the side, angle, side requirement.

19. **(5) $9090**
down payment:
$\frac{p}{46{,}500} = \frac{10}{100}$ $46{,}500 \times 10 = 465{,}000$
$465{,}000 \div 100 = 4{,}650$
$4650 + 1200 + 2790 + 450 = \$9090$

20. **(3) 6%**
$\frac{2{,}790}{46{,}500} = \frac{x}{100}$ $2{,}790 \times 100 = 279{,}000$
$279{,}000 \div 46{,}500 = 6\%$

21. **(5) Not enough information is provided.**

22. **(1) 20**
Area of rectangle: $A = lw$
$A = 25 \times 16 = 400$
Side of square: $400 = s^2$
$\sqrt{400} = s$
$20 = s$

23. **(5) Not enough information is provided.**

24. **(3) (0, −5)**
$y = \frac{2}{3}(0) - 5$
$y = 0 - 5 = -5$
Coordinates of the y-intercept are $(0, -5)$.

25. **(1) (7 × 92)/4**
men 4 + women 3 = total 7
$\frac{\text{men}}{\text{total}}$ $\frac{4}{7} = \frac{92}{x}$

$$4 \cdot x = 7 \times 92$$
$$x = \frac{7 \times 92}{4} \text{ or } (7 \times 92)/4$$

26. **(1) $w \leq 2$**
$$\begin{aligned} 3w + 1 &\leq 7(3 - w) \\ 3w + 1 &\leq 21 - 7w \\ +7w \quad &\quad +7w \\ 10w + 1 &\leq 21 \\ -1 \quad &\quad -1 \\ \frac{10w}{10} &\leq \frac{20}{10} \\ w &\leq 2 \end{aligned}$$

27. **(5) $x(x - 8)$**
$x^2 - 8x = x(x - 8)$

28. **(3) $6(a + 4b - 3)$**
$6a + 24b - 18 = 6(a + 4b - 3)$

29. **(4) $\frac{5}{6}$**
$\frac{1}{3} + \frac{1}{5} + \frac{3}{10} = \frac{25}{30} = \frac{5}{6}$

30. **(3) $624**
$$20\% + 10\% = 30\% \qquad \frac{p}{480} = \frac{30}{100} \qquad 480 \times 30 = 14{,}400$$
$$14{,}400 \div 100 = 144 \qquad \$480 + 144 = \$624$$

31. **(2) $249.60**
$$\frac{p}{624} = \frac{40}{100} \qquad 624 \times 40 = 24{,}960 \qquad 24{,}960.00 \div 100 = \$249.60$$

32. **(5) Not enough information is provided.**

33. **(4) 200**
$25 \div \frac{1}{8} =$
$25 \times 8 = 200$

34. **(3) 2.44**
$$7.00 - 4.56 = 2.44$$

35. **(4) D, A, B, E, C**
A: $2\frac{1}{2}$ ft = 30 in
B: $33\frac{1}{3}$ in
C: 3 ft $3\frac{1}{3}$ in = $39\frac{1}{3}$ in
D: 29 in
E: 3 ft = 36 in

36. **(2) 52**
$$\begin{aligned} a^2 &= b^2 + c^2 \\ a^2 &= 20^2 + 48^2 \\ a^2 &= 400 + 2304 \\ a^2 &= 2704 \\ a &= \sqrt{2704} \\ a &= 52 \end{aligned}$$

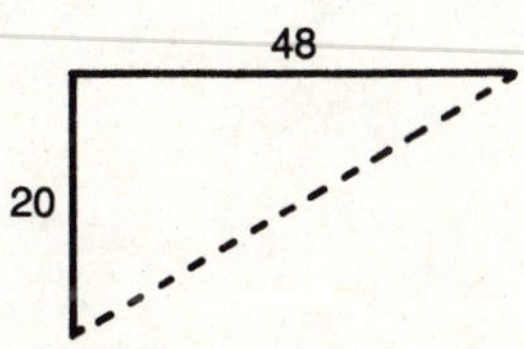

37. **(3) 6.75×10^7**
First remove the zeros and represent 67,500,000 as a number between 1 and 10. → 6.75
Then write a multiplication sign and 10 to the correct power. You are moving the decimal point seven places to the left, so 67,500,000 = 6.75×10^7.

38. **(4) (3, 7)**
$$\begin{aligned} y &= 3x - 2 \\ 7 &= 3(3) - 2 \\ 7 &= 9 - 2 \\ 7 &= 7 \end{aligned}$$

39. **(4) about twice the rate for 25- to 54-year-old males.**
$$20.3 \div 9.1 = 2 \qquad (18\,2,\ \text{remainder } 2\,1)$$

40. **(2) 1982**

41. **(4) $\frac{2}{9}$**
$$\frac{\text{favorable outcomes}}{\text{total outcomes}} \quad \frac{4}{18} = \frac{2}{9}$$

42. **(3) $5370**
Find the mean of the two middle prices.
$$\$4{,}950 + 5{,}790 = \$10{,}740 \qquad 10{,}740 \div 2 = \$5{,}370$$

43. **(4) 16**
height = $2x$
base = $3x$
$$\begin{aligned} A &= \tfrac{1}{2}bh \\ 192 &= (3x \cdot 2x)/2 \\ 192 &= 6x^2/2 \\ \frac{192}{3} &= \frac{3x^2}{3} \\ 64 &= x^2 \\ \sqrt{64} &= x \\ 8 &= x \\ 16 &= 2x \end{aligned}$$
AB is 16.

44. **(3) −4/3**
$$m = \frac{y_2 - y_1}{x_2 - x_1}$$
$$m = \frac{1 - 5}{2 - (-1)} = \frac{-4}{+3} = \frac{-4}{3}$$

45. **(1) 5**
$$\begin{aligned} \text{distance} &= \sqrt{(2 - (-1))^2 + (1 - 5)^2} \\ &= \sqrt{3^2 + 4^2} \\ &= \sqrt{25} \\ &= 5 \end{aligned}$$

46. **(2) B**

47. **(2) $7,600**
$$\$74 \times 200 = \$14{,}800 \qquad \$112 \times 200 = \$22{,}400 \qquad \$22{,}400 - 14{,}800 = \$7{,}600$$

48. **(1) The lower the cost, the higher the profit (net).**

49. **(1) B, A, D, C, E**
A = .5 kg
B = .485 kg
C = 1.1 kg
D = 0.65 kg
E = 1.2 kg

50. **(4) 70.2**
$$18 \times 3.9: \quad 16\,2 + 54 = 70.2$$

51. **(2) $300**
$100 \times \$7.50 = \$750$
$$\frac{p}{750} = \frac{40}{100} \qquad 750 \times 40 = 30{,}000 \qquad 30{,}000 \div 100 = \$300$$

52. **(3) 20**

$400 \times \$7.50 = \$3000$

$\frac{40}{100} \times \frac{\$3000}{1} = \$1200$

$\$240 = \frac{x}{100} \times \frac{\$1200}{1}$

$\$240 = 12x$

$20 = x$

53. **(4) the number of members in the band**

54. **(3) (.75 × 95,000) − 22,500**

(inventory × .75) − sales of previous month

55. **(3) 18 ft.**

area of rectangle: $\frac{1}{2} \times 180 = 90$ sq ft

area of triangle: $\frac{1}{2}bh = \frac{1}{2}b \times 10$

$90 = \frac{1}{2}b \times 10$

$\frac{90}{10} = \frac{\frac{1}{2}b \times 10}{10}$

$9 = \frac{1}{2}b$

$18 = b$

56. **(1) 5**

$\frac{50}{10{,}000} = \frac{x}{100}$ $\quad$ $10{,}000\overline{)50{,}000}$ = 5

Evaluation Chart

On the following chart, circle the number of any problem you got wrong. After each problem number, you will see the name of the section (or sections) where you can find the skills you need to solve the problem. When a problem involves more than one skill, the sections are separated by a slash (/).

This chart should help you decide which areas you need to review before you take the GED Test.

Posttest Evaluation Chart

Problem	Section	Starting Page
1	Perimeter	227
2, 3	Angle Relationships	254
4, 6, 8	Set-up Questions	42
5	Triangles	257
7	Area	238
9	Similarity	260
10	Solving Equations with Separated Unknowns	215
11	Volume	244
12	Using Algebra to Solve Word Problems	216
13	Ratio and Proportion	119
14, 43, 55	Using Algebra to Solve Geometry Problems	293
15, 16	Set-up Questions	42
17	Measurement Word Problems	158
18	Congruence	264
19, 20, 21	Proportion and Percent/Item Sets	136/70
22	Square Roots/Area Word Problems	236/238
23	When Not Enough Information Is Provided	152
24, 44	Slope and Intercepts	302
25	Proportion Word Problems/Set-up Questions	121/42
26	Solving Longer Inequalities	287
27, 28	Factoring	289
29	Adding Fractions	92
30, 31, 32	Proportion and Percent/Item Sets	136/70
33	Dividing Fractions	101
34	Subtracting Decimals	58
35, 49	Measurement Word Problems	158
36	Pythagorean Relationship	267
37	Scientific Notation	307
38	Linear Equations	300
39, 40	Tables	197
41	Probability	115
42	Mean and Median	34
45	Finding Distances on the Rectangular Coordinate System	297
46, 47, 48	Tables	197
50	Multiplying Decimals	62
51, 52	Percents/Item Sets	128/70
53	What More Do You Need to Know	170
54	Set-up Questions	42
56	Percents/Finding Necessary Information	128/22

Your score: _______ out of 56 correct.

If you got fewer than 40 items correct, go back and find the areas in which you had difficulty. Then review those pages that you need additional work in.

ANSWER KEY

CHAPTER 1

Whole Number Review Exercise 1

page 11

1. 40 20,000 3
2. 300,000 900 1000
3. 20 60,000,000 400
4. 400,000,000, 40,000 600,000
5. 1 × 10,000 = 10,000
   0 × 1000 = 0 (no thousands)
   4 × 100 = 400
   9 × 10 = 90
   6 × 1 = 6

Whole Number Review Exercise 2

page 12

	1	3	8	4	6	9	7	10	2	5
8	9	11	16	12	14	17	15	18	10	13
1	2	4	9	5	7	10	8	11	3	6
3	4	6	11	7	9	12	10	13	5	8
6	7	9	14	10	12	15	13	16	8	11
4	5	7	12	8	10	13	11	14	6	9
9	10	12	17	13	15	18	16	19	11	14
5	6	8	13	9	11	14	12	15	7	10
7	8	10	15	11	13	16	14	17	9	12
2	3	5	10	6	8	11	9	12	4	7
10	11	13	18	14	16	19	17	20	12	15

Whole Number Review Exercise 3

pages 12–13

	12	3	1	4	10	9	7	5	6	8	11	0	2
9	108	27	9	36	90	81	63	45	54	72	99	0	18
7	84	21	7	28	70	63	49	35	42	56	77	0	14
4	48	12	4	16	40	36	28	20	24	32	44	0	8
5	60	15	5	20	50	45	35	25	30	40	55	0	10
2	24	6	2	8	20	18	14	10	12	16	22	0	4
1	12	3	1	4	10	9	7	5	6	8	11	0	2
8	96	24	8	32	80	72	56	40	48	64	88	0	16
10	120	30	10	40	100	90	70	50	60	80	110	0	20
12	144	36	12	48	120	108	84	60	72	96	132	0	24
0	0	0	0	0	0	0	0	0	0	0	0	0	0
11	132	33	11	44	110	99	77	55	66	88	121	0	22
3	36	9	3	12	30	27	21	15	18	24	33	0	6
6	72	18	6	24	60	54	42	30	36	48	66	0	12

Whole Number Review Exercise 4

page 13

1. $5 \times 4 = 20$
2. $7 \times 8 = 56$
3. $3 \times 8 = 24$
4. $5 \times 9 = 45$
5. $6 \times 5 = 30$
6. $7 \times 6 = 42$
7. $9 \times 2 = 18$
8. $11 \times 11 = 121$
9. $12 \times 4 = 48$
10. $8 \times 5 = 40$
11. $6 \times 3 = 18$
12. $9 \times 6 = 54$
13. $8 \times 9 = 72$
14. $7 \times 4 = 28$
15. $9 \times 4 = 36$
16. $64 \div 8 = 8$
17. $42 \div 6 = 7$
18. $25 \div 5 = 5$
19. $24 \div 8 = 3$
20. $54 \div 9 = 6$
21. $36 \div 6 = 6$
22. $42 \div 7 = 6$
23. $81 \div 9 = 9$
24. $56 \div 8 = 7$
25. $49 \div 7 = 7$
26. $45 \div 5 = 9$
27. $32 \div 4 = 8$
28. $63 \div 9 = 7$
29. $48 \div 6 = 8$
30. $21 \div 3 = 7$

Whole Number Review Exercise 5

pages 17–18

1. 117,807
25,624
+ 92,183
117,807

2. 57,919
60,845
− 2,926
57,919

3. 41,231
48,005
− 6,774
41,231

4. 2373
83
2096
+ 194
2373

5. 34,248
5,708
× 6
34,248

6. 25,826
349
× 74
1 396
24 43
25,826

7. 356,648
872
× 409
7 848
348 80
356,648

8. 3,250,000
50,000
× 65
3,250,000

9. 26 r4
26 r4
17)446
34
106
102
4

10. 308
308
8)2464

11. 9,006
9,006
6)54,036

12. 14,459
30,045
− 15,586
14,459

13. 3418
194
8
2366
+ 850
3418

14. 48,000
500
× 96
48,000

15. 2,184,000
3,000,000
− 816,000
2,184,000

16. 203,406
29,058
× 7
203,406

17. 47
47
800)37,600
32 00
5 600
5 600
0

18. 592,000
8,000
× 74
592,000

19. 4,775,500
5,040,000
− 264,500
4,775,500

20. 306
306
35)10,710
10 5
21
0
210
210
0

21. 40,720
5,090
× 8
40,720

22. 423
423
12)5076
48
27
24
36
36
0

23. 6912
7000
− 88
6912

24. 391,084
400,300
− 9,216
391,084

25. 705

```
   705
4)2820
```

26. 155,700

```
    346
  × 450
 17 300
 138 4
155,700
```

27. 15,792

```
 2,900
   857
11,630
+  405
15,792
```

28. 8,060 r3

```
  8,060 r3
7)56,423
  56
   0 42
     42
      3
```

29. 492,869

```
 1,300,500
−  807,631
   492,869
```

30. 740

```
     740
19)14,060
   13 3
      76
      76
       0
```

CHAPTER 2

Problem-Solving Exercise 1

page 21

1. (3) Find the difference between the number of votes Mr. Sanders received and the number his opponent received.
2. (3) How much did the TV cost before it was on sale?
3. (2) Find Manny's total car payments for twelve months.
4. (3) Find out how far Pancho drove on one gallon of gasoline.
5. (2) How much did Gordon weigh at the end of December?
6. (2) How much do both Marilyn and Calvin take home in a month?
7. (3) Find the distance Carmen can drive with 13 gallons of gas.
8. (1) How much more does Gloria need to save for the TV?

Problem-Solving Exercise 2

pages 22–23

1. *Need:* 943 cars sold last month and 387 made in the U.S.
   Do not need: 198 cars made in Japan
2. *Need:* March 1985 unemployment of 10,549 and March 1984 unemployment of 8,248
   Do not need: December 1984 unemployment of 9,307
3. *Need* 132 miles and 3 hours
   Do not need: 22 miles on one gallon
4. *Need:* 132 miles and 22 miles on one gallon
   Do not need: 3 hours
5. *Need:* $6.50 for a child's ticket and 1230 children
   Do not need: $10.50 for an adult's ticket
6. *Need:* 2500 adults and 1250 high school graduates
   Do not need: 750 college graduates
7. *Need:* California pop. of 23,668,000 and New York pop. of 17,558,000
   Do not need: Texas pop. of 14,229,000
8. *Need:* $76.80 and 64 feet
   Do not need: 4 pounds a foot
9. *Need:* $285 for rent and $400 for food
   Do not need: $1100 income
10. *Need:* $2.49 a pound and 3 pounds of lamb chops
    Do not need: $1.89 a pound and 2 pounds of beef roast

Problem-Solving Exercise 3

pages 25–26

1. *Question:* How much older is father than son?
   Operation: subtract
2. *Question:* Find hourly wage.
   Operation: divide
3. *Question:* How much were monthly utilities?
   Operation: add
4. *Question:* Find how many dozen eggs are white.
   Operation: subtract
5. *Question:* Find net income.
   Operation: subtract
6. *Question:* Find year when the man was born.
   Operation: subtract
7. *Question:* Find daily pay.
   Operation: multiply
8. *Question:* Find monthly salary.
   Operation: divide
9. *Question:* Find number of minutes needed to type 1700-word letter.
   Operation: divide
10. *Question:* Find total amount in fund.
    Operation: add

Problem-Solving Exercise 4

page 27
(Exercise 1)

1. 4,695

```
  67,576
− 62,881
   4,695
```

2. $258

```
 $198
+  60
 $258
```

3. $2280

```
  $190
 ×  12
   380
  190
 $2280
```

4. **23 mi.**
   23
   20)460
   40
   60
   60

5. **211 lb.**
   178
   + 33
   211

6. **$1,850**
   $ 1,850
   12)$22,200

7. **312 mi.**
   24
   × 13
   72
   24
   312

8. **$95**
   $385
   − 290
   $ 95

(Exercise 2)

1. **556 cars**
   943
   − 387
   556

2. **2,301**
   10,549
   − 8,248
   2,301

3. **44 mi**
   44
   3)132

4. **6 gal**
   6
   22)132

5. **$7995**
   12 30
   × $6.50
   615 00
   7380
   $7995.00

6. **1250**
   2500
   − 1250
   1250

7. **6,110,000**
   23,668,000
   − 17,558,000
   6,110,000

8. **$1.20**
   $ 1.20
   64)$76.80
   64
   12 8
   12 8
   00

9. **$115**
   $400
   − 285
   $115

10. **$7.47**
   $2.49
   × 3
   $7.47

(Exercise 3)

1. **29 yrs.**
   37
   − 8
   29

2. **$6.50**
   $ 6.50
   40)$260.00
   240
   20 0
   20 0
   00

3. **$90**
   $12
   25
   + 53
   $90

4. **15 dozen**
   20
   − 5
   15

5. **$13,616**
   $18,296
   − 4,680
   $13,616

6. **1893**
   1985
   − 92
   1893

7. **$60**
   $ 7.50
   × 8
   $60.00

8. **$1365**
   $ 1,365
   12)$16,380

9. **20 min**
   20
   85)1700
   170
   00

10. **$112,000**
   $ 85,000
   + 27,000
   $112,000

Problem-Solving Exercise 5

page 29

1. 60
2. 230
3. 400
4. 6000
5. 8000
6. 7000
7. 25,000
8. 40,000
9. 1060
10. 28,100
11. 16,200
12. 1,480,000
13. 3000
14. 310
15. 8100

Problem-Solving Exercise 6

pages 30–31

1. *Round off wage to nearest dollar:* \$8
   Approximate answer: 9 × \$8 = **\$72**
   Actual answer: **\$74.25**
   $$\begin{array}{r} \$\ 8.25 \\ \times\quad 9 \\ \hline \$74.25 \end{array}$$

2. *Round off speed to nearest ten:* 50 mph
   Approximate answer: 50 × 6 = **300**
   Actual answer: **312 mi**
   $$\begin{array}{r} 52 \\ \times\ 6 \\ \hline 312 \end{array}$$

3. *Round off each attendance to nearest 100:* 500, 700, 500
   Approximate answer: 500 + 700 + 500 = **1700**
   Actual answer: **1672**
   $$\begin{array}{r} 487 \\ 662 \\ +\ 523 \\ \hline 1672 \end{array}$$

4. *Round off salary to a number easily divided by 12:* \$24,000
   Approximate answer: \$24,000 ÷ 12 = **\$2,000**
   Actual answer: **\$2080**
   $$\begin{array}{r} \$\ 2,080 \\ 12\overline{)\$24,960} \end{array}$$

5. *Round off year of construction to nearest ten:* 1900
   Round off 1988 to nearest ten: 1990
   Approximate answer: 1990 − 1900 = **90**
   Actual answer: **91 years**
   $$\begin{array}{r} 1988 \\ -\ 1897 \\ \hline 91 \end{array}$$

6. *Round off sale price and savings to nearest ten:* \$200 and \$60
   Approximate answer: \$200 + \$60 = **\$260**
   Actual answer: **\$256**
   $$\begin{array}{r} \$198 \\ +\ 58 \\ \hline \$256 \end{array}$$

7. *Round off price of one carton to nearest dollar:* \$2
   Approximate answer: 5 × \$2 = **\$10**
   Actual answer: **\$9.45**
   $$\begin{array}{r} \$1.89 \\ \times\ \ 5 \\ \hline \$9.45 \end{array}$$

8. *Round off each length to nearest ten:* 100 and 30
   Approximate answer: 100 − 30 = **70**
   Actual answer: **69 inches**
   $$\begin{array}{r} 96 \\ -\ 27 \\ \hline 69 \end{array}$$

Problem-Solving Exercise 7

pages 32–34

1. **a.** Find cost of educating 30 children in Upperville.
   b. Find cost of educating 30 children in Central City.
   c. Subtract answer to step b from answer to step a.
   Solution: **a.** $\begin{array}{r} \$7,800 \\ \times\quad 30 \\ \hline \$234,000 \end{array}$ **b.** $\begin{array}{r} \$2,100 \\ \times\quad 30 \\ \hline \$63,000 \end{array}$
   c. $\begin{array}{r} \$234,000 \\ -\ 63,000 \\ \hline \mathbf{\$171,000} \end{array}$

2. **a.** Find combined income for week.
   b. Find income for year. Multiply weekly income by 52. (1 year = 52 weeks)
   Solution: **a.** $\begin{array}{r} \$240 \\ +\ 200 \\ \hline \$440 \end{array}$ **b.** $\begin{array}{r} \$440 \\ \times\quad 52 \\ \hline 880 \\ 22\ 00 \\ \hline \mathbf{\$22,880} \end{array}$

3. **a.** Find total number of cases.
   b. Multiply number of cases by 20.
   Solution: **a.** $\begin{array}{r} 4 \\ 6 \\ +\ 10 \\ \hline 20 \end{array}$ **b.** $\begin{array}{r} 20 \\ \times\ 20 \\ \hline \textbf{400 cans} \end{array}$

4. **a.** Find the distance she drives in 5 hours.
   b. Find distance she drives in 3 hours.
   c. Add results from steps a. and b.
   Solution: **a.** $\begin{array}{r} 55 \\ \times\ 5 \\ \hline 275 \end{array}$ **b.** $\begin{array}{r} 35 \\ \times\ 3 \\ \hline 105 \end{array}$
   c. $\begin{array}{r} 275 \\ +\ 105 \\ \hline \textbf{380 miles} \end{array}$

5. **a.** Find the amount he makes in 35 hours.
   b. Find amount he makes in 6 hours.
   c. Add results from steps a. and b.
   Solution: **a.** $\begin{array}{r} \$8.50 \\ \times\quad 35 \\ \hline 42\ 50 \\ 255\ 0 \\ \hline \$297.50 \end{array}$ **b.** $\begin{array}{r} \$12.75 \\ \times\quad 6 \\ \hline \$76.50 \end{array}$
   c. $\begin{array}{r} \$297.50 \\ +\ 76.50 \\ \hline \mathbf{\$374.00} \end{array}$

6. **a.** Find total monthly payments.
   b. Add total monthly payments to down payment.
   Solution: **a.** $\begin{array}{r} \$45 \\ \times\ 18 \\ \hline \$810 \end{array}$ **b.** $\begin{array}{r} \$810 \\ +\ 150 \\ \hline \mathbf{\$960} \end{array}$

7. **a.** Find amount of last year's drive.
   b. Divide result from step a. by 5.
   Solution: **a.** $\begin{array}{r} \$30,400 \\ +\ 3,000 \\ \hline \$33,400 \end{array}$ **b.** $\begin{array}{r} \mathbf{\$\ 6,680} \\ 5\overline{)\$33,400} \end{array}$

8. **a.** Find total number of words.
b. Divide total by 75.

Solution: **a.** 1000 + 500 = 1500 **b.** $75\overline{)1500}$ = **20**; 150; 00

9. **a.** Find total cost of rent and food.
b. Subtract result from step a. from monthly income.

Solution: **a.** \$345 + 360 = \$705 **b.** \$1300 − 705 = **\$ 595**

10. **a.** Find cost of 2 station wagons.
b. Find cost of 3 sedans.
c. Add results from steps a. and b. to cost of truck.

Solution: **a.** \$11,500 × 2 = \$23,000 **b.** \$6,500 × 3 = \$19,500

c. \$23,000 + 19,500 + 24,800 = **\$67,300**

Problem-Solving Exercise 8

pages 35–36

1. **a.** Mean: **193**

353 + 19 + 207 = 579; $3\overline{)579}$ = 193

b. Median: **207**

19 207 353

2. **a.** Mean: **1995**

2046 + 971 + 3113 + 1850 = 7980; $4\overline{)7980}$ = 1995

b. Median: **1948**

971 1850 2046 3113

1850 + 2046 = 3896; $2\overline{)3896}$ = 1948

3. **a.** Mean: **253**

240 + 313 + 189 + 270 = 1012; $4\overline{)1012}$ = 253

b. Median: **255**

189 240 270 313

240 + 270 = 510; $2\overline{)510}$ = 255

4. **173 lb** 187 + 159 = 346; $2\overline{)346}$ = 173

5. **65°** 69 + 71 + 56 + 63 + 66 = 325; $5\overline{)325}$ = 65

6. **\$24,800** \$24,700 + 23,900 + 25,800 = \$74,400; $3\overline{)\$74,400}$ = \$24,800

7. **\$26.07** \$25.66 + 33.27 + 19.28 = \$78.21; $3\overline{)\$78.21}$ = \$26.07

8. **\$240** $450\overline{)\$108,000}$ = \$240; 90 0; 18 00; 18 00; 00

9. **\$1,850**
3 × \$1,650 = \$ 4,950
4 × \$1,875 = \$ 7,500
2 × \$2,100 = \$ 4,200
\$16,650

$9\overline{)\$16,650}$ = \$ 1,850

10. **47** 47 + 44 + 50 = 141; $3\overline{)141}$ = 47

Problem-Solving Exercise 9

pages 37–38

1. **96**

6 (×2) 12 (×2) 24 (×2) 48 (×2) 96

2. **96**

81 (+3) 84 (+3) 87 (+3) 90 (+3) 93 (+3) 96

3. **20**

32 (−3) 29 (−3) 26 (−3) 23 (−3) 20

4. **37**

13 (+4) 17 (+4) 21 (+4) 25 (+4) 29 (+4) 33 (+4) 37

5. **25**

10 (−5) 5 (+10) 15 (−5) 10 (+10) 20 (−5) 15 (+10) 25

6. **9**

100 (−19) 81 (−17) 64 (−15) 49 (−13) 36 (−11) 25 (−9) 16 (−7) 9

7. 65

5   9   17   33   65
+4   +8   +16   +32

8. 30

2   3   5   8   12   17   23   30
+1   +2   +3   +4   +5   +6   +7

9. 10

320   160   80   40   20   10
÷2   ÷2   ÷2   ÷2   ÷2

10. 49

4   8   7   14   13   26   25   50   49
×2   −1   ×2   −1   ×2   −1   ×2   −1

11. 60

72   68   64   60
−4   −4   −4

12. $0

March	April	May	June	July	August
$30,000	$24,000	$18,000	$12,000	$6,000	0
	−$6,000	−$6,000	−$6,000	−$6,000	−$6,000

13. 48,000

1,500   3,000   6,000   12,000   24,000   48,000
×2   ×2   ×2   ×2   ×2

Problem-Solving Exercise 10
pages 40–41

1. (4) 30 + 20
2. (2) 5(3 × 4)
3. (2) 9 + (7 + 4)
4. (5) (16 × 14) + (16 × 27)
5. (4) 3(15 − 6)

Problem-Solving Exercise 11
page 42

1. 0
2. 16
3. 4
4. 13
5. 4

Set-up Questions
page 43

1. (3) 5(45 + 30)
2. (2) 4(265 + 304)

Problem-Solving Exercise 12
pages 43–44

1. (1) 200(5 + 10)
2. (3) 9(3 + 4)
3. (4) (4 × 55) + (3 × 40)
4. (1) 12(50 − 10)
5. (2) (3 × 16) + (2 × 24)
6. (1) ($3.40 + $8.00 + $6.50 + $6.50)/4

Problem-Solving Exercise 13
pages 44–46

1. (4) Find the amount each program gets from the grant.

2. (4) 357
$$420 - 63 = 357$$

3. (3) 138,109
$$259{,}056 - 120{,}947 = 138{,}109$$

4. (3) Divide 81,480 by 84, then add 200.

5. (3) 36,000 6 is in the thousands place. The 4 indicates that you should leave 6 and put zeros to the right of 6.

6. (4) $25 10 × $1.29 = $12.90
$12.90 + 12.00 = $24.90 to the nearest dollar = $25

7. (5) $1040 $56 × 15: 280 + 56(0) = $840; $840 + 200 = $1040

8. (1) $14,200 7 × $1100 = $7700.
12 − 7 = 5 months left.
5 × $1300 = $6500.
$7700 + $6500 = $14,200

9. (5) 348 55 × 6 = 330
9 × 2 = 18
total = 348

10. (4) 34 27 + 36 + 28 + 39 + 40 = 170; 170 ÷ 5 = 34

11. (4) 1200 1024   1137   1263   1440
1137 + 1263 = 2400; 2400 ÷ 2 = 1200

12. (2) 9 30 ÷ 6 + (3 + 1)
5 + 4
9

13. (1) 63

1   3   7   15   31   63
+2   +4   +8   +16   +32

14. (2) 4(9 + 15)

15. (1) 12(9 + 6)

CHAPTER 4

Decimals Exercise 1

page 51

1. \$4 20¢ 0
2. 5¢ 40¢ 1¢

Decimals Exercise 2

page 52

1. a. three **b.** one **c.** two
2. a. none **b.** two **c.** five
3. a. none **b.** one **c.** four
4. a. three **b.** one **c.** six
5. a. hundredths or pennies
b. thousandths
c. tenths or dimes
6. a. ten thousandths
b. hundredths or pennies
c. hundred thousandths
7. a. hundredths or pennies
b. tens
c. hundreds
8. a. millionths
b. tenths or dimes
c. thousandths

Decimals Exercise 3

pages 52–53

1. .031
2. 2.12
3. .0780
4. 100.03
5. 71.5
6. 4.007
7. 3.0049
8. 13.13
9. 403.1
10. 10.20

Decimals Exercise 4

page 53

1. a. .067 or 0.067 (The zero in the units' place is often written to show that there is no whole number.)
b. 3.405 **c.** 8.0906
2. a. 80.025 **b.** 124.009 **c.** 7.5
3. a. 5 (Notice that you do not need to write the decimal point with a whole number.)
b. .3708 or 0.3708 **c.** 29.3
4. a. 6.3 **b.** .0023 or 0.0023 **c.** 60.0502

Decimals Exercise 5

page 54

1. .4 or 0.4
2. 6.3
3. .18 or 0.18
4. .009 or 0.009
5. .0002 or 0.0002
6. .013 or 0.013
7. 96.4
8. 5.016
9. 7500.8
10. .125 or 0.125
11. 84.009
12. 5602.028
13. .0312 or 0.0312
14. 204.03
15. 70.345

Decimals Exercise 6

page 55

1. a. 0.7 **b.** 0.62 **c.** 0.403
2. a. 0.0029 **b.** 0.101 **c.** 0.9
3. a. 0.8 **b.** 5.23 **c.** 0.4
4. a. 0.31 **b.** 1.68 **c.** 3.04
5. 0.062, 0.26, 0.602, 0.62
6. 0.0034, 0.34, 0.403, 0.43

Comparing and Ordering

page 57

Step 1. Give all lengths the same number of decimal places:
A—0.800
B—0.950
C—0.850
D—0.090
E—0.085

Step 2. Order the lengths from smallest to largest:
E—0.085
D—0.090
A—0.800
C—0.850
B—0.950

Step 3. Find this order in the answer choices:
(4) E, D, A, C, B

Decimals Exercise 7

pages 59–60

1. 1.5
$$\begin{array}{r} .6 \\ +\ .9 \\ \hline 1.5 \end{array}$$

2. 1.37
$$\begin{array}{r} .57 \\ +\ .8 \\ \hline 1.37 \end{array}$$

3. 195.909
$$\begin{array}{r} 16. \\ 9.24 \\ 170.3 \\ +\ .369 \\ \hline 195.909 \end{array}$$

4. 30.446
$$\begin{array}{r} 15.23 \\ 4. \\ 1.816 \\ +\ 9.4 \\ \hline 30.446 \end{array}$$

5. .54
$$\begin{array}{r} .80 \\ -\ .26 \\ \hline .54 \end{array}$$

6. .155
$$\begin{array}{r} .500 \\ -\ .345 \\ \hline .155 \end{array}$$

7. 17.68
$$\begin{array}{r} 18.00 \\ -\ .32 \\ \hline 17.68 \end{array}$$

8. .015
$$\begin{array}{r} .090 \\ -\ .075 \\ \hline .015 \end{array}$$

9. 233.316 12.3
.016
5.
+ 216.
233.316

10. 478.19 283.4
87.49
+ 107.3
478.19

11. 29.31 4.036
23.
2.19
+ .084
29.310 = 29.31

12. .206 .300
− .094
.206

13. 119.27 83.
+ 36.27
119.27

14. .0055 .0080
− .0025
.0055

15. .275 .600
− .325
.275

16. $16.32 $30.00
− 13.68
$16.32

17. $105.44 $84.
19.65
.23
+ 1.56
$105.44

18. .05084 .0075
.00128
.004
+ .03806
.05084

19. 15.37 20.00
− 4.63
15.37

20. .02075 .03000
− .00925
.02075

Decimals Exercise 8
pages 60–61

1. 0.22 meters 1.70 meters
− 1.48
0.22 meters

2. 11.75 pounds 16.25 pounds
− 4.50
11.75 pounds

3. 247.3 miles 85.4 miles
98.2
+ 63.7
247.3 miles

4. .301 .206
+ .095
.301

5. $.75 million
$1.35
.85
+ 1.05
$3.25 million

$4.00 million
− 3.25
$.75 million

6. 6.95 pounds
26.9
3.75
+ 12.4
43.05 pounds

50.00
− 43.05
6.95 pounds

7. .65 feet 10.65 feet
+ 0.2
10.85 feet

11.50
− 10.85
.65 feet

8. 1.1 million 239.0 million
− 226.8
12.2 million

12.2 million
− 11.1
1.1 million

9. (3) 5 − (2.45 + 0.95)

10. (4) Multiply 14 by .44, multiply 8 by 1.09, and add the two products.

Decimals Exercise 9
page 63

1. .56 .8 one place
× .7 one place
.56 two places

2. 6.18 20.6 one place
× 0.3 one place
6.18 two places

3. .836 2.09 two places
× .4 one place
.836 three places

4. 22.41 8.3 one place
× 2.7 one place
5 81
16 6
22.41 two places

5. .2268 0.378 three places
× 0.6 one place
.2268 four places

6. .0120 .0004 four places
× 30 no places
.0120 four places = .012

7. .008 0.8 one place
× 0.01 two places
.008 three places

8. 1.1512 1.439 three places
× 0.8 one place
1.1512 four places

9. **.114**
.076 three places
× 1.5 one place
380
76
.1140 four places = .114

10. **$104.16**
$24.80 two places
× 4.2 one place
4960
9920
$104.160 three places = $104.16

11. **$575**
$250 no places
× 2.3 one place
75 0
500
$575.0 one place = $575

12. **$1.45**
$1.16 two places
× 1.25 two places
580
232
1 16
$1.4500 four places = $1.45

13. 1.432 × 10 =
1.4.32 = **14.32**

14. 0.056 × 100 =
0.05.6 = **5.6**

15. 0.083 × 10 =
0.0.83 = **.83**

16. 19.7 × 1000 =
19.700. = **19,700**

17. **4.5**
15 no places
× 0.3 one place
4.5 one place

18. **357.5**
143 no places
× 2.5 one place
715
286
357.5 one place

19. **4.2874**
12.61 two places
× 0.34 two places
5044
3 783
4.2874 four places

20. **7950** 7.95 × 1000 =
7.950. = 7950

Decimals Exercise 10
page 65

1. **6.576**
8) 52.608

2. **84.3**
9) 758.7

3. **.75**
14) 10.50
9 8
70
70

4. **.013**
3) .039

5. **.095**
25) 2.375
2 25
125
125

6. **.0024**
7) .0168

7. **.4**
24) 9.6
9 6

8. **.04**
24) .96
96

9. **.15**
18) 2.70
1 8
90
90

10. **3.06**
12) 36.72

11. 42.3 ÷ 100 = .42.3 = **.423**

12. .04 ÷ 10 = .0.04 = **.004**

13. 19.5 ÷ 1000 = .019.5 = **.0195**

14. .65 ÷ 100 = .00.65 = **.0065**

Decimals Exercise 11
page 66

1. **42**
0.2.) 8.4.

2. **1 0.5**
.5.) 5.2.5

3. **2.6**
.13.) .33.8
26
7 8
7 8

4. **15**
.028.) .420.
28
140
140

5. **.27**
1.2.) .3.24
2 4
84
84

6. **20 90**
.04.) 83.60.

7. **4**
25.3.) 101.2.
101 2

8. 30 20
.03.)90.60.

9. 50
.134.)6.700.
6 70
00

10. 50
1.28.)64.00
64 0
00

11. 8 00
.12.)96.00
96
0 00

12. 60
4.35.)261.00
261 0
00

Decimal Exercise 12
pages 67–68

1. **$1** $.025
× 40
$1.000 = $1

2. **0.875 meter** 0.875 meter
4)3.500

3. **$6.40** $6.40
37.5.)$240.0.00

4. **(4)** Add 5.4 and 7.1, then divide by 2.

5. **$133.20** 18.5
× $7.20
3 700
129 5
$133.200 = $133.20

6. **$12.60** 12 − 3 = 9; $1.05 × 9 = $9.45; $ 9.45 + 3.15 = $12.60

7. **$55.90** $5.20 × 7 = $36.40; $7.80 × 2.5 = 3 900 + 15 60 = $19.500; $36.40 + 19.50 = $55.90

8. **222 miles** 2.5 × 60 = 150.0; 1.5 × 48 = 12 0 + 60 = 72.0; 150 + 72 = 222

9. **1.1 meters** 1.45 × 2 = 2.90; 4.0 − 2.9 = 1.1

10. **(3)** (1.3 + 3.5) ÷ 4

Decimals Exercise 13
pages 68–69

1. $.33
2. 6.5
3. 20.1
4. 8.038
5. .0083
6. **$15.58** $1.298
× 12
2 596
12 98
$15.576
to the nearest penny = $15.58

7. **160 miles** 4 5
× 3.5
22 5
135
157.5
to the nearest ten miles = 160 miles

8. **$294** $7.85
× 37.5
3 925
54 95
235 5
$294.375
to the nearest dollar = $294

9. **30 cm** 2.54
× 12
5 08
25 4
30.48
to the nearest centimeter = 30

10. **$8.38** $3.49
× 2.4
1 396
6 98
$8.376
to the nearest penny = $8.38

Decimals Exercise 14
page 69

1. **.03** .026
15).400
30
100
90
10
to the nearest hundredth = .03

2. **.01** .006
3).020
to the nearest hundredth = .01

3. **.1** .08
6).50
to the nearest tenth = .1

4. **.4** .37
.8.)3.00
to the nearest tenth = .4

Introducing Item Sets
pages 71–72

1. $180.24
Need: Price at store X of $1,199.75
Cash rebate at store X of $100
Price at store Y of $1,279.99
Solution:
```
$1,199.75      $1,279.99
−   100.     − 1,099.75
$1,099.75      $  180.24
```

2. $1788
Need: Down payment at store X of $150
Monthly payments at store X of $45.50
Number of months in three years, 36
Solution:
```
   $45.50    $1,638.00
  ×    36    + 150.00
   273 00    $1,788.00
  1 365
$1,638.00
```

3. $1510.40
Need: Down payment at store Y of $200
Monthly payments at store Y of $54.60
Number of months in two years, 24
Solution:
```
   $54.60    $1,310.40
  ×    24    + 200.00
   218 40    $1,510.40
 1 092 0
$1,310.40
```

4. $277.60
Need: Total cost at store X from problem 2, $1,788.00
Total cost at store Y from problem 3, $1,510.40
Solution:
```
  $1,788.00
− $1,510.40
  $  277.60
```

5. $688.25
Need: Price of furniture at store X of $1199.75
Cash rebate at store X of $100
Total installment price at store X from problem 2, $1788
Solution:
```
$1199.75      $1788
−   100     − 1099.75
$1099.75      $ 688.25
```

6. $340
Need: Assistant's wage of $8.50 and 40 hours
Solution:
```
   $8.50
 ×    40
 $340.00
```

7. $4160
Need: Assistant's total wages from problem 6, $340
Materials and expenses of $10,500
Price for job of $15,000
Solution:
```
$10,500      $15,000
+   340     − 10,840
$10,840      $ 4,160
```

8. $104
Need: Profit from problem 7, $4,160
Total number of hours, 40
Solution:
```
     $104
40)4,160
   4 0
     160
     160
       0
```

9. $3997.50
Need: Additional helper's wage of $6.50 an hour
Additional helper's total hours, 25
Profit from problem 7, $4160
Solution:
```
  $6.50     $4160.00
×    25    −  162.50
  32 50     $3997.50
 130 0
$162.50
```

Decimals Exercise 15
pages 72–75

1. hundredths
2. 70.016
3. 60.043
4. 408.0015
5. .06, .066, .6, .606
6. **42.995**
```
 36
  2.93
   .065
+ 4
 42.995
```
7. **.0737**
```
  .0800
− .0063
  .0737
```
8. **.016**
```
   3.2
× .005
 .0160 = .016
```
9. **.027**
```
    .027
15).405
    30
    105
    105
```
10. **200**
```
         200
.045.)9.000.
```
11. **(1) 10.18 million**
```
  6.79 million
+ 3.39
 10.18 million
```
12. **(4) 1.25 yards**
```
  3.00
− 1.75
  1.25
```
13. **(2) 18.6 miles**
```
     18.58   to the nearest
12)223.00    tenth = 18.6
   12
   103
    96
     7 0
     6 0
     1 00
       96
        4
```

14. (1) $1.77

$.034
× 52
68
1 70
$1.768 to the nearest penny = $1.77

15. (2) $25.78

100.0 − 56.3 = 43.7

43.7
× $.59
3 933
21 85
$25.783 to the nearest penny = $25.78

16. (5) .293

75)22.0000 = .2933 to the nearest thousandth = .293
15 0
7 00
6 75
250
225
250
225

17. (4) D, B, E, C, A

D = 0.125
B = 0.750
E = 1.125
C = 1.750
A = 1.800

18. (3) $60

$7.80 × 5 = $39.00

$8.40
× 2.5
4 200
16 80
$21.000

$39 + 21 = $60

19. (3) Multiply 3.6 by 3, then subtract 3.6.

20. (4) 5(125) + 3(190)

21. (5) 32.5 hours

13 + 4 + 15.5 = 32.5

22. (4) $96.60

$1.38 × 70 = $96.60

23. (4) 1.5 hours

17.0 − 15.5 = 1.5

24. (2) 48 mph

13)622.0 = 47.8 to the nearest unit = 48
52
102
91
11 0
10 4
6

CHAPTER 5

Fractions Exercise 1

page 77

1. $\frac{1}{3}$
2. $\frac{5}{6}$
3. $\frac{1}{2}$
4. $\frac{3}{8}$
5. $\frac{2}{5}$

Fractions Exercise 2

page 78

1. $2\frac{1}{3}$
2. $1\frac{1}{4}$
3. $1\frac{1}{4}$
4. $1\frac{5}{8}$
5. $5\frac{1}{6}$
6. $3\frac{1}{3}$

Fractions Exercise 3

page 79

1. $\frac{5}{3}$
2. $\frac{7}{4}$
3. $\frac{3}{2}$
4. $\frac{6}{6}$
5. $\frac{9}{4}$
6. $\frac{11}{8}$

Fractions Exercise 4

page 80

1. $\frac{1}{3}$
2. $\frac{5}{12}$
3. $\frac{199}{800}$
4. $\frac{7}{50}$
5. $\frac{137}{200}$
6. $\frac{11}{24}$
7. $\frac{2}{7}$

Fractions Exercise 5

page 81

In these answers, the number following the phrase *reduce by* tells what number to divide the numerator and denominator by in order to reduce to lowest terms in one step.

1. $\frac{6}{10} = \frac{3}{5}$ reduce by 2; $\frac{4}{6} = \frac{2}{3}$ reduce by 2; $\frac{14}{20} = \frac{7}{10}$ reduce by 2

$\frac{4}{32} = \frac{1}{8}$ reduce by 4; $\frac{18}{36} = \frac{1}{2}$ reduce by 18; $\frac{18}{24} = \frac{3}{4}$ reduce by 6

2. $\frac{25}{35} = \frac{5}{7}$ reduce by 5; $\frac{45}{50} = \frac{9}{10}$ reduce by 5; $\frac{15}{40} = \frac{3}{8}$ reduce by 5

$\frac{45}{60} = \frac{3}{4}$ reduce by 15; $\frac{30}{45} = \frac{2}{3}$ reduce by 15; $\frac{10}{25} = \frac{2}{5}$ reduce by 5

Fractions Exercise 6
page 82

$\frac{70}{100} = \frac{7}{10}$ reduce by 10 $\quad \frac{50}{60} = \frac{5}{6}$ reduce by 10 $\quad \frac{80}{160} = \frac{1}{2}$ reduce by 80

$\frac{40}{200} = \frac{1}{5}$ reduce by 40 $\quad \frac{30}{200} = \frac{3}{20}$ reduce by 10 $\quad \frac{50}{250} = \frac{1}{5}$ reduce by 50

Fractions Exercise 7
page 82

$\frac{7}{28} = \frac{1}{4}$ reduce by 7 $\quad \frac{9}{30} = \frac{3}{10}$ reduce by 3 $\quad \frac{27}{45} = \frac{3}{5}$ reduce by 9

$\frac{22}{33} = \frac{2}{3}$ reduce by 11 $\quad \frac{17}{34} = \frac{1}{2}$ reduce by 17 $\quad \frac{13}{39} = \frac{1}{3}$ reduce by 13

Fractions Exercise 8
pages 83–84

1. a. $\frac{6}{15} = \frac{2}{5}$ b. $\frac{9}{15} = \frac{3}{5}$ c. $\frac{3}{15} = \frac{1}{5}$
2. a. $\frac{18}{24} = \frac{3}{4}$ b. $\frac{6}{24} = \frac{1}{4}$
3. a. \$60 + \$30 = \$90 b. $\frac{30}{90} = \frac{1}{3}$ c. $\frac{60}{90} = \frac{2}{3}$
4. a. 175 + 125 = 300 mi. b. $\frac{175}{300} = \frac{7}{12}$ c. $\frac{125}{300} = \frac{5}{12}$
5. a. $\frac{300}{1200} = \frac{1}{4}$ b. $\frac{450}{1200} = \frac{3}{8}$ c. $\frac{150}{1200} = \frac{1}{8}$ d. $\frac{120}{1200} = \frac{1}{10}$

Fractions Exercise 9
page 85

1. $\frac{10}{4} = 4\overline{)10} = 2\frac{2}{4} = \mathbf{2\frac{1}{2}}$; $10 - 8 = 2$
2. $\frac{8}{6} = 6\overline{)8} = 1\frac{2}{6} = \mathbf{1\frac{1}{3}}$; $8 - 6 = 2$
3. $\frac{28}{8} = 8\overline{)28} = 3\frac{4}{8} = \mathbf{3\frac{1}{2}}$; $28 - 24 = 4$
4. $\frac{10}{3} = 3\overline{)10} = \mathbf{3\frac{1}{3}}$; $10 - 9 = 1$
5. $\mathbf{1\frac{2}{3}}$ $\quad 9\overline{)15}$: $15 - 9 = 6$, $1\frac{6}{9} = 1\frac{2}{3}$
   $\mathbf{4\frac{1}{2}}$ $\quad 8\overline{)36}$: $36 - 32 = 4$, $4\frac{4}{8} = 4\frac{1}{2}$
   $\mathbf{2\frac{3}{5}}$ $\quad 5\overline{)13}$: $13 - 10 = 3$, $2\frac{3}{5}$
   $\mathbf{5\frac{1}{6}}$ $\quad 6\overline{)31}$: $31 - 30 = 1$, $5\frac{1}{6}$
6. $\mathbf{3}$ $\quad 2\overline{)6}$: $6 - 6 = 0$, 3
   $\mathbf{1\frac{3}{4}}$ $\quad 16\overline{)28}$: $28 - 16 = 12$, $1\frac{12}{16} = 1\frac{3}{4}$
   $\mathbf{1}$ $\quad 15\overline{)15}$: $15 - 15 = 0$, 1
   $\mathbf{1\frac{1}{3}}$ $\quad 18\overline{)24}$: $24 - 18 = 6$, $1\frac{6}{18} = 1\frac{1}{3}$
7. $\mathbf{1\frac{3}{4}}$ $\quad 4\overline{)7}$: $7 - 4 = 3$, $1\frac{3}{4}$
   $\mathbf{7}$ $\quad 3\overline{)21}$: $21 - 21 = 0$, 7
   $\mathbf{4\frac{1}{6}}$ $\quad 12\overline{)50}$: $50 - 48 = 2$, $4\frac{2}{12} = 4\frac{1}{6}$
   $\mathbf{3\frac{1}{2}}$ $\quad 6\overline{)21}$: $21 - 18 = 3$, $3\frac{3}{6} = 3\frac{1}{2}$

Fractions Exercise 10
pages 86–87

1. $\mathbf{1}$: $\frac{2}{5} + \frac{3}{5} = \frac{5}{5} = 1$
   $\mathbf{\frac{2}{3}}$: $\frac{2}{9} + \frac{4}{9} = \frac{6}{9} = \frac{2}{3}$
   $\mathbf{1\frac{4}{15}}$: $\frac{8}{15} + \frac{11}{15} = \frac{19}{15} = 1\frac{4}{15}$
2. $\mathbf{1\frac{1}{4}}$: $\frac{5}{8} + \frac{5}{8} = \frac{10}{8} = 1\frac{2}{8} = 1\frac{1}{4}$
   $\mathbf{13\frac{2}{5}}$: $9\frac{3}{10} + 4\frac{1}{10} = 13\frac{4}{10} = 13\frac{2}{5}$
   $\mathbf{13}$: $10\frac{2}{3} + 2\frac{1}{3} = 12\frac{3}{3} = 13$
3. $\mathbf{14\frac{4}{9}}$: $6\frac{5}{9} + 7\frac{8}{9} = 13\frac{13}{9} = 14\frac{4}{9}$
   $\mathbf{22\frac{1}{4}}$: $1\frac{7}{8} + 20\frac{3}{8} = 21\frac{10}{8} = 22\frac{2}{8} = 22\frac{1}{4}$
   $\mathbf{\frac{13}{16}}$: $\frac{1}{16} + \frac{5}{16} + \frac{7}{16} = \frac{13}{16}$
4. $\mathbf{1\frac{3}{4}}$: $\frac{3}{4} + \frac{1}{4} + \frac{3}{4} = \frac{7}{4} = 1\frac{3}{4}$
   $\mathbf{16\frac{11}{12}}$: $2\frac{7}{12} + 4\frac{11}{12} + 9\frac{5}{12} = 15\frac{23}{12} = 16\frac{11}{12}$
   $\mathbf{13\frac{8}{9}}$: $5\frac{4}{9} + 4\frac{5}{9} + 3\frac{8}{9} = 12\frac{17}{9} = 13\frac{8}{9}$

Fractions Exercise 11
page 87

1. $\frac{1}{2}$ $\quad 6\frac{2}{3}$ $\quad 2\frac{1}{2}$

$$\frac{5}{8} - \frac{1}{8} = \frac{4}{8} = \frac{1}{2} \qquad 8\frac{5}{6} - 2\frac{1}{6} = 6\frac{4}{6} = 6\frac{2}{3} \qquad 4\frac{7}{8} - 2\frac{3}{8} = 2\frac{4}{8} = 2\frac{1}{2}$$

2. $2\frac{3}{5}$ $\quad \frac{3}{10}$ $\quad 4\frac{3}{5}$

$$10\frac{4}{5} - 8\frac{1}{5} = 2\frac{3}{5} \qquad \frac{19}{20} - \frac{13}{20} = \frac{6}{20} = \frac{3}{10} \qquad 6\frac{13}{15} - 2\frac{4}{15} = 4\frac{9}{15} = 4\frac{3}{5}$$

3. $\frac{2}{9}$ $\quad 7\frac{2}{3}$ $\quad \frac{2}{3}$

$$\frac{25}{36} - \frac{17}{36} = \frac{8}{36} = \frac{2}{9} \qquad 8\frac{5}{6} - 1\frac{1}{6} = 7\frac{4}{6} = 7\frac{2}{3} \qquad 4\frac{17}{18} - 4\frac{5}{18} = \frac{12}{18} = \frac{2}{3}$$

Fractions Exercise 12
page 89

1. $7\frac{1}{3}$ $\quad 7\frac{3}{8}$ $\quad 2\frac{5}{12}$

$$9 = 8\frac{3}{3};\ -1\frac{2}{3} = 1\frac{2}{3};\ 7\frac{1}{3} \qquad 11 = 10\frac{8}{8};\ -3\frac{5}{8} = 3\frac{5}{8};\ 7\frac{3}{8} \qquad 8 = 7\frac{12}{12};\ -5\frac{7}{12} = 5\frac{7}{12};\ 2\frac{5}{12}$$

2. $\frac{1}{2}$ $\quad 7\frac{3}{4}$ $\quad 9\frac{11}{16}$

$$7 = 6\frac{2}{2};\ -6\frac{1}{2} = 6\frac{1}{2};\ \frac{1}{2} \qquad 9 = 8\frac{4}{4};\ -1\frac{1}{4} = 1\frac{1}{4};\ 7\frac{3}{4} \qquad 12 = 11\frac{16}{16};\ -2\frac{5}{16} = 2\frac{5}{16};\ 9\frac{11}{16}$$

3. $2\frac{1}{2}$ $\quad 5\frac{1}{8} = 4\frac{1}{8} + \frac{8}{8} = 4\frac{9}{8}$; $-2\frac{5}{8} = 2\frac{5}{8}$; $2\frac{4}{8} = 2\frac{1}{2}$

$6\frac{3}{5}$ $\quad 8\frac{2}{5} = 7\frac{2}{5} + \frac{5}{5} = 7\frac{7}{5}$; $-1\frac{4}{5} = 1\frac{4}{5}$; $6\frac{3}{5}$

$\frac{1}{2}$ $\quad 3\frac{1}{4} = 2\frac{1}{4} + \frac{4}{4} = 2\frac{5}{4}$; $-2\frac{3}{4} = 2\frac{3}{4}$; $\frac{2}{4} = \frac{1}{2}$

4. $7\frac{3}{4}$ $\quad 24\frac{11}{16} = 23\frac{11}{16} + \frac{16}{16} = 23\frac{27}{16}$; $-16\frac{15}{16} = 16\frac{15}{16}$; $7\frac{12}{16} = 7\frac{3}{4}$

$1\frac{2}{3}$ $\quad 9\frac{1}{3} = 8\frac{1}{3} + \frac{3}{3} = 8\frac{4}{3}$; $-7\frac{2}{3} = 7\frac{2}{3}$; $1\frac{2}{3}$

$7\frac{2}{5}$ $\quad 12\frac{3}{10} = 11\frac{3}{10} + \frac{10}{10} = 11\frac{13}{10}$; $-4\frac{9}{10} = 4\frac{9}{10}$; $7\frac{4}{10} = 7\frac{2}{5}$

5. $2\frac{2}{3}$ $\quad 11\frac{7}{12} = 10\frac{7}{12} + \frac{12}{12} = 10\frac{19}{12}$; $-8\frac{11}{12} = 8\frac{11}{12}$; $2\frac{8}{12} = 2\frac{2}{3}$

$1\frac{1}{2}$ $\quad 15\frac{7}{24} = 14\frac{7}{24} + \frac{24}{24} = 14\frac{31}{24}$; $-13\frac{19}{24} = 13\frac{19}{24}$; $1\frac{12}{24} = 1\frac{1}{2}$

$12\frac{4}{9}$ $\quad 20\frac{5}{36} = 19\frac{5}{36} + \frac{36}{36} = 19\frac{41}{36}$; $-7\frac{25}{36} = 7\frac{25}{36}$; $12\frac{16}{36} = 12\frac{4}{9}$

Fractions Exercise 13
page 90

1. 10 12 36
2. 6 18 40
3. 18 24 20
4. 20 12 24
5. 18 40 36

Fractions Exercise 14
page 92

In these answers, the number following the phrase *raise by* tells you what number to multiply both the numerator and the denominator of the original fraction by. The letters *xp* stand for cross product.

1. $\frac{3}{5} = \frac{12}{20}$ raise by 4, $xp = 60$ $\quad \frac{5}{12} = \frac{10}{24}$ raise by 2, $xp = 120$ $\quad \frac{7}{8} = \frac{35}{40}$ raise by 5, $xp = 280$ $\quad \frac{3}{4} = \frac{12}{16}$ raise by 4, $xp = 48$

2. $\frac{5}{11} = \frac{20}{44}$ raise by 4, $xp = 220$ $\quad \frac{13}{20} = \frac{26}{40}$ raise by 2, $xp = 520$ $\quad \frac{7}{10} = \frac{42}{60}$ raise by 6, $xp = 420$ $\quad \frac{13}{20} = \frac{65}{100}$ raise by 5, $xp = 1300$

3. $\frac{3}{4} = \frac{75}{100}$ raise by 25, $xp = 300$ $\quad \frac{5}{9} = \frac{20}{36}$ raise by 4, $xp = 180$ $\quad \frac{1}{8} = \frac{3}{24}$ raise by 3, $xp = 24$ $\quad \frac{1}{3} = \frac{12}{36}$ raise by 12, $xp = 36$

Fractions Exercise 15

page 94

1. $1\frac{5}{12}$

$\frac{5}{6} = \frac{10}{12}$

$+\frac{7}{12} = \frac{7}{12}$

$\frac{17}{12} = 1\frac{5}{12}$

$1\frac{5}{12}$

$\frac{2}{3} = \frac{8}{12}$

$+\frac{3}{4} = \frac{9}{12}$

$\frac{17}{12} = 1\frac{5}{12}$

$2\frac{1}{2}$

$\frac{2}{3} = \frac{8}{12}$

$\frac{7}{12} = \frac{7}{12}$

$+\frac{5}{6} = \frac{10}{12}$

$\frac{25}{12} = 2\frac{1}{2}$

2. $14\frac{1}{6}$

$9\frac{2}{3} = 9\frac{4}{6}$

$+4\frac{1}{2} = 4\frac{3}{6}$

$13\frac{7}{6} = 14\frac{1}{6}$

$13\frac{11}{24}$

$10\frac{1}{8} = 10\frac{3}{24}$

$+3\frac{1}{3} = 3\frac{8}{24}$

$13\frac{11}{24}$

$25\frac{11}{20}$

$9\frac{1}{2} = 9\frac{10}{20}$

$8\frac{3}{4} = 8\frac{15}{20}$

$+7\frac{3}{10} = 7\frac{6}{20}$

$24\frac{31}{20} = 25\frac{11}{20}$

3. $7\frac{1}{3}$

$2\frac{5}{6} = 2\frac{5}{6}$

$+4\frac{1}{2} = 4\frac{3}{6}$

$6\frac{8}{6} = 7\frac{2}{6} = 7\frac{1}{3}$

$1\frac{7}{20}$

$\frac{1}{2} = \frac{10}{20}$

$\frac{1}{4} = \frac{5}{20}$

$+\frac{3}{5} = \frac{12}{20}$

$\frac{27}{20} = 1\frac{7}{20}$

$12\frac{19}{24}$

$4\frac{1}{3} = 4\frac{8}{24}$

$1\frac{5}{6} = 1\frac{20}{24}$

$+6\frac{5}{8} = 6\frac{15}{24}$

$11\frac{43}{24} = 12\frac{19}{24}$

4. $\frac{1}{5}$

$\frac{7}{10} = \frac{7}{10}$

$-\frac{1}{2} = \frac{5}{10}$

$\frac{2}{10} = \frac{1}{5}$

$\frac{11}{12}$

$2\frac{2}{3} = 1\frac{8}{12} + \frac{12}{12} = 1\frac{20}{12}$

$-1\frac{3}{4} = 1\frac{9}{12}$

$\frac{11}{12}$

$\frac{1}{40}$

$\frac{5}{8} = \frac{25}{40}$

$-\frac{3}{5} = \frac{24}{40}$

$\frac{1}{40}$

5. $4\frac{9}{20}$

$8\frac{1}{5} = 7\frac{4}{20} + \frac{20}{20} = 7\frac{24}{20}$

$-3\frac{3}{4} = 3\frac{15}{20}$

$4\frac{9}{20}$

$2\frac{1}{10}$

$11\frac{3}{5} = 11\frac{6}{10}$

$-9\frac{1}{2} = 9\frac{5}{10}$

$2\frac{1}{10}$

$3\frac{5}{6}$

$9\frac{1}{2} = 8\frac{3}{6} + \frac{6}{6} = 8\frac{9}{6}$

$-5\frac{2}{3} = 5\frac{4}{6}$

$3\frac{5}{6}$

6. $8\frac{5}{8}$

$16\frac{1}{4} = 15\frac{2}{8} + \frac{8}{8} = 15\frac{10}{8}$

$-7\frac{5}{8} = 7\frac{5}{8}$

$8\frac{5}{8}$

$4\frac{7}{24}$

$7\frac{11}{12} = 7\frac{22}{24}$

$-3\frac{5}{8} = 3\frac{15}{24}$

$4\frac{7}{24}$

$5\frac{3}{5}$

$8\frac{7}{20} = 7\frac{7}{20} + \frac{20}{20} = 7\frac{27}{20}$

$-2\frac{3}{4} = 2\frac{15}{20}$

$5\frac{12}{20} = 5\frac{3}{5}$

Fractions Exercise 16
pages 94–95

1. $1\frac{1}{3}$ **yards**
$$\begin{array}{r} 5 = 4\frac{3}{3} \\ -\ 3\frac{2}{3} = 3\frac{2}{3} \\ \hline 1\frac{1}{3} \end{array}$$

2. a. $6\frac{7}{8}$ **lb**
$$\begin{array}{r} 2\frac{1}{2} = 2\frac{4}{8} \\ 2\frac{5}{8} = 2\frac{5}{8} \\ +\ 1\frac{3}{4} = 1\frac{6}{8} \\ \hline 5\frac{15}{8} = 6\frac{7}{8} \end{array}$$

b. $13\frac{1}{4}$
$$\begin{array}{r} 10 \\ +\ 3\frac{1}{4} \\ \hline 13\frac{1}{4} \end{array}$$

c. **Yes**
$$\begin{array}{r} 6\frac{7}{8} = 6\frac{7}{8} \\ +\ 13\frac{1}{4} = 13\frac{2}{8} \\ \hline 19\frac{9}{8} = 20\frac{1}{8} \end{array}$$

Yes, the weight is more than 20 pounds.

3. $\frac{13}{16}$ Use 1 as the Robinsons' entire income.
$$\begin{array}{r} 1 = \frac{16}{16} \\ -\ \frac{3}{16} = \frac{3}{16} \\ \hline \frac{13}{16} \end{array}$$

4. $6\frac{1}{10}$ **miles**
$$\begin{array}{r} 2\frac{9}{10} = 2\frac{9}{10} \\ +\ 3\frac{1}{2} = 3\frac{5}{10} \\ \hline 5\frac{14}{10} = 6\frac{4}{10} \end{array}$$

$$\begin{array}{r} 12\frac{1}{2} = 12\frac{5}{10} \\ -\ 6\frac{2}{5} = 6\frac{4}{10} \\ \hline 6\frac{1}{10}\text{ miles} \end{array}$$

5. a. $61\frac{13}{16}$ **lb**
$$\begin{array}{r} 16\frac{3}{4} = 16\frac{12}{16} \\ 20\frac{1}{2} = 20\frac{8}{16} \\ +\ 24\frac{9}{16} = 24\frac{9}{16} \\ \hline 60\frac{29}{16} = 61\frac{13}{16} \end{array}$$

b. $38\frac{3}{16}$ **lb**
$$\begin{array}{r} 100 = 99\frac{16}{16} \\ -\ 61\frac{13}{16} = 61\frac{13}{16} \\ \hline 38\frac{3}{16} \end{array}$$

6. **(5)** Add $162\frac{1}{2}$ and 16.

7. a. $21\frac{1}{8}$ $19\frac{1}{8}$ $+\frac{4}{8}$ $19\frac{5}{8}$ $+\frac{4}{8}$ $20\frac{1}{8}$ $+\frac{4}{8}$ $20\frac{5}{8}$ $+\frac{4}{8}$ ——
$$\begin{array}{r} 20\frac{5}{8} \\ +\ \frac{4}{8} \\ \hline 20\frac{9}{8} = 21\frac{1}{8} \end{array}$$

b. $18\frac{5}{8}$
$$\begin{array}{r} 19\frac{1}{8} = 18\frac{1}{8} + \frac{8}{8} = 18\frac{9}{8} \\ -\ \frac{4}{8} = \frac{4}{8} \\ \hline 18\frac{5}{8} \end{array}$$

8. a. $28\frac{9}{16}$ **in.**
$$\begin{array}{r} 8\frac{5}{16} = 8\frac{5}{16} \\ 10\frac{1}{2} = 10\frac{8}{16} \\ +\ 9\frac{3}{4} = 9\frac{12}{16} \\ \hline 27\frac{25}{16} = 28\frac{9}{16} \end{array}$$

b. $2\frac{9}{16}$ **in.**
$$\begin{array}{r} 28\frac{9}{16} \\ -\ 26 \\ \hline 2\frac{9}{16} \end{array}$$

Fractions Exercise 17
page 96

1. $\frac{3}{4} \times \frac{5}{7} = \mathbf{\frac{15}{28}}$
$\frac{2}{3} \times \frac{1}{2} = \frac{2}{6} = \mathbf{\frac{1}{3}}$
$\frac{1}{10} \times \frac{5}{8} = \frac{5}{80} = \mathbf{\frac{1}{16}}$
$\frac{7}{8} \times \frac{1}{5} = \mathbf{\frac{7}{40}}$

2. $\frac{1}{4} \times \frac{5}{16} = \mathbf{\frac{5}{64}}$
$\frac{3}{10} \times \frac{3}{5} = \mathbf{\frac{9}{50}}$
$\frac{2}{9} \times \frac{3}{5} = \frac{6}{45} = \mathbf{\frac{2}{15}}$
$\frac{3}{5} \times \frac{1}{6} = \frac{3}{30} = \mathbf{\frac{1}{10}}$

Fractions Exercise 18
page 97

1. $\frac{3}{4} \times \frac{6}{7} = \mathbf{\frac{9}{14}}$
$\frac{14}{15} \times \frac{3}{7} = \mathbf{\frac{2}{5}}$
$\frac{4}{5} \times \frac{5}{6} = \mathbf{\frac{2}{3}}$
$\frac{4}{9} \times \frac{3}{8} = \mathbf{\frac{1}{6}}$

2. $\frac{5}{8} \times \frac{2}{15} = \mathbf{\frac{1}{12}}$
$\frac{9}{10} \times \frac{2}{3} = \mathbf{\frac{3}{5}}$
$\frac{3}{20} \times \frac{1}{3} = \mathbf{\frac{1}{20}}$
$\frac{4}{5} \times \frac{5}{24} = \mathbf{\frac{1}{6}}$

3. $\frac{2}{3} \times \frac{9}{20} = \mathbf{\frac{3}{10}}$
$\frac{6}{7} \times \frac{7}{8} \times \frac{4}{5} = \mathbf{\frac{3}{5}}$
$\frac{9}{10} \times \frac{1}{4} \times \frac{8}{9} = \mathbf{\frac{1}{5}}$
$\frac{3}{4} \times \frac{2}{9} \times \frac{15}{16} = \mathbf{\frac{5}{32}}$

Fractions Exercise 19

page 98

1. $\frac{1}{2} \times \frac{16}{1} = \frac{8}{1} = \mathbf{8}$
   $\frac{10}{1} \times \frac{2}{3} = \frac{20}{3} = \mathbf{6\frac{2}{3}}$
   $\frac{3}{8} \times \frac{12}{1} = \frac{9}{2} = \mathbf{4\frac{1}{2}}$
   $\frac{8}{1} \times \frac{7}{10} = \frac{28}{5} = \mathbf{5\frac{3}{5}}$
2. $\frac{2}{5} \times \frac{10}{1} = \frac{4}{1} = \mathbf{4}$
   $\frac{9}{1} \times \frac{11}{20} = \frac{99}{20} = \mathbf{4\frac{19}{20}}$
   $\frac{5}{6} \times \frac{9}{1} = \frac{15}{2} = \mathbf{7\frac{1}{2}}$
   $\frac{15}{1} \times \frac{7}{100} = \frac{21}{20} = \mathbf{1\frac{1}{20}}$
3. $\frac{18}{1} \times \frac{2}{3} = \frac{12}{1} = \mathbf{12}$
   $\frac{20}{1} \times \frac{4}{5} = \frac{16}{1} = \mathbf{16}$
   $\frac{4}{15} \times \frac{36}{1} = \frac{48}{5} = \mathbf{9\frac{3}{5}}$
   $\frac{24}{1} \times \frac{3}{10} = \frac{36}{5} = \mathbf{7\frac{1}{5}}$

Fractions Exercise 20

page 99

1. $\frac{6}{3} + \frac{2}{3} = \mathbf{\frac{8}{3}}$
   $\frac{8}{8} + \frac{5}{8} = \mathbf{\frac{13}{8}}$
   $\frac{40}{5} + \frac{2}{5} = \mathbf{\frac{42}{5}}$
   $\frac{12}{4} + \frac{1}{4} = \mathbf{\frac{13}{4}}$
   $\frac{18}{6} + \frac{5}{6} = \mathbf{\frac{23}{6}}$
2. $\frac{35}{7} + \frac{4}{7} = \mathbf{\frac{39}{7}}$
   $\frac{6}{2} + \frac{1}{2} = \mathbf{\frac{7}{2}}$
   $\frac{21}{3} + \frac{1}{3} = \mathbf{\frac{22}{3}}$
   $\frac{54}{9} + \frac{2}{9} = \mathbf{\frac{56}{9}}$
   $\frac{30}{3} + \frac{1}{3} = \mathbf{\frac{31}{3}}$
3. $\frac{60}{5} + \frac{3}{5} = \mathbf{\frac{63}{5}}$
   $\frac{36}{4} + \frac{1}{4} = \mathbf{\frac{37}{4}}$
   $\frac{39}{3} + \frac{2}{3} = \mathbf{\frac{41}{3}}$
   $\frac{45}{3} + \frac{1}{3} = \mathbf{\frac{46}{3}}$
   $\frac{32}{8} + \frac{3}{8} = \mathbf{\frac{35}{8}}$

Fractions Exercise 21

page 100

1. $1\frac{2}{3} \times 3\frac{1}{2} =$
   $\frac{5}{3} \times \frac{7}{2} = \frac{35}{6} = \mathbf{5\frac{5}{6}}$
   $1\frac{1}{2} \times \frac{2}{3} =$
   $\frac{3}{2} \times \frac{2}{3} = \mathbf{1}$
   $6\frac{1}{2} \times 3\frac{1}{4} =$
   $\frac{13}{2} \times \frac{13}{4} = \frac{169}{8} = \mathbf{21\frac{1}{8}}$
2. $\frac{5}{8} \times 2\frac{2}{9} =$
   $\frac{5}{8} \times \frac{20}{9} = \frac{25}{18} = \mathbf{1\frac{7}{18}}$
   $\frac{2}{3} \times 3\frac{3}{4} =$
   $\frac{2}{3} \times \frac{15}{4} = \frac{5}{2} = \mathbf{2\frac{1}{2}}$
   $2\frac{5}{8} \times 1\frac{1}{3} =$
   $\frac{21}{8} \times \frac{4}{3} = \frac{7}{2} = \mathbf{3\frac{1}{2}}$
3. $1\frac{1}{2} \times \frac{4}{5} \times 2\frac{1}{4} =$
   $\frac{3}{2} \times \frac{4}{5} \times \frac{9}{4} = \frac{27}{10} = \mathbf{2\frac{7}{10}}$
   $\frac{5}{8} \times 4 \times 1\frac{2}{5} =$
   $\frac{5}{8} \times \frac{4}{1} \times \frac{7}{5} = \frac{7}{2} = \mathbf{3\frac{1}{2}}$
   $2\frac{1}{2} \times 1\frac{3}{5} \times 6 =$
   $\frac{5}{2} \times \frac{8}{5} \times \frac{6}{1} = \frac{24}{1} = \mathbf{24}$
4. $1\frac{7}{20} \times 5\frac{1}{3} \times 1\frac{1}{4} =$
   $\frac{27}{20} \times \frac{16}{3} \times \frac{5}{4} = \frac{9}{1} = \mathbf{9}$
   $1\frac{7}{8} \times \frac{3}{4} \times 2\frac{2}{9} =$
   $\frac{15}{8} \times \frac{3}{4} \times \frac{20}{9} = \frac{25}{8} = \mathbf{3\frac{1}{8}}$
   $\frac{3}{8} \times 2\frac{5}{6} \times 1\frac{1}{3} =$
   $\frac{3}{8} \times \frac{17}{6} \times \frac{4}{3} = \frac{17}{12} = \mathbf{1\frac{5}{12}}$

Fractions Exercise 22

page 102

1. $\frac{1}{3} \div \frac{1}{6} =$
   $\frac{1}{3} \times \frac{6}{1} = \frac{2}{1} = \mathbf{2}$
   $5 \div \frac{5}{6} =$
   $\frac{5}{1} \times \frac{6}{5} = \frac{30}{5} = \mathbf{6}$
   $4\frac{1}{2} \div \frac{3}{4} =$
   $\frac{9}{2} \times \frac{4}{3} = \frac{6}{1} = \mathbf{6}$
   $2\frac{2}{3} \div \frac{2}{15} =$
   $\frac{8}{3} \times \frac{15}{2} = \frac{20}{1} = \mathbf{20}$
2. $\frac{1}{3} \div \frac{2}{3} =$
   $\frac{1}{3} \times \frac{3}{2} = \mathbf{\frac{1}{2}}$
   $\frac{5}{7} \div \frac{5}{14} =$
   $\frac{5}{7} \times \frac{14}{5} = \mathbf{2}$
   $4 \div \frac{3}{8} =$
   $\frac{4}{1} \times \frac{8}{3} = \frac{32}{3} = \mathbf{10\frac{2}{3}}$
   $\frac{5}{9} \div \frac{3}{4} =$
   $\frac{5}{9} \times \frac{4}{3} = \mathbf{\frac{20}{27}}$
3. $5\frac{5}{6} \div \frac{7}{8} =$
   $\frac{35}{6} \times \frac{8}{7} = \frac{20}{3} = \mathbf{6\frac{2}{3}}$
   $\frac{9}{10} \div \frac{3}{5} =$
   $\frac{9}{10} \times \frac{5}{3} = \frac{3}{2} = \mathbf{1\frac{1}{2}}$
   $10 \div \frac{5}{6} =$
   $\frac{10}{1} \times \frac{6}{5} = \mathbf{12}$
   $3\frac{1}{3} \div \frac{1}{3} =$
   $\frac{10}{3} \times \frac{3}{1} = \mathbf{10}$

Fractions Exercise 23

page 103

1. $10 \div 1\frac{1}{2} =$
   $\frac{10}{1} \div \frac{3}{2} =$
   $\frac{10}{1} \times \frac{2}{3} = \frac{20}{3} = \mathbf{6\frac{2}{3}}$
   $1\frac{1}{3} \div 3\frac{1}{5} =$
   $\frac{4}{3} \div \frac{16}{5} =$
   $\frac{4}{3} \times \frac{5}{16} = \mathbf{\frac{5}{12}}$
   $6 \div 1\frac{1}{3} =$
   $\frac{6}{1} \div \frac{4}{3} =$
   $\frac{6}{1} \times \frac{3}{4} = \frac{9}{2} = \mathbf{4\frac{1}{2}}$

2. $2\frac{1}{2} \div 3\frac{1}{4} =$

$\frac{5}{2} \div \frac{13}{4} =$

$\frac{5}{2} \times \frac{4}{13} = \mathbf{\frac{10}{13}}$

$21 \div 4\frac{1}{5} =$

$\frac{21}{1} \div \frac{21}{5} =$

$\frac{21}{1} \times \frac{5}{21} = \frac{5}{1} = \mathbf{5}$

$2\frac{2}{9} \div 2 =$

$\frac{20}{9} \div \frac{2}{1} =$

$\frac{20}{9} \times \frac{1}{2} = \frac{10}{9} = \mathbf{1\frac{1}{9}}$

3. $\frac{9}{10} \div 3 =$

$\frac{9}{10} \div \frac{3}{1} =$

$\frac{9}{10} \times \frac{1}{3} = \mathbf{\frac{3}{10}}$

$1\frac{3}{4} \div 7 =$

$\frac{7}{4} \div \frac{7}{1} =$

$\frac{7}{4} \times \frac{1}{7} = \mathbf{\frac{1}{4}}$

$5\frac{5}{6} \div 7 =$

$\frac{35}{6} \div \frac{7}{1} =$

$\frac{35}{6} \times \frac{1}{7} = \mathbf{\frac{5}{6}}$

4. $\frac{3}{4} \div 3\frac{1}{5} =$

$\frac{3}{4} \div \frac{16}{5} =$

$\frac{3}{4} \times \frac{5}{16} = \mathbf{\frac{15}{64}}$

$3\frac{3}{4} \div 4 =$

$\frac{15}{4} \div \frac{4}{1} =$

$\frac{15}{4} \times \frac{1}{4} = \mathbf{\frac{15}{16}}$

$10 \div 1\frac{2}{3} =$

$\frac{10}{1} \div \frac{5}{3} =$

$\frac{10}{1} \times \frac{3}{5} = \frac{6}{1} = \mathbf{6}$

Multiplying and Dividing with Fractions

page 104

(4) 60

Answer should be greater than 45 because the fabric was cut into pieces smaller than one inch. *Divide* to get an answer that is greater than the dividend. 45 (the amount being divided up) comes first.

$45 \div \frac{3}{4} = 45 \times \frac{4}{3} = 60$

Fractions Exercise 24

pages 105–106

1. Answer should be more than $62\frac{1}{2}$ pounds. Multiplication.

$3 \times 62\frac{1}{2} =$

$\frac{3}{1} \times \frac{125}{2} = \frac{375}{2} = \mathbf{187\frac{1}{2}}$

2. \$38 comes first. It is being divided up. Answer should be less than \$38.

$38 \div 9\frac{1}{2} =$

$\frac{38}{1} \div \frac{19}{2} =$

$\frac{38}{1} \times \frac{2}{19} = \frac{4}{1} = \mathbf{4}$

3. $17\frac{1}{2}$ comes first. Answer should be less than 17. You are dividing by something larger than one.

$17\frac{1}{2} \div 1\frac{3}{4} =$

$\frac{35}{2} \div \frac{7}{4} =$

$\frac{35}{2} \times \frac{4}{7} = \frac{10}{1} = \mathbf{10}$

4. Answer should be more than \$8.20. Multiplication.

$8\frac{1}{2} \times 8.20 =$

$\frac{17}{2} \times \frac{8.20}{1} = \frac{69.70}{1} = \mathbf{69.70}$

5. 90 comes first. It is being divided up. Answer should be less than 90. You are dividing by something larger than one.

$90 \div 1\frac{1}{2} =$

$\frac{90}{1} \div \frac{3}{2} =$

$\frac{90}{1} \times \frac{2}{3} = \frac{60}{1} = \mathbf{60}$

6. a. Answer should be less than 45,000. Multiplication.

$\frac{2}{5} \times 45{,}000 =$

$\frac{2}{5} \times \frac{45{,}000}{1} = \frac{18{,}000}{1} = \mathbf{18{,}000}$

b. $\begin{array}{r} 45{,}000 \\ -\ 18{,}000 \\ \hline 27{,}000 \end{array}$

7. $\$1\frac{1}{2}$ million comes first. Answer should be less than $\$1\frac{1}{2}$ million.

$1\frac{1}{2} \div 6 =$

$\frac{3}{2} \div \frac{6}{1} =$

$\frac{3}{2} \times \frac{1}{6} = \mathbf{\frac{1}{4}}$

8. $6\frac{3}{4}$ comes first. It is being divided up. Answer should be less than $6\frac{3}{4}$.

$6\frac{3}{4} \div 12 =$

$\frac{27}{4} \div \frac{12}{1} =$

$\frac{27}{4} \times \frac{1}{12} = \mathbf{\frac{9}{16}}$

9. $\$600$

$$\begin{array}{rl} \$21,500 & \\ +\ \$14,500 & \\ \hline \$36,000 & \text{yearly income} \end{array}$$

$$\begin{array}{rl} \$\ 3,000 & \text{monthly income} \\ 12\overline{)\$36,000} & \end{array}$$

$\frac{1}{5} \times \frac{3000}{1} = 600$

10. $\mathbf{\frac{29}{40}}$

$$\begin{array}{rcl} \frac{1}{5} & = & \frac{8}{40} \\ \frac{2}{5} & = & \frac{16}{40} \\ +\frac{1}{8} & = & \frac{5}{40} \\ \hline & & \frac{29}{40} \end{array}$$

11. 4500 $\frac{1}{8} \times \frac{36,000}{1} = 4,500$

12. $\$40$ First find $\frac{2}{5}$ of monthly income.

$\frac{2}{5} \times \frac{3000}{1} = 1200$

Divide $\$1200$ by 30.

$\$1200 \div 30 = \$40$ each day.

13. $\$3625$ $\frac{1}{4} \times \frac{14,500}{1} = \$3,625$

Fractions Exercise 25

page 107

1. $.6 = \frac{6}{10} = \mathbf{\frac{3}{5}}$ $\quad .5 = \frac{5}{10} = \mathbf{\frac{1}{2}}$

   $.45 = \frac{45}{100} = \mathbf{\frac{9}{20}}$ $\quad .80 = \frac{80}{100} = \mathbf{\frac{4}{5}}$

2. $.125 = \frac{125}{1000} = \mathbf{\frac{1}{8}}$ $\quad .065 = \frac{65}{1000} = \mathbf{\frac{13}{200}}$

   $.15 = \frac{15}{100} = \mathbf{\frac{3}{20}}$ $\quad .96 = \frac{96}{100} = \mathbf{\frac{24}{25}}$

3. $.024 = \frac{24}{1000} = \mathbf{\frac{3}{125}}$ $\quad .0002 = \frac{2}{10,000} = \mathbf{\frac{1}{5000}}$

   $.010 = \frac{10}{1000} = \mathbf{\frac{1}{100}}$ $\quad .34 = \frac{34}{100} = \mathbf{\frac{17}{50}}$

Fractions Exercise 26

page 108

1. $\begin{array}{r}\mathbf{.75}\\ 4\overline{)3.00}\end{array}$ $\quad \begin{array}{r}\mathbf{.33\frac{1}{3}}\\ 3\overline{)1.00}\end{array}$ $\quad \begin{array}{r}\mathbf{.7}\\ 10\overline{)7.0}\end{array}$ $\quad \begin{array}{r}.625 \text{ or } \mathbf{.62\frac{1}{2}}\\ 8\overline{)5.000}\end{array}$

2. $\begin{array}{r}\mathbf{.5}\\ 2\overline{)1.0}\end{array}$ $\quad \begin{array}{r}\mathbf{.05}\\ 20\overline{)1.00}\end{array}$ $\quad \begin{array}{r}\mathbf{.83\frac{1}{3}}\\ 6\overline{)5.00}\end{array}$ $\quad \begin{array}{r}\mathbf{.44\frac{4}{9}}\\ 9\overline{)4.00}\end{array}$

3. $\begin{array}{r}\mathbf{.08\frac{1}{3}}\\ 12\overline{)1.00}\end{array}$ $\quad \begin{array}{r}\mathbf{.66\frac{2}{3}}\\ 3\overline{)2.00}\end{array}$ $\quad \begin{array}{r}\mathbf{.12}\\ 25\overline{)3.00}\end{array}$ $\quad \begin{array}{r}\mathbf{.16\frac{2}{3}}\\ 6\overline{)1.00}\end{array}$

Fractions Exercise 27

page 109

Fraction	Decimal
$\frac{1}{4}$	.25
$\frac{1}{2}$	.5
$\frac{3}{4}$	.75
$\frac{1}{8}$	.125 or $.12\frac{1}{2}$
$\frac{1}{5}$	.2
$\frac{3}{5}$	.6
$\frac{4}{5}$	.8
$\frac{1}{10}$	.1
$\frac{9}{10}$	.9
$\frac{1}{3}$	$.33\frac{1}{3}$
$\frac{2}{3}$	$.66\frac{2}{3}$

Decimals and Fractions in Word Problems

page 110

(1) $\mathbf{6\frac{1}{3}}$

Step 1: Change decimal to a fraction:

$9.5 = 9\frac{5}{10} = 9\frac{1}{2}$

Step 2: Add three fractions:

$$\begin{array}{rcl} 9\frac{1}{2} & = & 9\frac{3}{6} \\ 4\frac{1}{3} & = & 4\frac{2}{6} \\ 5\frac{1}{6} & = & 5\frac{1}{6} \\ \hline & & 18\frac{6}{6} = 19 \end{array}$$

Step 3: Divide by 3 to find average:

$$\begin{array}{r} 6\frac{1}{3} \\ 3\overline{)19} \end{array}$$

Fractions Exercise 28

pages 110–11

1. a. $$\begin{array}{rcl} 8\frac{1}{2} & = & 8.50 \\ -\ 3.75 & = & 3.75 \\ \hline & & \mathbf{4.75} \end{array}$$ **feet**

   b. $$\begin{array}{rcl} 8\frac{1}{2} & = 7\frac{2}{4} + \frac{4}{4} = & 7\frac{6}{4} \\ -\ 3\frac{3}{4} & = & 3\frac{3}{4} \\ \hline & & \mathbf{4\frac{3}{4}} \end{array}$$ **feet**

2. $$\begin{array}{rcl} 2\frac{1}{2} & = & 2.5 \\ 5\frac{3}{4} & = & 5.75 \\ +\ 6.3 & = & 6.3 \\ \hline & & 14.55 \end{array}$$ $\begin{array}{r} 4.85 \\ 3\overline{)14.55} \end{array}$ to the nearest tenth = **4.9 pound**

3. **(2)** $.75(2.50) + 1.5(3.90)$

4. **(4)** $10 - 3.5 - .75$

Fractions Exercise 29

pages 111–12

1. a. $\frac{1}{2} = \frac{9}{18}$ and $\frac{4}{9} = \frac{8}{18}$

   $\mathbf{\frac{1}{2}}$ **is larger.**

   b. $\frac{2}{3} = \frac{10}{15}$ and $\frac{3}{5} = \frac{9}{15}$

   $\mathbf{\frac{2}{3}}$ **is larger.**

   c. $\frac{3}{4} = \frac{9}{12}$ and $\frac{5}{6} = \frac{10}{12}$

   $\mathbf{\frac{5}{6}}$ **is larger.**

2. a. $\frac{13}{20} = \frac{13}{20}$ and $\frac{7}{10} = \frac{14}{20}$

   $\mathbf{\frac{7}{10}}$ **is larger.**

   b. $\frac{5}{8} = \frac{25}{40}$ and $\frac{3}{5} = \frac{24}{40}$

   $\mathbf{\frac{5}{8}}$ **is larger.**

   c. $\frac{1}{6} = \frac{3}{18}$ and $\frac{2}{9} = \frac{4}{18}$

   $\mathbf{\frac{2}{9}}$ **is larger.**

3. a. $\frac{5}{6} = \frac{25}{30}$ and $\frac{7}{10} = \frac{21}{30}$

   $\mathbf{\frac{5}{6}}$ **is larger.**

   b. $\frac{5}{12} = \frac{15}{36}$ and $\frac{5}{9} = \frac{20}{36}$

   $\mathbf{\frac{5}{9}}$ **is larger.**

   c. $\frac{5}{8} = \frac{25}{40}$ and $\frac{11}{20} = \frac{22}{40}$

   $\mathbf{\frac{5}{8}}$ **is larger.**

4. $\frac{5}{8} = \frac{15}{24}, \frac{5}{6} = \frac{20}{24}, \frac{7}{12} = \frac{14}{24}$

   from smallest to largest: $\mathbf{\frac{7}{12}, \frac{5}{8}, \frac{5}{6}}$

5. $\frac{7}{10} = \frac{14}{20}, \frac{4}{5} = \frac{16}{20}, \frac{3}{4} = \frac{15}{20}, \frac{13}{20} = \frac{13}{20}$

   from smallest to largest: $\mathbf{\frac{13}{20}, \frac{7}{10}, \frac{3}{4}, \frac{4}{5}}$

6. **(5)** D, A, C, B

   $A = \frac{9}{16}$

   $B = .75 = \frac{75}{100} = \frac{3}{4} = \frac{12}{16}$

   $C = \frac{5}{8} = \frac{10}{16}$

   $D = 0.5 = \frac{5}{10} = \frac{1}{2} = \frac{8}{16}$

7. **E, D, A, B, C**

   $A = 5\frac{1}{2} = 5\frac{60}{120}$

   $B = 5.30 = 5\frac{3}{10} = 5\frac{36}{120}$

   $C = 5.2 = 5\frac{2}{10} = 5\frac{24}{120}$

   $D = 5\frac{2}{3} = 5\frac{80}{120}$

   $E = 5\frac{7}{8} = 5\frac{105}{120}$

Fractions Exercise 30

pages 112–14

1. $\frac{11}{20}$

2. $\frac{48}{60} = \mathbf{\frac{4}{5}}$

3. $\frac{52}{8} = 6\frac{4}{8} = \mathbf{6\frac{1}{2}}$

4. $\mathbf{15\frac{7}{8}}$

$$\begin{array}{r} 7\frac{3}{8} \\ 3\frac{5}{8} \\ +\ 4\frac{7}{8} \\ \hline 14\frac{15}{8} = 15\frac{7}{8} \end{array}$$

5. $\mathbf{3\frac{2}{5}}$

$$\begin{array}{rcl} 12\frac{1}{5} & = & 11\frac{1}{5} + \frac{5}{5} = 11\frac{6}{5} \\ -\ 8\frac{4}{5} & = & 8\frac{4}{5} \\ \hline & & 3\frac{2}{5} \end{array}$$

6. $\mathbf{20\frac{19}{20}}$

$$\begin{array}{rcl} 5\frac{3}{4} & = & 5\frac{15}{20} \\ 6\frac{1}{2} & = & 6\frac{10}{20} \\ +\ 8\frac{7}{10} & = & 8\frac{14}{20} \\ \hline & & 19\frac{39}{20} = 20\frac{19}{20} \end{array}$$

7. $\mathbf{4\frac{3}{4}}$

$$\begin{array}{rcl} 8\frac{5}{12} & = & 7\frac{5}{12} + \frac{12}{12} = 7\frac{17}{12} \\ -\ 3\frac{2}{3} & = & 3\frac{8}{12} \\ \hline & & 4\frac{9}{12} = 4\frac{3}{4} \end{array}$$

8. $\frac{3}{8} \times \frac{1}{6} \times \frac{4}{5} = \mathbf{\frac{1}{20}}$

9. $5 \times \frac{3}{10} =$

   $\frac{5}{1} \times \frac{3}{10} = \frac{3}{2} = \mathbf{1\frac{1}{2}}$

10. $\frac{21}{3} + \frac{2}{3} = \frac{23}{3}$

11. $3\frac{3}{4} \times 3\frac{1}{3} =$

    $\frac{15}{4} \times \frac{10}{3} = \frac{25}{2} = \mathbf{12\frac{1}{2}}$

12. $\frac{3}{4} \div \frac{1}{8} =$

    $\frac{3}{4} \times \frac{8}{1} = \frac{6}{1} = \mathbf{6}$

13. $24\frac{1}{2} \div 3\frac{1}{2} =$

    $\frac{49}{2} \div \frac{7}{2} = \frac{49}{2} \times \frac{2}{7} = \frac{7}{1} = \mathbf{7}$

14. $.65 = \frac{65}{100} = \mathbf{\frac{13}{20}}$

15. $$\begin{array}{r} \mathbf{.24} \\ 25\overline{)6.00} \end{array}$$

16. $\frac{5}{9} = \frac{5}{9}$ and $\frac{2}{3} = \frac{6}{9}$

    $\mathbf{\frac{5}{9}}$ **is smaller.**

17. **(4) 371**

$$\begin{array}{r} 524.7 \\ 208.6 \\ +\ 380.9 \\ \hline 1114.2 \end{array}$$

$$\begin{array}{r} 371.4 \\ 3\overline{)1114.2} \end{array}$$ to the nearest mile $= 371$

18. (3) Multiply $2\frac{3}{4}$ by \$7.20, then subtract the product from \$35.00.

19. (1) $\frac{23}{40}$

$$\begin{array}{rcl} \frac{1}{5} & = & \frac{8}{40} \\ \frac{1}{8} & = & \frac{5}{40} \\ +\frac{1}{10} & = & \frac{4}{40} \\ \hline & & \frac{17}{40} \end{array} \qquad \begin{array}{rcl} 1 & = & \frac{40}{40} \\ -\frac{17}{40} & = & \frac{17}{40} \\ \hline & & \frac{23}{40} \end{array}$$

20. (5) \$72 $\frac{1}{5} \times \frac{360}{1} = \frac{72}{1} = \$72$

21. (2) \$69

Take-home pay $= \frac{23}{40} \times \frac{360}{1} = \frac{207}{1} = 207$

Food $= \frac{1}{3} \times \frac{207}{1} = \frac{69}{1} = \$69$

22. (4) \$9.20 $41.40 \div 4\frac{1}{2} =$

$41.40 \div 4.5$

$$\begin{array}{r} \$9.20 \\ 4.5\overline{)\,41.40} \\ \underline{40\;5} \\ 90 \\ \underline{90} \\ 0 \end{array}$$

23. (1) $21\frac{7}{8}$

$$\begin{array}{rcl} 28\frac{3}{8} = 27\frac{3}{8} + \frac{8}{8} = & & 27\frac{11}{8} \\ -\;6\frac{1}{2} = & & 6\frac{4}{8} \\ \hline & & \mathbf{21\frac{7}{8}} \end{array}$$

24. (2) 200

$$\begin{array}{r} 2\;00 \\ .45\overline{)\,90.00} \end{array}$$

25. (1) $(3 \times 12.50) - 5$

CHAPTER 6

Probability Exercise 1

page 117

1. $\frac{1}{6}$
2. $\frac{2}{6} = \frac{1}{3}$
3. $\frac{4}{6} = \frac{2}{3}$
4. $\frac{1}{6}$
5. $\frac{3}{6} = \frac{1}{2}$
6. $\frac{3}{12} = \frac{1}{4}$

Probability Exercise 2

pages 117–18

1. $\frac{4}{52} = \mathbf{\frac{1}{13}}$
2. $\frac{16}{20} = \mathbf{\frac{4}{5}}$
3. $\mathbf{\frac{2}{3}}$

$$\begin{array}{r} 12 \text{ blue} \\ +\;6 \text{ white} \\ \hline 18 \text{ total} \end{array}$$

$\frac{12}{18} = \frac{2}{3}$

4. $\frac{2}{1000} = \mathbf{\frac{1}{500}}$
5. $\mathbf{\frac{3}{500}}$ 2 + 3 + 1 = 6 tickets sold to the Robinsons.

$\frac{6}{1000} = \frac{3}{500}$

6. $\frac{15}{200} = \mathbf{\frac{3}{40}}$
7. $\mathbf{\frac{1}{5}}$
8. $\mathbf{\frac{1}{5}}$
9. Together 1 and 5 appear twice. $\mathbf{\frac{2}{5}}$
10. $\mathbf{\frac{3}{5}}$
11. $\mathbf{\frac{1}{8}}$

$$\begin{array}{r} 1 \text{ white} \\ 3 \text{ gray} \\ +\;4 \text{ green} \\ \hline 8 \text{ total cans} \end{array}$$

$\frac{\text{white}}{\text{total}} = \frac{1}{8}$

12. $\mathbf{\frac{1}{7}}$ Total is down to 7 cans.

$\frac{\text{white}}{\text{total}} = \frac{1}{7}$

13. $\mathbf{\frac{2}{3}}$ Now there are 1 white can, 1 gray can, and 4 green cans.

$\frac{4}{6} = \frac{2}{3}$

14. $\mathbf{\frac{4}{13}}$ Total is 12 + 8 + 6 = 26.

$\frac{8}{26} = \frac{4}{13}$

15. $\mathbf{\frac{1}{5}}$ Total is down to 25. Number of green is down to 5.

$\frac{5}{25} = \frac{1}{5}$

16. $\mathbf{\frac{1}{2}}$ Total is down to 24.

$\frac{12}{24} = \mathbf{\frac{1}{2}}$

Ratio and Proportion Exercise 1

pages 120–21

1. $\frac{12}{15} = \mathbf{\frac{4}{5}}$ **or 4:5**

2. a. $\frac{\text{women}}{\text{total}}$ $\frac{12}{20} = \mathbf{\frac{3}{5}}$ **or 3:5**

   b. $\frac{\text{men}}{\text{total}}$ $\frac{20 - 12}{20} = \frac{8}{20} = \mathbf{\frac{2}{5}}$ **or 2:5**

   c. $\frac{\text{men}}{\text{women}}$ $\frac{8}{12} = \mathbf{\frac{2}{3}}$ **or 2:3**

   d. $\frac{\text{women}}{\text{men}}$ $\frac{12}{8} = \mathbf{\frac{3}{2}}$ **or 3:2**

3. a. $\frac{\text{union}}{\text{total}}$ $\frac{105}{105 + 45} = \frac{105}{150} = \frac{21}{30} =$

   $\mathbf{\frac{7}{10}}$ **or 7:10**

   b. $\frac{\text{non-union}}{\text{total}}$ $\frac{45}{150} = \frac{9}{30} = \mathbf{\frac{3}{10}}$ **or 3:10**

c. $\frac{\text{union}}{\text{non-union}}$ $\frac{105}{45} = \frac{21}{9} = \mathbf{\frac{7}{3}}$ **or 7:3**

d. $\frac{\text{total}}{\text{union}}$ $\frac{150}{105} = \frac{30}{21} = \mathbf{\frac{10}{7}}$ **or 10:7**

4. $\frac{\text{\$ for education}}{\text{\$ not for education}}$

$\frac{\$3{,}000{,}000}{\$18{,}000{,}000 - \$3{,}000{,}000} =$

$\frac{\$\ 3{,}000{,}000}{\$15{,}000{,}000} = \mathbf{\frac{1}{5}}$ **or 1:5**

5. $\frac{\text{fractions and decimals}}{\text{total number of problems}}$

$\frac{15 + 5}{50} = \frac{20}{50} = \mathbf{\frac{2}{5}}$ **or 2:5**

6. a. $\frac{\text{boys}}{\text{girls}}$ $\frac{80 - 48}{48} = \frac{32}{48} = \mathbf{\frac{2}{3}}$ **or 2:3**

b. $\frac{\text{girls}}{\text{total}}$ $\frac{48}{80} = \mathbf{\frac{3}{5}}$ **or 3:5**

c. $\frac{\text{girls}}{\text{boys}}$ $\frac{48}{32} = \mathbf{\frac{3}{2}}$ **or 3:2**

d. $\frac{\text{boys}}{\text{total}}$ $\frac{32}{80} = \mathbf{\frac{2}{5}}$ **or 2:5**

7. $\frac{\text{games won}}{\text{games played}}$ $\frac{62}{62 + 20} = \frac{62}{82} = \mathbf{\frac{31}{41}}$ **or 31:41**

8. $\frac{\text{games won}}{\text{games played}}$ $\frac{82 - 41}{82} = \frac{41}{82} = \mathbf{\frac{1}{2}}$ **or 1:2**

Ratio and Proportion Exercise 2

page 122

1. **4** $10 \times 6 = 60$ $\quad 60 \div 15 = 4$

$3\frac{3}{5}$ $3 \times 6 = 18$ $\quad 18 \div 5 = 3\frac{3}{5}$

$1\frac{1}{3}$ $4 \times 3 = 12$ $\quad 12 \div 9 = 1\frac{3}{9} = 1\frac{1}{3}$

$3\frac{1}{2}$ $7 \times 4 = 28$ $\quad 28 \div 8 = 3\frac{4}{8} = 3\frac{1}{2}$

2. **$1\frac{2}{3}$** $1 \times 5 = 5$ $\quad 5 \div 3 = 1\frac{2}{3}$

$2\frac{1}{2}$ $3 \times 5 = 15$ $\quad 15 \div 6 = 2\frac{3}{6} = 2\frac{1}{2}$

22 $11 \times 4 = 44$ $\quad 44 \div 2 = 22$

36 $8 \times 9 = 72$ $\quad 72 \div 2 = 36$

3. **$5\frac{1}{3}$** $\frac{4}{e} = \frac{6}{8}$ $\quad 4 \times 8 = 32$ $\quad 32 \div 6 = 5\frac{2}{6} = 5\frac{1}{3}$

$9\frac{1}{3}$ $\frac{3}{7} = \frac{4}{y}$ $\quad 7 \times 4 = 28$ $\quad 28 \div 3 = 9\frac{1}{3}$

$22\frac{1}{2}$ $\frac{15}{40} = \frac{x}{60}$ $\quad 15 \times 60 = 900$ $\quad 900 \div 40 = 22\frac{20}{40} = 22\frac{1}{2}$

40 $\frac{30}{a} = \frac{12}{16}$ $\quad 16 \times 30 = 480$ $\quad 480 \div 12 = 40$

Using Proportions

page 125

(3) 21

$\frac{\text{acre}}{\text{bushels}}$ $\frac{35}{3150} = \frac{?}{1890}$ $\quad 1890 \times 35 = 66{,}150$

$66150 \div 3150 = 21$ acres

Ratio and Proportion Exercise 3

page 125

1. **\$5.10**

$\frac{\text{feet}}{\$}$ $\frac{6}{3.40} = \frac{9}{x}$ $\quad 3.40 \times 9 = 30.60$ $\quad 30.60 \div 6 = \$5.10$

2. **$4\frac{1}{3}$ inches**

$\frac{\text{inches}}{\text{miles}}$ $\frac{2}{150} = \frac{x}{325}$ $\quad 325 \times 2 = 650$

$650 \div 150 = 4\frac{50}{150} = 4\frac{1}{3}$ inches

3. **125 eggs**

$\frac{\text{chickens}}{\text{eggs}}$ $\frac{30}{75} = \frac{50}{x}$ $\quad 75 \times 50 = 3750$

$3750 \div 30 = 125$ eggs

4. **$5\frac{1}{3}$ hrs.**

$\frac{\text{miles}}{\text{hours}}$ $\frac{450}{2} = \frac{1200}{x}$ $\quad 1200 \times 2 = 2400$

$2400 \div 450 = 5\frac{150}{450} = 5\frac{1}{3}$ hours

$2400 - 2250 = 150$

5. **35 gallons**

$\frac{\text{blue}}{\text{white}}$ $\frac{5}{2} = \frac{x}{14}$ $\quad 14 \times 5 = 70$

$70 \div 2 = 35$ gallons

6. $3\frac{2}{3}$ **in.** $\frac{\text{inches}}{\text{degrees}}$ $\frac{2\frac{1}{2}}{75} = \frac{x}{110}$

$\frac{5}{2} \times \frac{110}{1} = 275$

$3\frac{50}{75} = 3\frac{2}{3}$ inches

$75\overline{)275}$
225
50

7. (3) $\frac{\text{width}}{\text{length}}$ $\frac{3}{5} = \frac{x}{12}$ $\frac{3 \times 12}{5}$ or $(3 \cdot 12)/5$

8. (1) $\frac{\text{parts}}{\text{hours}}$ $\frac{16}{2} = \frac{100}{x}$ $\frac{100 \times 2}{16}$ or $(2 \times 100)/16$

9. (4) $\frac{\text{sugar}}{\text{flour}}$ $\frac{2}{3} = \frac{x}{12}$ $\frac{2 \times 12}{3}$ or $(2 \times 12)/3$

10. (2) $\frac{\text{cost}}{\text{number}}$ $\frac{90}{12} = \frac{x}{8}$ $\frac{90 \times 8}{12}$ or $(90 \times 8)/12$

Ratio and Proportion Exercise 4

pages 126–27

1. $1500 take-home 10
withheld + 3
total 13

$\frac{\text{take-home}}{\text{total}}$ $\frac{10}{13} = \frac{t}{1950}$

1950 × 10 = 19,500

$1,500
$13\overline{)19,500}$

2. 20 games won 5
lost + 3
total 8

$\frac{\text{won}}{\text{total}}$ $\frac{5}{8} = \frac{w}{32}$ 32 × 5 = 160

20 games won
$8\overline{)160}$

3. 80 women men 7
women + 2
total 9

$\frac{\text{women}}{\text{total}}$ $\frac{2}{9} = \frac{w}{360}$ 360 × 2 = 720

80 women
$9\overline{)720}$

4. 500 parts good 20
defective + 1
total 21

$\frac{\text{defective}}{\text{total}}$ $\frac{1}{21} = \frac{d}{10,500}$

10,500 × 1 = 10,500

500 parts
$21\overline{)10,500}$

5. 250 people passed 5
failed + 1
total 6

$\frac{\text{passed}}{\text{total}}$ $\frac{5}{6} = \frac{p}{300}$

300 × 5 = 1500

250 people passed
$6\overline{)1500}$

6. 216 workers strikers 3
non-strikers + 2
total 5

$\frac{\text{strikers}}{\text{total}}$ $\frac{3}{5} = \frac{s}{360}$

360 × 3 = 1080

216 voted to strike
$5\overline{)1080}$

7. $6\frac{1}{2}$ **feet** $\frac{\text{ground}}{\text{side}}$ $\frac{40}{13} = \frac{20}{x}$

20
× 13
60
20
260

6.5
$40\overline{)260}$
240
200
200

CHAPTER 7

Percent Exercise 1

page 129

1. 60% 99% 6.5% 2% $83\frac{1}{3}$%
2. 110% 240% 650%
3. $\frac{1}{4}$% 0.9% .65% 0.1%

Percent Exercise 2

page 130

1. .09
2. .24
3. 1
4. .003
5. $.87\frac{1}{2}$
6. $.08\frac{1}{3}$
7. .0015
8. 2.75
9. .027
10. .0395
11. .57
12. 10

Percent Exercise 3

page 131

1. 81%
2. $37\frac{1}{2}$%
3. 50%
4. .4%
5. .09%
6. 21.7%
7. 3%
8. $33\frac{1}{3}$%
9. 210%
10. 485%
11. 392.4%
12. 1.5%

Percent Exercise 4

page 133

1. $\frac{1}{5} \times \frac{100\%}{1} = \mathbf{20\%}$
2. $\frac{5}{6} \times \frac{100\%}{1} = \frac{250}{3} = \mathbf{83\frac{1}{3}\%}$
3. $\frac{3}{8} \times \frac{100\%}{1} = \frac{75}{2} = \mathbf{37\frac{1}{2}\%}$
4. $\frac{2}{3} \times \frac{100\%}{1} = \frac{200}{3} = \mathbf{66\frac{2}{3}\%}$
5. $\frac{14}{8} = \frac{7}{4} \qquad \frac{7}{4} \times \frac{100\%}{1} = \mathbf{175\%}$
6. $\frac{9}{10} \times \frac{100\%}{1} = \mathbf{90\%}$
7. $\frac{5}{12} \times \frac{100\%}{1} = \frac{125}{3} = \mathbf{41\frac{2}{3}\%}$
8. $\frac{6}{7} \times \frac{100\%}{1} = \frac{600}{7} = \mathbf{85\frac{5}{7}\%}$
9. $\frac{1}{6} \times \frac{100\%}{1} = \frac{50}{3} = \mathbf{16\frac{2}{3}\%}$
10. $\frac{10}{5} = \frac{2}{1} \qquad \frac{2}{1} \times \frac{100\%}{1} = \mathbf{200\%}$
11. $\frac{1}{12} \times \frac{100\%}{1} = \frac{25}{3} = \mathbf{8\frac{1}{3}\%}$
12. $\frac{2}{11} \times \frac{100\%}{1} = \frac{200}{11} = \mathbf{18\frac{2}{11}\%}$

Percent Exercise 5

page 134

1. $\frac{45}{100} = \mathbf{\frac{9}{20}}$
2. $37\frac{1}{2} \div 100 = \frac{75}{2} \times \frac{1}{100} = \mathbf{\frac{3}{8}}$
3. $6\frac{2}{3} \div 100 = \frac{20}{3} \times \frac{1}{100} = \mathbf{\frac{1}{15}}$
4. $\frac{8}{100} = \mathbf{\frac{2}{25}}$
5. $\frac{2}{100} = \mathbf{\frac{1}{50}}$
6. $83\frac{1}{3} \div 100 = \frac{250}{3} \times \frac{1}{100} = \mathbf{\frac{5}{6}}$
7. $\frac{24}{100} = \mathbf{\frac{6}{25}}$
8. $33\frac{1}{3} \div 100 = \frac{100}{3} \times \frac{1}{100} = \mathbf{\frac{1}{3}}$
9. $28\frac{4}{7} \div 100 = \frac{200}{7} \times \frac{1}{100} = \mathbf{\frac{2}{7}}$
10. $\frac{80}{100} = \mathbf{\frac{4}{5}}$
11. $\frac{150}{100} = 1\frac{50}{100} = \mathbf{1\frac{1}{2}}$
12. $12\frac{1}{2} \div 100 = \frac{25}{2} \times \frac{1}{100} = \mathbf{\frac{1}{8}}$
13. $1.5\% = .015 = \frac{15}{1000} = \mathbf{\frac{3}{200}}$
14. $.09\% = .0009 = \mathbf{\frac{9}{10{,}000}}$
15. $.6\% = .006 = \frac{6}{1000} = \mathbf{\frac{3}{500}}$
16. $\mathbf{\frac{1}{100}}$

Percent Exercise 6

page 135

Fraction	Decimal	Percent
$\frac{1}{4}$	.25	25%
$\frac{1}{2}$	.50	50%
$\frac{3}{4}$	.75	75%
$\frac{1}{8}$	.125 or $.12\frac{1}{2}$	$12\frac{1}{2}\%$ or 12.5%
$\frac{1}{5}$	.2	20%
$\frac{2}{5}$	.4	40%
$\frac{3}{5}$	.6	60%
$\frac{4}{5}$	.8	80%
$\frac{1}{10}$	.1	10%
$\frac{3}{10}$	.3	30%
$\frac{7}{10}$	.7	70%
$\frac{9}{10}$	.9	90%
$\frac{1}{3}$	$.33\frac{1}{3}$	$33\frac{1}{3}\%$
$\frac{2}{3}$	$.66\frac{2}{3}$	$66\frac{2}{3}\%$

Percent Exercise 7

page 136

1. $\frac{18}{36} = \frac{50}{100}$
2. $\frac{20}{25} = \frac{80}{100}$
3. $\frac{45}{450} = \frac{10}{100}$
4. $\frac{4}{25} = \frac{16}{100}$
5. $\frac{60}{200} = \frac{30}{100}$
6. $\frac{13}{65} = \frac{20}{100}$
7. $\frac{107}{107} = \frac{100}{100}$
8. $\frac{50}{100} = \frac{50}{100}$

Percent Exercise 8

page 137

1. **25%** $\quad \frac{9}{36} = \frac{x}{100} \quad 9 \times 100 = 900 \quad 900 \div 36 = 25$
2. **40%** $\quad \frac{14}{35} = \frac{x}{100} \quad 14 \times 100 = 1400 \quad 1400 \div 35 = 40$
3. **$66\frac{2}{3}\%$** $\quad \frac{50}{75} = \frac{x}{100} \quad 50 \times 100 = 5000 \quad 5000 \div 75 = 66\frac{2}{3}$
4. **250%** $\quad \frac{40}{16} = \frac{x}{100} \quad 40 \times 100 = 4000 \quad 4000 \div 16 = 250$
5. **75%** $\quad \frac{120}{160} = \frac{x}{100} \quad 120 \times 100 = 12{,}000 \quad 12{,}000 \div 160 = 75$

Percent Exercise 9

page 138

1. **27** $\frac{p}{90} = \frac{30}{100}$ $\begin{array}{r} 90 \\ \times\ 30 \\ \hline 2700 \end{array}$ $\begin{array}{r} 27 \\ 100\overline{)2700} \end{array}$

2. **91** $\frac{p}{140} = \frac{65}{100}$ $\begin{array}{r} 140 \\ \times\ 65 \\ \hline 9100 \end{array}$ $\begin{array}{r} 91 \\ 100\overline{)9100} \end{array}$

3. **12** $\frac{p}{72} = \frac{16\frac{2}{3}}{100}$ $72 \times 16\frac{2}{3} = \frac{72}{1} \times \frac{50}{3}$
$= 1200$
$\begin{array}{r} 12 \\ 100\overline{)1200} \end{array}$

4. **42.5** $\frac{p}{500} = \frac{8.5}{100}$ $\begin{array}{r} 8.5 \\ \times\ 500 \\ \hline 4250.0 \end{array}$ $\begin{array}{r} 42.5 \\ 100\overline{)4250.0} \end{array}$

5. **180** $\frac{p}{60} = \frac{300}{100}$ $\begin{array}{r} 60 \\ \times\ 300 \\ \hline 18{,}000 \end{array}$ $\begin{array}{r} 180 \\ 100\overline{)18{,}000} \end{array}$

Percent Exercise 10

page 139

1. **36** $\frac{18}{w} = \frac{50}{100}$ $\begin{array}{r} 18 \\ \times\ 100 \\ \hline 1800 \end{array}$ $\begin{array}{r} 36 \\ 50\overline{)1800} \end{array}$

2. **160** $\frac{24}{w} = \frac{15}{100}$ $\begin{array}{r} 24 \\ \times\ 100 \\ \hline 2400 \end{array}$ $\begin{array}{r} 160 \\ 15\overline{)2400} \end{array}$

3. **2000** $\frac{90}{w} = \frac{4.5}{100}$ $\begin{array}{r} 90 \\ \times\ 100 \\ \hline 9000 \end{array}$ $\begin{array}{r} 200\ 0 \\ 4.5.\overline{)9000.0.} \end{array}$

4. **16** $\frac{2}{w} = \frac{12\frac{1}{2}}{100}$ $\begin{array}{r} 2 \\ \times\ 100 \\ \hline 200 \end{array}$
$200 \div 12\frac{1}{2} = \frac{200}{1} \div \frac{25}{2} = \frac{200}{1} \times \frac{2}{25} = 16$

5. **43** $\frac{86}{w} = \frac{200}{100}$ $\begin{array}{r} 86 \\ \times\ 100 \\ \hline 8600 \end{array}$ $\begin{array}{r} 43 \\ 200\overline{)8600} \end{array}$

Percent Exercise 11

page 140

1. **50% percent**
$\frac{16}{32} = \frac{x}{100}$ $\begin{array}{r} 16 \\ \times\ 100 \\ \hline 1600 \end{array}$ $\begin{array}{r} 50\% \\ 32\overline{)1600} \end{array}$

2. **72 part**
$\frac{p}{90} = \frac{80}{100}$ $\begin{array}{r} 80 \\ \times\ 90 \\ \hline 7200 \end{array}$ $\begin{array}{r} 72 \\ 100\overline{)7200} \end{array}$

3. **50 whole**
$\frac{30}{w} = \frac{60}{100}$ $\begin{array}{r} 30 \\ \times\ 100 \\ \hline 3000 \end{array}$ $\begin{array}{r} 50 \\ 60\overline{)3000} \end{array}$

4. **36 part**
$\frac{p}{800} = \frac{4\frac{1}{2}}{100}$
$4\frac{1}{2} \times 800 = \frac{9}{2} \times \frac{800}{1} = 3600$
$\begin{array}{r} 36 \\ 100\overline{)3600} \end{array}$

5. **28% percent**
$\frac{14}{50} = \frac{x}{100}$ $\begin{array}{r} 14 \\ \times\ 100 \\ \hline 1{,}400 \end{array}$ $\begin{array}{r} 28 \\ 50\overline{)1{,}400} \end{array}$

6. **120 whole**
$\frac{15}{w} = \frac{12\frac{1}{2}}{100}$ $\begin{array}{r} 15 \\ \times\ 100 \\ \hline 1{,}500 \end{array}$
$1{,}500 \div 12\frac{1}{2} = \frac{1{,}500}{1} \div \frac{25}{2} =$
$\frac{1{,}500}{1} \times \frac{2}{25} =$
$\frac{3{,}000}{25} = 120$

7. **$33\frac{1}{3}$% percent**
$\frac{15}{45} = \frac{x}{100}$ $\begin{array}{r} 15 \\ \times\ 100 \\ \hline 1500 \end{array}$
$\begin{array}{r} 33\frac{15}{45} \\ 45\overline{)1500} \end{array} = 33\frac{1}{3}\%$

8. **32.4 part**
$\frac{p}{900} = \frac{3.6}{100}$ $\begin{array}{r} 3.6 \\ \times\ 900 \\ \hline 3240.0 \end{array}$ $\begin{array}{r} 32.4 \\ 100\overline{)3240.0} \end{array}$

9. **150% percent**
$\frac{120}{80} = \frac{x}{100}$ $\begin{array}{r} 120 \\ \times\ 100 \\ \hline 12{,}000 \end{array}$ $\begin{array}{r} 150 \\ 80\overline{)12{,}000} \end{array}$

10. **135 whole**
$\frac{45}{w} = \frac{33\frac{1}{3}}{100}$ $\begin{array}{r} 45 \\ \times\ 100 \\ \hline 4{,}500 \end{array}$
$4{,}500 \div 33\frac{1}{3} = \frac{4{,}500}{1} \div \frac{100}{3} =$
$\frac{4{,}500}{1} \times \frac{3}{100} = 135$

11. **17.2 part**
$\frac{p}{200} = \frac{8.6}{100}$ $\begin{array}{r} 8.6 \\ \times\ 200 \\ \hline 1720.0 \end{array}$ $\begin{array}{r} 17.2 \\ 100\overline{)1720.0} \end{array}$

12. **38.7 part**
$\frac{p}{38.7} = \frac{100}{100}$ $\begin{array}{r} 38.7 \\ \times\ 100 \\ \hline 3870.0 \end{array}$ $\begin{array}{r} 38.7 \\ 100\overline{)3870.0} \end{array}$

Percent Exercise 12

page 141

1. **$6 part**

$\frac{p}{40} = \frac{15}{100}$ $\quad 15 \times 40 = 600 \quad 600 \div 100 = \$6$

2. **75% percent**

$\frac{6{,}000}{8{,}000} = \frac{x}{100}$ $\quad 6{,}000 \times 100 = 600{,}000 \quad 600{,}000 \div 8{,}000 = 75$

3. **$30 part**

$\frac{p}{250} = \frac{12}{100}$ $\quad 250 \times 12$: 500, 250, 3,000 $\quad 3{,}000 \div 100 = \$30$

4. **215 pounds whole**

$\frac{172}{w} = \frac{80}{100}$ $\quad 172 \times 100 = 17{,}200 \quad 17{,}200 \div 80 = 215$ pounds

5. **25% percent**

$\frac{150}{600} = \frac{x}{100}$ $\quad 150 \times 100 = 15{,}000 \quad 15{,}000 \div 600 = 25$

6. **$14.40 part**

$\frac{p}{240} = \frac{6}{100}$ $\quad 240 \times 6 = 1{,}440 \quad 1{,}440.00 \div 100 = \$14.40$

7. **24 people whole**

$\frac{18}{w} = \frac{75}{100}$ $\quad 18 \times 100 = 1{,}800 \quad 1{,}800 \div 75 = 24$ people

8. **30% percent**

$\frac{6}{20} = \frac{x}{100}$ $\quad 100 \times 6 = 600 \quad 600 \div 20 = 30$

9. **$345.60 part**

$\frac{p}{3840} = \frac{9}{100}$ $\quad 3{,}840 \times 9 = 34{,}560 \quad 34{,}560.00 \div 100 = \$345.60$

10. **60% percent**

$\frac{90}{150} = \frac{x}{100}$ $\quad 90 \times 100 = 9{,}000 \quad 9{,}000 \div 150 = 60$

11. **12% percent**

$\frac{30}{250} = \frac{x}{100}$ $\quad 30 \times 100 = 3{,}000 \quad 3{,}000 \div 250 = 12$

12. **150 children whole**

$\frac{60}{w} = \frac{40}{100}$ $\quad 60 \times 100 = 6{,}000 \quad 6{,}000 \div 40 = 150$ children

Percent Exercise 13

page 143

1. **$17.40**

First find the markup.

$\frac{p}{12} = \frac{45}{100}$ $\quad 45 \times 12 = 540 \quad 540.00 \div 100 = \$5.40$

Then find the new price.

original price	$12.00
markup	+ 5.40
selling price	$17.40

2. **$1,440**

First find the assessed value of the farm.

$\frac{p}{120{,}000} = \frac{60}{100}$ $\quad 120{,}000 \times 60 = 7{,}200{,}000 \quad 7{,}200{,}000 \div 100 = \$72{,}000$

Find the taxes on $72,000.

$\frac{p}{72{,}000} = \frac{2}{100}$ $\quad 72{,}000 \times 2 = 144{,}000 \quad 144{,}000 \div 100 = \$1{,}440$

3. **$325**

First find yearly rent.

$\frac{p}{15{,}600} = \frac{25}{100}$ $\quad 15{,}600 \times 25 = 390{,}000 \quad 390{,}000 \div 100 = \$3{,}900$

Find the monthly rent.

$3900 \div 12 = \$325$

4. **$2070**

First find the down payment.

$\frac{p}{1800} = \frac{15}{100}$ $\quad 1{,}800 \times 15$: 9 000, 18 00, 27,000 $\quad 27{,}000 \div 100 = \$270$

Next find the total monthly payments.

$36 \times \$50 = \$1800$

Finally, find the total.

monthly payments	$1800
down payment	+ 270
total	$2070

5. **$175**

Find 65% of the total.

$\frac{p}{500} = \frac{65}{100}$ $\quad 65 \times 500 = 32{,}500 \quad 32{,}500 \div 100 = \$325$

Then find out what they need.

want	$500
have	− 325
still need	$175

6. 852.80

total deductions = 9% + 5% + 4% = 18%

Find 18% of his salary.

$\frac{p}{1040} = \frac{18}{100}$ 1,040 × 18 = 8 320 + 10 40 = 18,720 $\quad$ 100)18,720.00 = $ 187.20

Subtract deductions from gross salary.

gross salary	$1040.00
deductions	− $87.20
net salary	$ 852.80

7. 4950

First find 125% increase.

$\frac{p}{2200} = \frac{125}{100}$ 125 × 2200 = 25 000 + 250 = 275,000 $\quad$ 100)275,000 = 2,750

Then add the increase to the 1970 figure.

1970	2200
increase	+ 2750
1980	4950

8. 18%

total discount = 10% + 5% + 3% = 18%

9. $11.70

$\frac{p}{65} = \frac{18}{100}$ 65 × 18 = 1170 $\quad$ 100)1170.00 = $ 11.70

10. $5330

original price per desk	$65.00
discount	− 11.70
discount price per desk	$53.30
	$53.30
	× 100
total for 100 desks	$5330.00

Percent Exercise 14

page 145

1. $7\frac{1}{7}$%

90 − 84 = 6 $\quad \frac{6}{84} = \frac{x}{100}$ $\quad$ 100 × 6 = 600 $\quad$ 84)600 = $7\frac{12}{84} = 7\frac{1}{7}$ (588, remainder 12)

2. 25%

1,200 − 900 = 300 $\quad \frac{300}{1200} = \frac{x}{100}$ $\quad$ 300 × 100 = 30,000

1,200)30,000 = 25

3. 20%

12 − 10 = 2 $\quad \frac{2}{10} = \frac{x}{100}$ $\quad$ 100 × 2 = 200 $\quad$ 10)200 = 20

4. $8\frac{1}{3}$%

300 − 275 = 25 $\quad \frac{25}{300} = \frac{x}{100}$ $\quad$ 25 × 100 = 2,500 $\quad$ 300)2,500 = $8\frac{1}{3}$

5. 160%

65,000 − 25,000 = 40,000 $\quad \frac{40,000}{25,000} = \frac{x}{100}$

40,000 × 100 = 4,000,000

25,000)4,000,000 = 160

6. 8%

3,600 − 3,312 = 288 $\quad \frac{288}{3,600} = \frac{x}{100}$ $\quad$ 288 × 100 = 28,800

3,600)28,800 = 8%

7. 12%

1,400 − 1,250 = 150 $\quad \frac{150}{1,250} = \frac{x}{100}$ $\quad$ 150 × 100 = 15,000

1,250)15,000 = 12%

8. 20%

90 − 75 = 15 $\quad \frac{15}{75} = \frac{x}{100}$ $\quad$ 15 × 100 = 1,500 $\quad$ 75)1,500 = 20

Percent Exercise 15

page 146

1. $1500

$\frac{6,000}{w} = \frac{80}{100}$ $\quad$ 6000 × 100 = 600,000 $\quad$ 80)600,000 = 7,500

original price	$7,500
current value	− 6,000
depreciation	$1,500

2. 3200 registered voters

$\frac{4,800}{w} = \frac{60}{100}$ $\quad$ 4,800 × 100 = 480,000 $\quad$ 60)480,000 = 8,000

registered	8,000
went to polls	− 4,800
did not go to polls	3,200

3. **$38**

$\frac{342}{w} = \frac{90}{100}$ $\quad 342 \times 100 = 34{,}200 \quad 34{,}200 \div 90 = 380$

original price $380
sale price − 342
savings $ 38

4. **$2300**

$\frac{6{,}900}{w} = \frac{75}{100}$ $\quad 6{,}900 \times 100 = 690{,}000 \quad 690{,}000 \div 75 = 9{,}200$

total $9,200
saved − 6,900
need $2,300

5. **300 people**

$\frac{1{,}800}{w} = \frac{120}{100}$ $\quad 1{,}800 \times 100 = 180{,}000 \quad 180{,}000 \div 120 = 1{,}500$

this year 1,800
last year − 1,500
new members 300

Percent Exercise 16
page 147

1. **$36.90**

First find the 18% discount.

$\frac{x}{50} = \frac{18}{100}$ $\quad 18 \times 50 = 900 \quad 900 \div 100 = \$9$

Subtract the discount from the marked price.

50
− 9
41

Then find 10% of the discounted price.

$\frac{x}{41} = \frac{10}{100}$ $\quad 41 \times 10 = 410 \quad 410.00 \div 100 = \$4.10$

Subtract this amount to find the total cost.

$41.00
− 4.10
$36.90

2. **$12.45**

12% + 5% = 17%

$\frac{17}{100} = \frac{x}{15}$ $\quad 17 \times 15 = 255$

$255 \div 100 = 2.55 \quad 15.00 - 2.55 = 12.45$

3. **22 students**

30% + 25% = 55%

$\frac{55}{100} = \frac{x}{40}$ $\quad 55 \times 40 = 2{,}200 \quad 2{,}200 \div 100 = 22$

4. **147 people**

First find the number of no-shows.

$\frac{14}{100} = \frac{x}{1{,}400}$ $\quad 1{,}400 \times 14 = 19{,}600 \quad 19{,}600 \div 100 = 196$

Then find the number of no-shows from the Toronto Tour Group.

$\frac{75}{100} = \frac{x}{196}$

$196 \times 75 = 14{,}700 \quad 14{,}700 \div 100 = 147$

5. **21,000 accidents**

First find the number of accidents caused by carelessness.

$\frac{25}{100} = \frac{x}{240{,}000}$ $\quad 240{,}000 \times 25 = 6{,}000{,}000 \quad 6{,}000{,}000 \div 100 = 60{,}000$

Then find the number of accidents caused by carelessness *and* involving children.

$\frac{35}{100} = \frac{x}{60{,}000}$ $\quad 60{,}000 \times 35 = 2{,}100{,}000 \quad 2{,}100{,}000 \div 100 = 21{,}000$

Working with Item Sets
page 149

	Lincoln	Milwaukee
cost	$47,700	$58,900
down payment	12%	8%
home improvements	$6200	$2945

1. **(2) $1012**

12% of $47,700 = $5724
8% of $58,900 = $4712

5724 − 4712 = $1012

2. **(3) five**

$\frac{\$2{,}945}{\$58{,}900} = \frac{x}{100}$ $\quad 2945 \times 100 = 294{,}500$

$294{,}500 \div 58{,}900 = .05 = 5\%$

3. **(4) $53,900**

$47,700
+ 6,200
$53,900

Percent Exercise 17
page 151

1. **$375**

$\frac{3000}{1} \times \frac{12.5}{100} = 375$

2. $842

$\frac{800}{1} \times \frac{5\frac{1}{4}}{100} = \$42$

principal	$800
interest	+ 42
total	$842

3. $900

$\frac{5,000}{1} \times \frac{9}{100} \times \frac{2}{1} = 900$

4. $36

9 months $= \frac{9}{12} = \frac{3}{4}$ year

$\frac{800}{1} \times \frac{6}{100} \times \frac{3}{4} = 36$

5. $51.75

6 months $= \frac{6}{12} = \frac{1}{2}$ **year**

$\frac{900}{1} \times \frac{11.5}{100} \times \frac{1}{2} = \frac{103.50}{2} = 51.75$

6. $105

1 year and 6 months $= 1\frac{6}{12} = 1\frac{1}{2}$ years

$\frac{500}{1} \times \frac{14}{100} \times 1\frac{1}{2} =$

$\frac{500}{1} \times \frac{14}{100} \times \frac{3}{2} = 105$

7. $2300

2 years and 6 months $= 2\frac{6}{12} = 2\frac{1}{2}$ years

$\frac{6}{100} \times 2\frac{1}{2} \times \frac{2,000}{1} =$

$\frac{6}{100} \times \frac{5}{2} \times 2,000 = 300$

principal	$2,000
interest	+ 300
total	$2,300

8. $1095

1 year and 8 months $= 1\frac{8}{12} = 1\frac{2}{3}$ years

$\frac{900}{1} \times \frac{13}{100} \times 1\frac{2}{3} =$

$\frac{900}{1} \times \frac{13}{100} \times \frac{5}{3} = 195$

principal	$ 900
interest	+ 195
total	$1,095

9. $933.33

2 years and 4 months $= 2\frac{4}{12} = 2\frac{1}{3}$ years

$\frac{4000}{1} \times \frac{10}{100} \times 2\frac{1}{3} =$

$\frac{4000}{1} \times \frac{10}{100} \times \frac{7}{3} = \frac{2,800}{3}$

to the nearest penny = $933.33

10. $1332.75

nine months $= \frac{9}{12} = \frac{3}{4}$ year

$\frac{1,200}{1} \times \frac{14.75}{100} \times \frac{3}{4} = 132.75$

principal	$1,200.00
interest	+ 132.75
total	$1,332.75

When Not Enough Information Is Provided

pages 153–54

1. (4) $16,125

Find Fred's raise after 3 months.

$\frac{p}{1,250} = \frac{10}{100}$ $1,250 \times 10 = 12,500$ $12,500 \div 100 = 125$

Then find his new monthly salary.

$1,250 + 125 = $1,375

Finally, find his yearly salary.

$1,250 × 3 months = $3,750 $1,375 × 9 months = $12,375

$12,375 + 3,750 = $16,125

2. (3) $187.50

$\frac{p}{1250} = \frac{15}{100}$ 1250×15: 6 250 + 12 50 = 18,750 $18,750.00 \div 100 = \$187.50$

3. (5) Not enough information is provided. You do not know what percent of gross salary goes for social security.

4. (2) $1485

$\frac{p}{1375} = \frac{8}{100}$ $1375 \times 8 = 11,000$ $11,000 \div 100 = 110$ $1375 + 110 = $1485

Percent Exercise 18

page 154

1. (3) $270

$\frac{p}{250} = \frac{8}{100}$ $250 \times 8 = 2000$ $2000 \div 100 = 20$ $250 + 20 = $270

2. (1) $275

$\frac{p}{250} = \frac{10}{100}$ $250 \times 10 = 2500$ $2500 \div 100 = 25$ $250 + 25 = $275

3. (2) $297

$\frac{p}{270} = \frac{10}{100}$ $270 \times 10 = 2700$ $2700 \div 100 = 27$ $270 + 27 = $297

Percent Exercise 19

pages 155–57

1. $\frac{18}{24} = \frac{3}{4}$ **or 3:4**

2. $\frac{20}{36} = \frac{5}{9}$ **or 5:9**

3. $4\frac{1}{2}$ $\frac{s}{15} = \frac{6}{20}$

$\begin{array}{r} 15 \\ \times\ 6 \\ \hline 90 \end{array}$ $\quad 20\overline{)90}^{\,4\frac{10}{20}} = 4\frac{1}{2}$

4. 35 $\frac{5}{12} = \frac{x}{84}$ $\quad \begin{array}{r} 84 \\ \times\ 5 \\ \hline 420 \end{array}$ $\quad 12\overline{)420}^{\,35}$

5. .95 = $\underset{\smile}{.95.}$ = **95%**

6. 3.2% = $\underset{\smile}{.03}.2$ = **.032**

7. $\frac{5}{8} \times \frac{100\%}{1} = \frac{125}{2} =$ **62.5%**

8. $\frac{37\frac{1}{2}}{100} = 37\frac{1}{2} \div 100 = \frac{75}{2} \times \frac{1}{100} = \mathbf{\frac{3}{8}}$

9. 9.72 $\frac{p}{360} = \frac{2.7}{100}$ $\quad \begin{array}{r} 360 \\ \times\ 2.7 \\ \hline 972.0 \end{array}$ $\quad 100\overline{)972.00}^{\,9.72}$

10. 60 $\frac{21}{35} = \frac{x}{100}$ $\quad \begin{array}{r} 21 \\ \times\ 100 \\ \hline 2{,}100 \end{array}$ $\quad 35\overline{)2100}^{\,60}$

11. 375 $\frac{75}{w} = \frac{20}{100}$ $\quad \begin{array}{r} 75 \\ \times\ 100 \\ \hline 7{,}500 \end{array}$ $\quad 20\overline{)7{,}500}^{\,375}$

12. 32 $\frac{48}{w} \times \frac{150}{100}$ $\quad \begin{array}{r} 48 \\ \times\ 100 \\ \hline 4{,}800 \end{array}$ $\quad 150\overline{)4{,}800}^{\,32}$

13. $27 9 months = $\frac{9}{12} = \frac{3}{4}$ year

$\frac{450}{1} \times \frac{8}{100} \times \frac{3}{4} = \frac{54}{2} = 27$

14. (4) $\frac{1}{3}$ 2 + 5 + 4 + 1 = 12 socks $\frac{4}{12} = \frac{1}{3}$

15. (3) 3:5

won	12
lost	+ 8
played	20

$\frac{12}{20} = \frac{3}{5}$ or 3:5

16. (3) $64 $\frac{\text{hours}}{\$} \quad \frac{6}{38.40} = \frac{10}{x}$

$\begin{array}{r} 38.40 \\ \times\ 10 \\ \hline 384.00 \end{array}$ $\quad 6\overline{)384}^{\,\$64}$

17. (5) (2 × 215)/5 $\frac{\text{total } \$}{\text{food } \$} \quad \frac{5}{2} = \frac{215}{x}$

18. (2) 14,250

voted	3
stayed home	+ 5
total	8

$\frac{\text{voted}}{\text{total}} \quad \frac{3}{8} = \frac{x}{38{,}000}$

$\begin{array}{r} 38{,}000 \\ \times\ 3 \\ \hline 114{,}000 \end{array}$ $\quad 8\overline{)114{,}000}^{\,14{,}250}$

19. (5) $2300

$\frac{115}{x} = \frac{5}{100}$ $\quad \begin{array}{r} 115 \\ \times\ 100 \\ \hline 11{,}500 \end{array}$ $\quad 5\overline{)11{,}500}^{\,\$2{,}300}$

20. (4) $83\frac{1}{3}\%$ $\frac{1{,}000}{1{,}200} = \frac{x}{100}$ $\quad \begin{array}{r} 1{,}000 \\ \times\ 100 \\ \hline 100{,}000 \end{array}$

$1{,}200\overline{)100{,}000}^{\,83\frac{1}{3}\%}$

21. (1) $12\frac{1}{2}\%$
First find the discount.

$\begin{array}{r} \$1.20 \\ -\ 1.05 \\ \hline \$\ .15 \end{array}$

Then find the percent of decrease.

$\frac{.15}{1.20} = \frac{x}{100}$ $\quad \begin{array}{r} .15 \\ \times\ 100 \\ \hline 15.00 \end{array}$ $\quad 1.2.\overline{)15.0.}^{\,12\frac{1}{2}\%}$

22. (4) $296.80
First find the discount.

$\frac{p}{350} = \frac{20}{100}$ $\quad \begin{array}{r} 350 \\ \times\ 20 \\ \hline 7{,}000 \end{array}$ $\quad 100\overline{)7{,}000}^{\,\$\ 70}$

The selling price is $350 − 70 = $280.
Now find the tax.

$\frac{p}{280} = \frac{6}{100}$ $\quad \begin{array}{r} 280 \\ \times\ 6 \\ \hline 1680 \end{array}$ $\quad 100\overline{)1{,}680.00}^{\,\$\ 16.80}$

The final price is $\begin{array}{r} \$280.00 \\ +\ 16.80 \\ \hline \$296.80 \end{array}$

23. (2) $786.70

$\frac{p}{695} = \frac{6}{100}$ $\quad \begin{array}{r} 695 \\ \times\ 6 \\ \hline 4{,}170 \end{array}$ $\quad 100\overline{)4{,}170.00}^{\,\$\ 41.70}$

price	$695.00
tax	41.70
shipping	+ 50.00
	$786.70

24. (5) Not enough information is provided. You do not know the cost of tolls.

25. (2) $75

$\frac{p}{749} = \frac{10}{100}$ $\quad \begin{array}{r} 749 \\ \times\ 10 \\ \hline 7{,}490 \end{array}$

$100\overline{)\$7{,}490.00}^{\,\$\ 74.90}$ to the nearest dollar = $75

CHAPTER 8

Measurements Exercise 1

page 61

1. $2\frac{1}{2}$ lb

$\frac{1 \text{ lb}}{16 \text{ oz}} = \frac{x}{40 \text{ oz}}$ $\quad \begin{array}{r} 40 \\ \times\ 1 \\ \hline 40 \end{array}$ $\quad 16\overline{)40}^{\,2\frac{8}{16}} = 2\frac{1}{2}$ lb

$\begin{array}{r} 32 \\ \hline 8 \end{array}$

2. 6,000 lb

$$\frac{1\text{ T}}{2000\text{ lb}} = \frac{3\text{ T}}{x} \qquad \begin{array}{r}2000\\ \times\ \ 3\\ \hline 6000\end{array} \qquad \begin{array}{r}6000\text{ lb}\\ 1\overline{)\,6000}\end{array}$$

3. 2 yd

$$\frac{1\text{ yd}}{3\text{ ft}} = \frac{x}{6\text{ ft}} \qquad \begin{array}{r}6\\ \times\ 1\\ \hline 6\end{array} \qquad \begin{array}{r}2\text{ yd}\\ 3\overline{)\,6}\end{array}$$

4. $3\frac{2}{7}$ wk

$$\frac{1\text{ wk}}{7\text{ da}} = \frac{x}{23\text{ da}} \qquad \begin{array}{r}23\\ \times\ 1\\ \hline 23\end{array} \qquad \begin{array}{r}3\frac{2}{7}\text{ wk}\\ 7\overline{)\,23}\end{array}$$

5. $\frac{7}{8}$ lb

$$\frac{1\text{ lb}}{16\text{ oz}} = \frac{x}{14\text{ oz}} \qquad \begin{array}{r}14\\ \times\ 1\\ \hline 14\end{array} \qquad \frac{14}{16} = \frac{7}{8}\text{ lb}$$

Since 14 is less than 16, you can put the 14 over 16 and reduce.

6. 24 qt

$$\frac{1\text{ gal}}{4\text{ qt}} = \frac{6\text{ gal}}{x} \qquad \begin{array}{r}6\\ \times\ 4\\ \hline 24\end{array} \qquad \begin{array}{r}24\\ 1\overline{)\,24}\end{array}$$

7. $\frac{1}{2}$ mi

$$\frac{1\text{ mi}}{5280\text{ ft}} = \frac{x}{2640\text{ ft}} \qquad \begin{array}{r}2640\\ \times\ 1\\ \hline 2640\end{array} \qquad \frac{2640}{5280} = \frac{1}{2}\text{ mi}$$

8. $2\frac{7}{9}$ yd

$$\frac{1\text{ yd}}{36\text{ in}} = \frac{x}{100\text{ in}} \qquad \begin{array}{r}100\\ \times\ 1\\ \hline 100\end{array} \qquad \begin{array}{r}2\frac{28}{36}\\ 36\overline{)\,100}\\ 72\\ \hline 28\end{array} = 2\frac{7}{9}\text{ yd}$$

9. $2\frac{1}{2}$ cups

$$\frac{1\text{ cup}}{8\text{ oz}} = \frac{x}{20\text{ oz}} \qquad \begin{array}{r}20\\ \times\ 1\\ \hline 20\end{array} \qquad \begin{array}{r}2\frac{4}{8}\\ 8\overline{)\,20}\end{array} = 2\frac{1}{2}\text{ cups}$$

10. 60 pt

$$\frac{1\text{ qt}}{2\text{ pt}} = \frac{30\text{ qt}}{x} \qquad \begin{array}{r}30\\ \times\ 2\\ \hline 60\end{array} \qquad \begin{array}{r}60\text{ pt}\\ 1\overline{)\,60}\end{array}$$

11. 15,840 ft

$$\frac{1\text{ mi}}{5280\text{ ft}} = \frac{3\text{ mi}}{x} \qquad \begin{array}{r}5280\\ \times\ \ 3\\ \hline 15{,}840\end{array} \qquad \begin{array}{r}15{,}840\text{ ft}\\ 1\overline{)\,15{,}840}\end{array}$$

12. 6 ft

$$\frac{1\text{ yd}}{3\text{ ft}} = \frac{2\text{ yd}}{x} \qquad \begin{array}{r}3\\ \times\ 2\\ \hline 6\end{array} \qquad \begin{array}{r}6\text{ ft}\\ 1\overline{)\,6}\end{array}$$

13. $32\frac{1}{2}$ gal

$$\frac{1\text{ gal}}{4\text{ qt}} = \frac{x}{130\text{ qt}} \qquad \begin{array}{r}130\\ \times\ 1\\ \hline 130\end{array} \qquad \begin{array}{r}32\frac{2}{4}\\ 1\overline{)\,130}\end{array} = 32\frac{1}{2}\text{ gal}$$

14. 72 hr

$$\frac{1\text{ da}}{24\text{ hr}} = \frac{3\text{ da}}{x} \qquad \begin{array}{r}24\\ \times\ 3\\ \hline 72\end{array} \qquad \begin{array}{r}72\text{ hr}\\ 1\overline{)\,72}\end{array}$$

15. $\frac{3}{4}$ hr

$$\frac{1\text{ hr}}{60\text{ min}} = \frac{x}{45\text{ min}} \qquad \begin{array}{r}45\\ \times\ 1\\ \hline 45\end{array} \qquad \frac{45}{60} = \frac{3}{4}\text{ hr}$$

16. 2 mi

$$\frac{1\text{ mi}}{5280\text{ ft}} = \frac{x}{10{,}560\text{ ft}} \qquad \begin{array}{r}10{,}560\\ \times\ \ \ 1\\ \hline 10{,}560\end{array}$$

$$\begin{array}{r}2\text{ mi}\\ 5280\overline{)\,10{,}560}\end{array}$$

17. $3\frac{1}{3}$ hr

$$\frac{60\text{ min}}{1\text{ hr}} = \frac{200\text{ min}}{x} \qquad \begin{array}{r}200\\ \times\ 1\\ \hline 200\end{array} \qquad \begin{array}{r}3\frac{20}{60}\\ 60\overline{)\,200}\\ 180\\ \hline 20\end{array} = 3\frac{1}{3}\text{ hr}$$

18. 12 trips

$$\frac{1\text{ T}}{2000\text{ lb}} = \frac{3\text{ T}}{6000\text{ lb}} \qquad \begin{array}{r}12\\ 500\overline{)\,6000}\end{array}$$

Measurements Exercise 2

page 164

1. .915 L

$$\frac{1\text{ L}}{1000\text{ mL}} = \frac{x}{915\text{ mL}} \qquad \begin{array}{r}915\\ \times\ 1\\ \hline 915\end{array} \qquad \begin{array}{r}.915\text{ L}\\ 1000\overline{)\,915.000}\end{array}$$

2. .5 m

$$\frac{1\text{ m}}{100\text{ cm}} = \frac{x}{50\text{ cm}} \qquad \begin{array}{r}50\\ \times\ 1\\ \hline 50\end{array} \qquad \begin{array}{r}.5\text{ m}\\ 100\overline{)\,50.0}\end{array}$$

3. 2400 m

$$\frac{1\text{ km}}{1000\text{ m}} = \frac{2.4\text{ km}}{x} \qquad \begin{array}{r}2.4\\ \times\ 1000\\ \hline 2400.0\end{array} \qquad \begin{array}{r}2400\text{ m}\\ 1\overline{)\,2400}\end{array}$$

4. 10,000 mg

$$\frac{1\text{ g}}{1000\text{ mg}} = \frac{10\text{ g}}{x} \qquad \begin{array}{r}1000\\ \times\ \ 10\\ \hline 10{,}000\end{array} \qquad \begin{array}{r}10{,}000\text{ mg}\\ 1\overline{)\,10{,}000}\end{array}$$

5. 50 cL

$$\frac{1\text{ L}}{100\text{ cL}} = \frac{.5\text{ L}}{x} \qquad \begin{array}{r}100\\ \times\ .5\\ \hline 50.0\end{array} \qquad \begin{array}{r}50\text{ cL}\\ 1\overline{)\,50}\end{array}$$

6. .2 km

$\frac{1\text{ km}}{1000\text{ m}} = \frac{x}{200\text{ m}}$ 200 × 1 = 200

$1000\overline{)200.0}$ = .2 km

7. 1.25g

$\frac{1\text{ g}}{100\text{ cg}} = \frac{x}{125\text{ cg}}$ 125 × 1 = 125

$100\overline{)125.00}$ = 1.25 g

8. 3000 g

$\frac{1\text{ kg}}{1000\text{ g}} = \frac{3\text{ kg}}{x}$ 1000 × 3 = 3000

$1\overline{)3000}$ = 3000 g

9. 48 jars

$\frac{1\text{ L}}{10\text{ dL}} = \frac{4.8\text{ L}}{x}$ 4.8 × 10 = 48.0

$1\overline{)48}$ = 48 dL

10. 8.75 m

$\frac{1\text{ m}}{1000\text{ mm}} = \frac{x}{8750\text{ mm}}$ 8750 × 1 = 8750

$1000\overline{)8750.0}$ = 8.75 m

Comparing and Ordering Measurements

page 166

(2) E, D, C, A, B

Board A: 2 feet = 24 in
Board B: 20 inches = 20 in
Board C: 1 yd 2 in = 38 in
Board D: 40 inches = 40 in
Board E: 1 yd 2 ft = 60 in

From bottom to top: E, D, C, A, B

Measurements Exercise 3

page 167

1. **1 ft 7 in**
   $12\overline{)19}$; 12; 7

2. **1 min 15 sec**
   $60\overline{)75}$; 60; 15

3. **2 lb 4 oz**
   $16\overline{)36}$; 32; 4

4. **$8\frac{1}{3}$ yd**
   $3\overline{)25}$; 24; 1

5. **3 hr 20 min**
   $60\overline{)200}$; 180; 20

6. **2 T 500 lb**
   $2000\overline{)4500}$; 4000; 500

7. **$3\frac{3}{4}$ gal**
   $4\overline{)15}$; 12; 3

8. **2 yd 3 in**
   $36\overline{)75}$; 72; 3

9. **3 mi**
   $5280\overline{)15{,}840}$; 15,840; 0

10. **$6\frac{4}{16} = 6\frac{1}{4}$ lb**
    $16\overline{)100}$; 96; 4

11. **4 qt 1 pt**
    $2\overline{)9}$; 8; 1

12. **5 wk 5 days**
    $7\overline{)40}$; 35; 5

What More Do You Need to Know?

page 170

(2) her regular hourly wage

Measurements Exercise 4

pages 170–71

1. **7 ft 1 in.** 4 ft 9 in + 2 ft 4 in = 6 ft 13 in = 7 ft 1 in

2. **2 min 45 sec**
   4 min 20 sec = 3 min 80 sec
   − 1 min 35 sec = − 1 min 35 sec
   2 min 45 sec

3. **43 yd 1 ft** 8 yd 2 ft × 5 = 40 yd 10 ft = 43 yd 1 ft

4. **3 pt 6 oz** 2 pt 12 oz + 10 oz = 2 pt 22 oz = 3 pt 6 oz

5. **3 min 25 sec**
   $3\overline{)10\text{ min }15\text{ sec}}$
   9 min
   1 min = 60 sec
   75 sec

6. 1 mile 4780 ft

```
 2 miles = 1 mile 5280 ft
− 500 ft = −        500 ft
          1 mile 4780 ft
```

7. 11 yd 2 ft

```
  2 yd  5 ft  8 in
×             3
6 yd 15 ft 24 in = 11 yd 2 ft
```

8. 12 T 600 lb

```
  2 T  500 lb
  1 T  900 lb
+ 8 T 1200 lb
 11 T 2600 lb = 12 T 600 lb
```

9. (2) The number of skirts she ironed and the number she altered.

10. 2 gal 1 qt

```
  4 gal 2 qt
+ 3 gal 3 qt
  7 gal 5 qt = 8 gal 1 qt

 10 gal 2 qt
− 8 gal 1 qt
  2 gal 1 qt
```

11. 1 day 21 hr 45 min

```
  3 days 10 hr 15 min
− 1 day 12 hr 30 min
  1 day 21 hr 45 min
```

12. 5 oz

```
1 lb 9 oz = 25 oz      5
                   5)25
```

13. 1 yd 1 ft 2 in

```
  8 yd 2 ft  8 in
− 1 yd 2 ft 10 in
  6 yd 2 ft 10 in

   1 yd 1 ft  2 in
5)6 yd 2 ft 10 in
```

14. 10 jars

```
  2 gal =   1 gal 4 qt
− 3 qt  = −       3 qt
            1 gal 1 qt
```

1 gal 1 qt =
5 qt =
10 pints or 10 jars

15. 2 ft

```
   1 yd 1 ft
8)10 yd 2 ft
   8 yd
   2 yd = 6 ft
          8 ft

  1 yd 1 ft
−      2 ft
       2 ft
```

16. $8.10 2 lb 4 oz = $2\frac{1}{4}$ lb

$2\frac{1}{4} \times \$3.60 =$

$\frac{9}{4} \times \frac{3.60}{1} = \$8.10$

Measurement Exercise 5

pages 172–73

1. 3.245 L

```
  2.75  L
+  .495 L
  3.245 L
```

2. 9.4 km

```
 10.0 km
−  .6 km
  9.4 km
```

3. 7.35 km

```
  2.45 km
×    3
  7.35 km
```

4.

```
    400 g
5)2 kg = 2000 g
```

5.

```
  2 kg 500 g
2)5 kg
  4 kg
  1 kg = 1000 g
```

6. 4.8 g or 4800 mg

```
    5 g   =   4 g 1000 mg
− 200 mg  = −      200 mg
              4 g  800 mg =
                   4.8 g =
                  4800 mg
```

7.

```
       25 cm
16)4 m = 400 cm
          32
          80
```

8. (5) Subtract 25 centimeters from 2 meters and divide by 6.

9. 115 mL or .115 L

```
  .9 L = 900 mL
−        785 mL
         115 mL or .115 L
```

10. $18.80

```
  2.35 L
×   20
 47.00 L = 47 L

   .40
 × 47
   280
  160
$18.80
```

11. 3550 m or 3.550 km

```
  3 km
       450 m
+      100 m
  3 km 550 m = 3550 m or 3.550 km
```

12. .009 kg

```
  .0087 or .009 kg
4).035
    32
    30
```

Measurement Exercise 6

page 173

1. 7.568 L $\frac{1\text{ qt}}{.946\text{ L}} = \frac{8\text{ qt}}{?}$

```
  .946
×    8
 7.568 L
```

2. 3937 in $\frac{1\text{ m}}{39.37} = \frac{100\text{ m}}{?}$

39.37 × 100 = 3937

3. 80 oz $\frac{1 \text{ oz}}{28.4 \text{ g}} = \frac{?}{2272}$

$28.4\overline{)22720}$ = 80.
2272

4. 60.96 cm $\frac{2.54 \text{ cm}}{1} = \frac{?}{24}$

2.54
× 24
10 16
50 8
60.96

Measurement Exercise 7
pages 176–77

1. J $\frac{7}{8}$ in
K $1\frac{1}{4}$ in
L $3\frac{1}{2}$ in
M $4\frac{1}{8}$ in
N $5\frac{5}{8}$ in

2. P 2.5 cm
Q 4 cm
R 6.3 cm
S 9.2 cm
T 13.9 cm

3. 36 amperes

4. 14 volts

5. 9 seconds

Measurement Exercise 8
pages 177–79

1. 5 kg .65 kg
1.2
+ 3.15
5.00 = 5 kg

2. 14 hr 20 min 3 hr 45 min
× 2
6 hr 90 min = 7 hr 30 min

7 hr 30 min
4 hr 20 min
+ 2 hr 30 min
13 hr 80 min = 14 hr 20 min

3. $25.20 18 ft 8 in
× 3
54 ft 24 in

56
× .45
280
224
$25.20

4. 4 lb 6 oz 14 oz
× 5
70 oz = 4 lb 6 oz

5. $4.69 $4.5\overline{)21.11}$ = 4.691 to nearest cent = $4.69

6. 36 L 29.3 L
36.5
42.5
108.0 L

$3\overline{)108}$ = 36 L

7. 13 gal 3 gal 1 qt
× 4
12 gal 4 qt = 13 gal

8. 55 cm 1.15 m
× 3
3.45 m

4 m
− 3.45 m
.55 m or 55 cm

9. (2) The number of shelves he cut

10. 4 3 yd 9 in = 3.25 yd

$3.25\overline{)1500}$ = 4.615

Gianni can make 4 whole suits. The remainder (fraction or decimal) does not count in this problem.

11. 7:20 A.M. 10 hr 20 min
− 3 hr
7 hr 20 min = 7:20 A.M.

12. 4:05 P.M. 1 hr 5 min
+ 3 hr
4 hr 5 min = 4:05 P.M.

13. 5 hr 45 min

Change arrival time to hours and minutes.
1:05 = 12 hr (noon)
+ 1 hr 5 min
13 hr 5 min

Subtract departure time from arrival time.
13 hr 5 min
− 7 hr 20 min
5 hr 45 min

14. longest to shortest: C, D, E, A, B
A — 19 in
B — 15 in
C — 27 in
D — 26 in
E — 21 in

15. 4 pieces 10 × 36 = 360 inches *or* 30 feet
Nick needs 4 pieces, each 8 ft long
8 × 4 = 32 ft

16. heaviest to lightest: B, C, D, A, E
A — 1000 lb
B — 2200 lb
C — 1800 lb
D — 1200 lb
E — 750 lb

17. 1700 ft

18. 15,100 cu ft

CHAPTER 9

Graphs Exercise 1

pages 183–84

1. 100 houses
2. 550 single family homes
3. **200 houses** $\begin{array}{r} 800 \\ -\ 600 \\ \hline 200 \end{array}$
4. $\frac{400}{800} = \mathbf{\frac{1}{2}}$
5. 1984-1985
6. **75%**
   $\frac{x}{100} = \frac{6}{8}$ $\begin{array}{r} 100 \\ \times\ 6 \\ \hline 600 \end{array}$ $600 \div 8 = 75\%$
7. $\begin{array}{r} 80{,}000 \\ \times\ \ 400 \\ \hline \mathbf{\$32{,}000{,}000} \end{array}$
8. **(2)** The number decreased each year.
9. **1,000,000 domestic cars**
10. **$4\frac{1}{2}$ times** $9 \div 2 = 4\frac{1}{2}$
11. 1976
12. $\frac{500{,}000}{2{,}000{,}000} = \frac{1}{4} = \mathbf{25\%}$
13. $\begin{array}{r} 9{,}000{,}000 \\ \times\ \ 900 \\ \hline 8{,}100{,}000{,}000 \end{array}$ $\begin{array}{r} 2{,}000{,}000 \\ \times\ \ 400 \\ \hline 800{,}000{,}000 \end{array}$ $\begin{array}{r} 8{,}100{,}000{,}000 \\ +\ 800{,}000{,}000 \\ \hline \mathbf{\$8{,}900{,}000{,}000} \end{array}$
14. **(3)** $\frac{1{,}000{,}000\,(7 + 9)}{2}$
    There are 7 striped symbols next to year 1975 and 9 next to 1977. Each symbol is worth 1,000,000. Divide by 2 to find the average.

Graphs Exercise 2

page 186

1. 5 categories
2. 17.8%
3. $\begin{array}{r} 27.0\% \\ +\ 17.8 \\ \hline \mathbf{44.8\%} \end{array}$
4. **27 families**
   $\frac{13.6}{100} = \frac{p}{200}$ $\begin{array}{r} 13.6 \\ \times\ 200 \\ \hline 2720.0 \end{array}$
   $2720.0 \div 100 = 27.2$ to the nearest unit = 27 families
5. **10%** $\begin{array}{r} 27\% \\ -\ 17 \\ \hline 10\% \end{array}$
6. **(3)** $\mathbf{\frac{3}{10}}$
   29.2% is close to 30%. $30\% = \frac{30}{100} = \frac{3}{10}$.

Graphs Exercise 3

pages 187–88

1. 48%
2. $\frac{8}{48} = \frac{1}{6}$
3. $\mathbf{\frac{1}{3}}$ You know that $33\frac{1}{3}\% = \frac{1}{3}$. 33% is close to $33\frac{1}{3}\%$.
4. **3 times** $33 \div 11 = 3$
5. **$48 billion**
   $\frac{8}{100} = \frac{p}{600 \text{ billion}}$ $\begin{array}{r} 600 \\ \times\ 8 \\ \hline 4800 \end{array}$ $4800 \div 100 = \$\,48$ billion
6. **(1)** Together individual income tax and social security tax make up over 80% of the federal budget. (48% + 33% = 81%)
7. 1970 and 1982
8. natural gas
9. nuclear power
10. **$6\frac{1}{2}$ times** $13 \div 2 = 6\frac{1}{2}$
11. $\begin{array}{r} 53\% \\ +\ 7 \\ \hline 60\% \end{array}$ $\frac{60\%}{100\%} = \mathbf{\frac{3}{5}}$

Graphs Exercise 4

pages 190–91

1. approximately 230
2. Northeast
3. **(3)** 495
   $300{,}000 \div 100{,}000 = 3$ $\begin{array}{r} 165 \\ \times\ 3 \\ \hline 495 \end{array}$
4. **(5) 444,000**
   $240{,}000{,}000 \div 100{,}000 = 2{,}400$ $\begin{array}{r} 185 \\ \times\ 2400 \\ \hline 74\ 000 \\ 370 \\ \hline 444{,}000 \end{array}$
5. **(5) 35**
   50,000 is $\frac{1}{2}$ of the 100,000 population the graph shows. The Northeast is about 230. Find $\frac{1}{2}$ of 230.
   $\frac{1}{2} \times 230 = 115$
   The North Central is about 160. Find $\frac{1}{2}$ of 160.
   $\frac{1}{2} \times 160 = 80$
   $115 - 80 = 35$
6. **(4) 25%**
   $\begin{array}{rl} 200 & \text{West} \\ -\ 160 & \text{South} \\ \hline 40 & \end{array}$
   You are comparing 40 to 161 which is approximately 160.
   $\frac{40}{160} = \frac{x}{100}$ $\begin{array}{r} 40 \\ \times\ 100 \\ \hline 4000 \end{array}$ $4000 \div 160 = 25\%$

Graphs Exercise 5
page 192

1. **a.** approximately \$255,000
   b. approximately \$175,000
   c. approximately \$100,000
2. Gas service stations and hardware stores
3. **20%** $\frac{50}{250} = \frac{x}{100}$ $\quad 50 \times 100 = 5000 \quad 5000 \div 250 = 20\%$
4. **$\frac{1}{2}$** $\frac{50}{100} = \frac{x}{100}$ $\quad 50 \times 100 = 5000 \quad 5000 \div 100 = 50\%$
5. **Car dealers**
   Clothing stores in 1985 was approximately \$60,000. Eating and drinking places in 1985 was approximately \$110,000. The total, \$170,000 is almost equal to the sales of car dealers in 1985.
6. **\$33,000**
   $\$100{,}000 \div 3 = \$33{,}333\frac{1}{3}$ to the nearest thousand = \$33,000

Reading a Graph Carefully
page 195

(5) Not enough information is provided.
The question asks for information about North America. The graph provides information only for the United States.

Graphs Exercise 6
page 196

1. **(3)** 1975
2. **(5)** Not enough information is given. 1968 is not included on the graph.
3. **(1)** 1979
4. **(4)** **2 times** $\quad 6 \div 3 = 2$

Graphs Exercise 7
page 198

1. **a.** 572,000
   b. 87,000
   c. 249,000
2. Communication Workers of America; Service Employees International Union; and American Federation of State, County, and Municipal Employees.
3. United Steel Workers
4. International Typographical Union (They lost over 50% of membership. While the Steelworkers lost a larger *number* of members, the percent decrease was only **46%**.)
5. **10 times**
   $997 \div 99$: 99 into 997 = 10, 997 − 990 = 07
6. **(5)** 1000(1260 + 1150)/2

Graphs Exercise 8
pages 199–201

1. **a.** 273,700
   b. 1,804,200
   c. 15,400
   d. 72,900
2. 22%
3. **15 times** $\quad 90 \div 6 = 15$
4. 6% + 1% + 3% = **10%**
5. **(4)** 72,900
6. **(3) 1,011,700**
   1258.0 − 246.3 = 1011.7 $\quad$ 1011.7 × 1000 = 1,011,700
7. **(4) \$172.75**
   \$845.00 + 32.50 + 680.00 + 170.00 = 1727.50 $\quad$ 10% of 1727.50 = \$172.75
8. **(3) 4 times**
   $680 \div 170 = 4$ (680 − 680 = 0)
9. **(2) \$5.00**
   34 hours × 4 weeks = 136 hours $\quad$ $680 \div 136 = \$5.00$ (680 − 680 = 0)
10. **(3) \$36.60**
    First find amount saved.
    $\frac{?}{1727.50} = \frac{4}{100}$ $\quad$ 1727.50 × 4 = 6910.00 $\quad$ $6910.00 \div 100 = 69.10$

    Then subtract overtime amount from saved amount.
    69.10 − 32.50 = 36.60

CHAPTER 10

Basics of Algebra Exercise 1
pages 203–04

1. $x + 9$ or $9 + x$
2. $7x$
3. $5x$
4. $x + 8$ or $8 + x$
5. $x - 10$
6. $x - 1$
7. $\frac{3}{x}$
8. $x - 20$
9. $\frac{x}{15}$
10. $x + \frac{1}{2}$ or $\frac{1}{2} + x$
11. $\frac{x}{2}$
12. $x + 4$ or $4 + x$

Basics of Algebra Exercise 2
pages 204–05

1. $w + 2$
2. $s - 5$
3. $3c$
4. $\frac{1}{5}i$
5. $\frac{x}{4}$
6. $\frac{5}{6}t$
7. $l - 2$
8. $.06m$
9. $y + 10$
10. $p - 15$

Basics of Algebra Exercise 3
page 207

1. $f + 20 = 57$; $-20 \quad -20$; $\mathbf{f = 37}$
2. $42 = t + 7$; $-7 \quad -7$; $\mathbf{35 = t}$
3. $n + 36 = 60$; $-36 \quad -36$; $\mathbf{n = 24}$
4. $41 = c + 18$; $-18 \quad -18$; $\mathbf{23 = c}$
5. $b - 19 = 28$; $+19 \quad +19$; $\mathbf{b = 47}$
6. $33 = k - 8$; $+8 \quad +8$; $\mathbf{41 = k}$
7. $c - 4 = 27$; $+4 \quad +4$; $\mathbf{c = 31}$
8. $18 = d - 6$; $+6 \quad +6$; $\mathbf{24 = d}$

Basics of Algebra Exercise 4
page 208

1. $8y = 96$
$$\frac{8y}{8} = \frac{96}{8}$$
$\mathbf{y = 12}$

2. $11 = 2d$
$$\frac{11}{2} = \frac{2d}{2}$$
$\mathbf{5\frac{1}{2} = d}$

3. $15p = 75$
$$\frac{15p}{15} = \frac{75}{15}$$
$\mathbf{p = 5}$

4. $25z = 100$
$$\frac{25z}{25} = \frac{100}{25}$$
$\mathbf{z = 4}$

5. $\frac{x}{3} = 9$
$$3 \cdot \frac{x}{3} = 9 \cdot 3$$
$\mathbf{x = 27}$

6. $9 = \frac{m}{4}$
$$4 \cdot 9 = \frac{m}{4} \cdot 4$$
$\mathbf{36 = m}$

7. $\frac{a}{5} = 8$
$$5 \cdot \frac{a}{5} = 8 \cdot 5$$
$\mathbf{a = 40}$

8. $\frac{c}{9} = 2$
$$9 \cdot \frac{c}{9} = 2 \cdot 9$$
$\mathbf{c = 18}$

Basics of Algebra Exercise 5
pages 209–10

1. e
2. f
3. i
4. g
5. b
6. h

7. $x - 9 = 15$; $+9 \quad +9$; $\mathbf{x = 24}$

8. $6x = 27$
$$\frac{6x}{6} = \frac{27}{6}$$
$\mathbf{x = 4\frac{3}{6} = 4\frac{1}{2}}$

9. $\frac{x}{5} = 50$
$$5 \cdot \frac{x}{5} = 50 \cdot 5$$
$\mathbf{x = 250}$

10. $3 = x - 5$; $+5 \quad +5$; $\mathbf{8 = x}$

11. $60 = \frac{3}{4}x$
$$\frac{4}{3} \cdot 60 = \frac{3}{4}x \cdot \frac{4}{3}$$
$\mathbf{80 = x}$

12. $7x = 84$
$$\frac{7x}{7} = \frac{84}{7}$$
$\mathbf{x = 12}$

13. **$330**
$x - 40 = 290$; $+40 \quad +40$; $x = 330$

14. **$2,925**
$$\frac{x}{15} = 195$$
$$15 \cdot \frac{x}{15} = 195 \cdot 15$$
$x = 2,925$

15. $2080

$\frac{1}{4}x = 520$

$4 \cdot \frac{1}{4}x = 520 \times 4$

$x = \$2080$

16. 120 members

$.75x = 90$

$\frac{.75x}{.75} = \frac{90}{.75}$

$x = 120$

Basics of Algebra Exercise 6

page 211

1. e
2. g
3. d
4. f
5. h
6. a
7. c
8. b
9. $2x + 6$
10. $8x - 5$
11. $2x + 4x$ or $4x + 2x$
12. $\frac{1}{2}x + 8$ or $8 + \frac{1}{2}x$
13. $\frac{1}{3}x - 1$
14. $15 - 3x$
15. $\frac{x+7}{4}$ or $(x + 7)/4$
16. $10(x + 12)$ or $10(12 + x)$
17. $9(x - 4)$

Basics of Algebra Exercise 7

page 213

1. $7m - 2 = 54$

$+2 \quad +2$

$7m = 56$

$\frac{7m}{7} = \frac{56}{7}$

$\mathbf{m = 8}$

2. $\frac{a}{3} + 5 = 9$

$-5 \quad -5$

$\frac{a}{3} = 4$

$3 \cdot \frac{a}{3} = 4 \cdot 3$

$\mathbf{a = 12}$

3. $7 = \frac{c}{2} + 3$

$-3 \quad -3$

$4 = \frac{c}{2}$

$2 \cdot 4 = \frac{c}{2} \cdot 2$

$\mathbf{8 = c}$

4. $82 = 9d + 10$

$-10 \quad -10$

$72 = 9d$

$\frac{72}{9} = \frac{9d}{9}$

$\mathbf{8 = d}$

5. $25c - 17 = 183$

$+17 \quad +17$

$25c = 200$

$\frac{25c}{25} = \frac{200}{25}$

$\mathbf{c = 8}$

6. $\frac{w}{2} - 7 = 3$

$+7 \quad +7$

$\frac{w}{2} = 10$

$2 \cdot \frac{w}{2} = 10 \cdot 2$

$\mathbf{w = 20}$

7. $2 = 6x - 10$

$+10 \quad +10$

$12 = 6x$

$\frac{12}{6} = \frac{6x}{6}$

$\mathbf{2 = x}$

8. $\frac{1}{3}p + 8 = 11$

$-8 \quad -8$

$\frac{1}{3}p = 3$

$3 \cdot \frac{1}{3}p = 3 \cdot 3$

$\mathbf{p = 9}$

9. $40 = 13z + 14$

$-14 \quad -14$

$26 = 13z$

$\frac{26}{13} = \frac{13z}{13}$

$\mathbf{2 = z}$

10. $\frac{n}{2} + 3 = 7$

$-3 \quad -3$

$\frac{n}{2} = 4$

$2 \cdot \frac{n}{2} = 4 \cdot 2$

$\mathbf{n = 8}$

11. $\frac{3}{4}y - 3 = 12$

$+3 \quad +3$

$\frac{3}{4}y = 15$

$\frac{4}{3} \cdot \frac{3}{4}y = 15 \cdot \frac{4}{3}$

$\mathbf{y = 20}$

12. $39 = 16k - 9$

$+9 \quad +9$

$48 = 16k$

$\frac{48}{16} = \frac{16k}{16}$

$\mathbf{3 = k}$

13. $10 = 6a + 7$

$-7 \quad -7$

$3 = 6a$

$\frac{3}{6} = \frac{6a}{6}$

$\mathbf{\frac{1}{2} = a}$

14.
$$\begin{aligned} 9r + 15 &= 18 \\ -15 &\quad -15 \\ 9r &= 3 \\ \frac{9r}{9} &= \frac{3}{9} \\ \boldsymbol{r} &= \boldsymbol{\frac{1}{3}} \end{aligned}$$

Basics of Algebra Exercise 8
page 214

1.
$$\begin{aligned} 3(x + 4) &= 27 \\ 3x + 12 &= 27 \\ 3x &= 27 - 12 \\ 3x &= 15 \\ \boldsymbol{x} &= \boldsymbol{5} \end{aligned}$$

2.
$$\begin{aligned} 10 &= 5(a - 3) \\ 10 &= 5a - 15 \\ 25 &= 5a \\ \boldsymbol{5} &= \boldsymbol{a} \end{aligned}$$

3.
$$\begin{aligned} 36 &= 4(c + 4) \\ 36 &= 4c + 16 \\ 20 &= 4c \\ \boldsymbol{5} &= \boldsymbol{c} \end{aligned}$$

4.
$$\begin{aligned} 28 &= 7m - 32 \\ 60 &= 7m \\ \frac{60}{7} &= m \\ \boldsymbol{8\tfrac{4}{7}} &= \boldsymbol{m} \end{aligned}$$

5.
$$\begin{aligned} 3(y - 2) &= 3 \\ 3y - 6 &= 3 \\ 3y &= 9 \\ \boldsymbol{y} &= \boldsymbol{3} \end{aligned}$$

6.
$$\begin{aligned} 80 &= 8(s + 7) \\ 80 &= 8s + 56 \\ 24 &= 8s \\ \boldsymbol{3} &= \boldsymbol{s} \end{aligned}$$

7.
$$\begin{aligned} 3(t + 10) &= 90 \\ 3t + 30 &= 90 \\ 3t &= 60 \\ \boldsymbol{t} &= \boldsymbol{20} \end{aligned}$$

8.
$$\begin{aligned} 7(m - 2) &= 28 \\ 7m - 14 &= 28 \\ 7m &= 42 \\ \boldsymbol{m} &= \boldsymbol{6} \end{aligned}$$

Basics of Algebra Exercise 9
page 216

1.
$$\begin{aligned} 5y - y &= 19 + 9 \\ 4y &= 28 \\ \frac{4y}{4} &= \frac{28}{4} \\ \boldsymbol{y} &= \boldsymbol{7} \end{aligned}$$

2.
$$\begin{aligned} 6t + 8 + 4t &= 58 \\ 8 + 10t &= 58 \\ -8 &\quad -8 \\ 10t &= 50 \\ \frac{10t}{10} &= \frac{50}{10} \\ \boldsymbol{t} &= \boldsymbol{5} \end{aligned}$$

3.
$$\begin{aligned} 9c &= 44 - 2c \\ +2c &= \quad +2c \\ 11c &= 44 \\ \frac{11c}{11} &= \frac{44}{11} \\ \boldsymbol{c} &= \boldsymbol{4} \end{aligned}$$

4.
$$\begin{aligned} 8m &= 2m + 30 \\ -2m &\quad -2m \\ 6m &= 30 \\ \frac{6m}{6} &= \frac{30}{6} \\ \boldsymbol{m} &= \boldsymbol{5} \end{aligned}$$

5.
$$\begin{aligned} 4a + 55 &= 9a \\ -4a &\quad -4a \\ 55 &= 5a \\ \frac{55}{5} &= \frac{5a}{5} \\ \boldsymbol{11} &= \boldsymbol{a} \end{aligned}$$

6.
$$\begin{aligned} 4p &= p + 18 \\ -p &\quad -p \\ 3p &= 18 \\ \frac{3p}{3} &= \frac{18}{3} \\ \boldsymbol{p} &= \boldsymbol{6} \end{aligned}$$

7.
$$\begin{aligned} 6f &= 14 - f \\ +f &\quad +f \\ 7f &= 14 \\ \frac{7f}{7} &= \frac{14}{7} \\ \boldsymbol{f} &= \boldsymbol{2} \end{aligned}$$

8.
$$\begin{aligned} 3 &= y + 8y \\ 3 &= 9y \\ \frac{3}{9} &= \frac{9y}{9} \\ \boldsymbol{\frac{1}{3}} &= \boldsymbol{y} \end{aligned}$$

9.
$$\begin{aligned} 8r + 17 &= 5r + 32 \\ -5r &\quad -5r \\ 3r + 17 &= 32 \\ -17 &\quad -17 \\ \frac{3r}{3} &= \frac{15}{3} \\ \boldsymbol{r} &= \boldsymbol{5} \end{aligned}$$

10.
$$\begin{aligned} 7n - 9 &= 3n + 7 \\ -3n &\quad -3n \\ 4n - 9 &= 7 \\ +9 &\quad +9 \\ \frac{4n}{4} &= \frac{16}{4} \\ \boldsymbol{n} &= \boldsymbol{4} \end{aligned}$$

11.
$$\begin{aligned} 6z + 11 &= 5z + 20 \\ -5z \quad & \quad -5z \\ z + 11 &= 20 \\ -11 \quad & \quad -11 \\ z &= 9 \end{aligned}$$

12.
$$\begin{aligned} 5y - 4 &= 2y + 77 \\ +4 \quad & \quad +4 \\ 5y &= 2y + 81 \\ -2y \quad & \quad -2y \\ 3y &= 81 \\ y &= 27 \end{aligned}$$

Using Algebra to Solve Word Problems

page 217

(2) 8

Bill's age = x

Stan's age = $2x + 1$

$$\begin{aligned} 2x + 1 &= 17 \\ 2x &= 16 \\ x &= 8 \end{aligned}$$

Basics of Algebra Exercise 10

page 218

1. **13 and 14**

First number = x

Next number = $x + 1$

(Consecutive means one following another.)

$$\begin{aligned} x + x + 1 &= 27 \\ 2x + 1 &= 27 \\ -1 \quad & \quad -1 \\ 2x &= 26 \\ \frac{2x}{2} &= \frac{26}{2} \\ x &= 13 \\ x + 1 &= 14 \end{aligned}$$

2. **18, 20, and 22**

First number = x

Second number = $x + 2$

Third number = $x + 2 + 2 = x + 4$

$$\begin{aligned} x + x + 2 + x + 4 &= 60 \\ 3x + 6 &= 60 \\ -6 \quad & \quad -6 \\ 3x &= 54 \\ \frac{3x}{3} &= \frac{54}{3} \\ x &= 18 \\ x + 2 &= 20 \\ x + 4 &= 22 \end{aligned}$$

3. **24 union workers**

Nonunion workers = x

Union workers = $6x$

$$\begin{aligned} 6x - 8 &= 4x \\ +8 \quad & \quad +8 \\ 6x &= 4x + 8 \\ 6x - 4x &= 8 \\ 2x &= 8 \\ x &= 4 \\ 6x &= 24 \end{aligned}$$

4. **6 men and 18 women**

Number of men = x

Number of women = $3x$

$$\begin{aligned} 3x - 2 &= x + 10 \\ -x \quad & \quad -x \\ 2x - 2 &= 10 \\ +2 \quad & \quad +2 \\ 2x &= 12 \\ \frac{2x}{2} &= \frac{12}{2} \\ x &= 6 \\ 3x &= 3(6) = 18 \end{aligned}$$

5. **$270**

Juan's wage = x

Felipe's wage = $3x$

$$\begin{aligned} x + 3x &= 360 \\ 4x &= 360 \\ \frac{4x}{4} &= \frac{360}{4} \\ x &= 90 \\ 3x &= 270 \end{aligned}$$

6. **(2) $\frac{294}{3}$**

Mrs. Migliaccio's salary = x

Mr. Migliaccio's salary = $2x$

$$\begin{aligned} 3x &= 294 \\ x &= \frac{294}{3} \end{aligned}$$

7. **24 hours**

Jerry's hours = x

Jeff's hours = $2x$

Paul's hours = $2x + 6$

$$\begin{aligned} x + 2x + 2x + 6 &= 51 \\ 5x + 6 &= 51 \\ -6 \quad & \quad -6 \\ 5x &= 45 \\ \frac{5x}{5} &= \frac{45}{5} \\ x &= 9 \\ 2x + 6 &= 24 \end{aligned}$$

8. **16,500 people**

Voters who voted = x

People who did not vote = $3x$

$$\begin{aligned} x + 3x &= 22{,}000 \\ 4x &= 22{,}000 \\ \frac{4x}{4} &= \frac{22{,}000}{4} \\ x &= 5{,}500 \\ 3x &= 16{,}500 \end{aligned}$$

9. **(4) $(4 \times 18) + 2$**

$$\begin{aligned} x &= \text{Laila's age} \\ 4x &= \text{David's age} \\ 4x + 2 &= \text{Nancy's age} \end{aligned}$$

$(4 \times 18) + 2$

10. **(1) $(\frac{1}{2} \times 40) + 5$**

x = height of house

$\frac{1}{2}x$ = height of ladder

Basics of Algebra Exercise 11

page 221

1. 419 mph
$$d = r \cdot t$$
$$1{,}676 = r \cdot 4$$
$$\frac{1{,}676}{4} = \frac{r \cdot 4}{4}$$
$$419 = r$$

2. 89 hammers
$$c = n \cdot r$$
$$\$111.25 = n \cdot \$1.25$$
$$\frac{111.25}{1.25} = \frac{n \cdot 1.25}{1.25}$$
$$89 = n$$

3. 68 degrees
$$F = (\tfrac{9}{5})C + 32$$
$$F = (\tfrac{9}{5}) \cdot 20 + 32$$
$$F = 36 + 32$$
$$F = 68$$

4. $1,300
$$i = prt$$
$$\$32.50 = p \cdot 5\% \cdot \frac{1}{2}$$
$$\$32.50 = p \cdot \frac{5}{100} \cdot \frac{1}{2}$$
$$\$32.50 = p \cdot \frac{5}{200}$$
$$\$32.50 = p \cdot \frac{1}{40}$$
$$\$32.50 \cdot 40 = p$$
$$\$1{,}300 = p$$

Basics of Algebra Exercise 12

pages 222–23

1. $x + 11$
Remember, you can use x or any other letter to stand for the unknown.)

2. $3a + \frac{1}{3}a$

3. $4c$

4. $\frac{4}{5}s$

5. $9x = 108$
$\mathbf{x = 12}$

6. $\frac{c}{4} = 8$
$\mathbf{c = 32}$

7.
$$14 + x = 3x - 8$$
$$+8 \qquad\qquad +8$$
$$22 + x = 3x$$
$$22 = 2x$$
$$\mathbf{11 = x}$$

8.
$$\frac{21}{p} + 1 = 8$$
$$\frac{21}{p} = 7$$
$$\mathbf{p = 3}$$

9.
$$\frac{1}{8}y = 24$$
$$\mathbf{y = 192}$$

10.
$$12.5 = a + 1.5$$
$$\mathbf{11 = a}$$

11.
$$96 + 2x = x + 101$$
$$2x = x + 5$$
$$\mathbf{x = 5}$$

12.
$$20 - 4z = 5z - 35$$
$$55 = 9z$$
$$\frac{55}{9} = z$$
$$6\tfrac{1}{9} = z$$

13.
$$x - 18 = 1$$
$$\mathbf{x = 19}$$

14.
$$\frac{4}{5}b = 40$$
$$b = 40 \cdot \frac{5}{4}$$
$$\mathbf{b = 50}$$

15.
$$9y - 5 = 6y + 7$$
$$3y = 12$$
$$\mathbf{y = 4}$$

16. (2) 4.25
$$\begin{array}{r} 7.50 \\ -\ 3.25 \\ \hline 4.25 \end{array}$$

17. (4) $840
$$.10 \times t = 84$$
$$t = \frac{84}{.10}$$
$$t = \$840$$

18. (3) 70
Let x = number of men
$x + 25$ = number of women
$$x + x + 25 = 115$$
$$2x + 25 = 115$$
$$2x = 90$$
$$x = 45$$
$$x + 25 = 70$$

19. (2) $4.50
Let x = Joe's wage
$x - 2$ = Sam's wage
$$40(x + x - 2) = \$440$$
$$40(2x - 2) = 440$$
$$80x - 80 = 440$$
$$80x = 520$$
$$x = \$6.50$$
$$x - 2 = \$4.50$$

20. (4) 26
Let x = Ed's miles
$x + 6$ = Nina's miles
$$x + x + 6 = 46$$
$$2x + 6 = 46$$
$$2x = 40$$
$$x = 20$$
$$x + 6 = 26$$

21. (4) $46

$46 \times \$1.00 = \$46$

22. (2) Ed's rate of walking

23. (3) $3(46 \times \$.50)$

24. (4) $1020

$$\begin{aligned} \text{Let } x &= \text{Kathleen's wages} \\ x + \$280 &= \text{Mel's wages} \\ x + x + 280 &= 1760 \\ 2x + 280 &= 1760 \\ 2x &= 1480 \\ x &= 740 \\ x + 280 &= \$1020 \end{aligned}$$

25. (1) 12

$$\begin{aligned} 10n - 7 &= 101 + n \\ 9n &= 108 \\ n &= 12 \end{aligned}$$

CHAPTER 11

Geometry Exercise 1

pages 226–27

1. parallel
2. square
3. perpendicular
4. angle
5. vertical
6. horizontal
7. diameter
8. radius
9. sphere
10. right angle
11. rectangle
12. triangle
13. cube
14. rectangular solid
15. cylinder

Geometry Exercise 2

pages 229–30

1. 178 ft

$P = 2l + 2w$

$P = (2 \times 52) + (2 \times 37)$

$P = 104 + 74 = 178$ ft

2. 57 cm

$P = a + b + c$

$P = 19 + 19 + 19 = 57$ cm

3. 32 yd

$P = 4s$

$P = 4 \times 8 = 32$ yd

4. $8\frac{1}{2}$ in

$P = 2l + 2w$

$P = 2(2\frac{3}{4}) + 2(1\frac{1}{2})$

$P = (\frac{2}{1} \times \frac{11}{4}) + (\frac{2}{1} \times \frac{3}{2})$

$P = \frac{11}{2} + 3$

$P = 5\frac{1}{2} + 3 = 8\frac{1}{2}$ in

5. 2 in

$P = 4s$

$P = 4(\frac{1}{2}) = 2$ in

6. $27\frac{1}{2}$ in

$P = a + b + c$

$$\begin{aligned} 7\tfrac{3}{4} &= 7\tfrac{3}{4} \\ 9\tfrac{1}{4} &= 9\tfrac{1}{4} \\ +\ 10\tfrac{1}{2} &= 10\tfrac{2}{4} \\ & \quad 26\tfrac{6}{4} = 27\tfrac{1}{2} \text{ in} \end{aligned}$$

7. 23 mi

$P = 2l + 2w$

$P = 2(6.3) + 2(5.2)$

$P = 12.6 + 10.4 = 23$ mi

8. 7 m

$P = 4s$

$P = 4(1.75) = 7.00 = 7$ m

9. 35 ft

$P = a + b + c$

$P = 13 + 13 + 9 = 35$ ft

10. 31.4 yd

$C = \pi d$

$C = 3.14 \times 10 = 31.4$ yd

11. 132 in

$d = 2 \times 21 = 42$ in

$C = \pi d$

$C = \frac{22}{7} \times \frac{42}{1} = 132$ in

12. 4.7 cm

$C = \pi d$

$C = 3.14 \times 1.5 =$

$$\begin{array}{r} 3.14 \\ \times\ 1.5 \\ \hline 1\,570 \\ 3\,14 \\ \hline 4.710 \end{array}$$

4.710 to the nearest tenth = 4.7 cm

13. 44 yd

$C = \pi d$

$C = \frac{22}{7} \times \frac{14}{1} = 44$ yd

Drawing Pictures

page 231

(5) 152

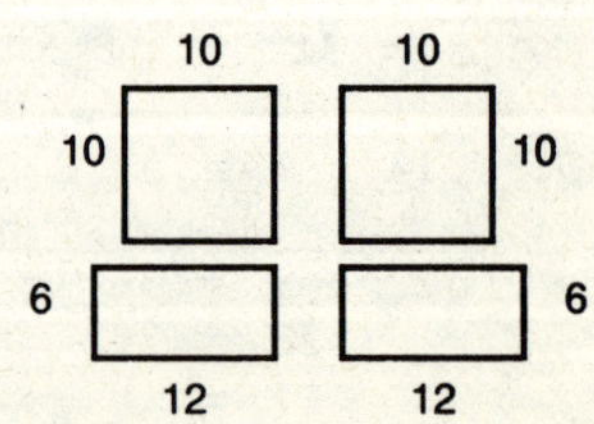

perimeter of each square:

$P = 4s =$

$4 \times 10 = 40$

perimeter of each rectangle:

$P = 2l + 2w =$

$(2 \times 12) + (2 \times 6)$

$24 + 12$

36

$40 + 40 + 36 + 36 = 152$

Geometry Exercise 3
pages 232–33

1. **36 in**
   $P = 2l + 2w$
   $P = (2 \times 8) + (2 \times 10)$
   $P = 16 + 20 = 36$ in

2. **78 m**
   width $= \frac{1}{2} \times 26 = 13$ m
   $P = 2l + 2w$
   $P = (2 \times 26) + (2 \times 13)$
   $P = 52 + 26 = 78$ m

3. **25 in**
   To find the side, divide the perimeter by 4.
   $100 \div 4 = 25$ in

4. **9 ft 2 in**
   $c = \pi d$
   $\frac{22}{7} \times \frac{35}{1} = 110$ in $= 9$ ft 2 in

5. **156 in**
   $P = 2l + 2w$
   $P = 2 \times 36 + 2 \times 42$
   $P = 72 + 84 = 156$ in

6. **78 ft**
   $156 \times 6 = 936$ $\frac{1\text{ ft}}{12\text{ in}} = \frac{x}{936}$ $936 \times 1 = 936$ in $936 \div 12 = 78$ ft

7. **$46.80**
   $78 \times \$.60 = \$46.80$

8. **12 yd**
   $P = a + b + c$
   $P = 12 + 15 + 9 = 36$ ft
   $\frac{1\text{ yd}}{3\text{ ft}} = \frac{x}{36}$ $36 \times 1 = 36$ $36 \div 3 = 12$ yd

9. **210 bricks**
   First find the circumference of the pool in feet.
   $C = \pi d$
   $C = 3.14 \times 50 = 157$ ft
   Change 157 feet to inches.
   $\frac{1\text{ ft}}{12\text{ in}} = \frac{157}{x}$ 157×12: 314, 157, $= 1884$ $1884 \div 1 = 1884$ in
   Divide the perimeter by the length of one brick.
   $1884 \div 9 = 209\frac{1}{3}$ or 210 bricks

10. **7 feet**
    First change 30 inches to feet.
    $\frac{1}{12} = \frac{x}{30}$ $30 \div 12 = 2\frac{1}{2}$ ft
    Then substitute the information into the perimeter formula.
    $P = 2l + 2w$
    $19 = 2l + 2(2\frac{1}{2})$
    $19 = 2l + 5$
    Use inverse operations to solve for *P*.
    $19 = 2l + 5$
    $19 - 5 = 2l$
    $14 = 2l$
    $7 = l$

11. **(3)** $(2 \times 2.3) + (2 \times 1.4)$
12. **(1)** $2(1.6) + 2.4$

Geometry Exercise 4
page 236

1. $6 \times 6 =$ **36**
2. $1 \times 1 \times 1 =$ **1**
3. $5 \times 5 \times 5 =$ **125**
4. $\frac{3}{5} \times \frac{3}{5} = \mathbf{\frac{9}{25}}$
5. $3 \times 3 \times 3 \times 3 \times 3 =$ **243**
6. $5 \times 5 = 25$
   $3 \times 3 = 9$
   $25 + 9 =$ **34**
7. $5 \times 5 \times 5 = 125$
   $10 \times 10 = 100$
   $125 - 100 =$ **25**
8. $.06 \times .06 =$ **.0036**
9. **1**
   Any number to the zero power is 1.
10. $25 \times 25 =$ **625**
11. $11 \times 11 = 121$
    $4 \times 4 = 16$
    $121 - 16 =$ **105**
12. $.005 \times .005 =$ **.000025**
13. $1 + 10 + 100 =$ **111**
14. $20 \times 20 \times 20 =$ **8000**
15. **13**
    Any number to the first power is that number.
16. $\frac{1}{4} \times \frac{1}{4} \times \frac{1}{4} = \mathbf{\frac{1}{64}}$
17. $10 \times 10 \times 10 \times 10 =$ **10,000**
18. $8 \times 8 = 64$
    $2 \times 2 \times 2 \times 2 = 16$
    $64 - 16 =$ **48**
19. $3 \times 3 \times 3 = 27$
    $7 \times 7 = 49$
    $27 + 49 =$ **76**
20. $0.1 \times 0.1 \times 0.1 =$ **0.001**

Geometry Exercise 5

page 238

1. $\sqrt{196} = 14$

Guess 12 because 12 × 12 = 144.

```
   16      16      14
12)196   + 12    2)28
   12      28
   76
   72
    4
```

2. $\sqrt{441} = 21$

Guess 20 because 20 × 20 = 400.

```
   22      22      21
20)441   + 20    2)42
   40      42
   41
   40
    1
```

3. $\sqrt{1024} = 32$

Guess 30 because 30 × 30 = 900.

```
   34      34      32
30)1024  + 30    2)64
   90      64
   124
   120
     4
```

4. $\sqrt{1521} = 39$

Guess 40 because 40 × 40 = 1600.

```
   38      38      39
40)1521  + 40    2)78
   120     78
   321
   320
     1
```

5. $\sqrt{361} = 19$

Guess 20 because 20 × 20 = 400.

```
   18      18      19
20)361   + 20    2)38
   20      38
   161
   160
     1
```

6. $\sqrt{676} = 26$

Guess 30 because 30 × 30 = 900.

```
   22      22      26
30)676   + 30    2)52
   60      52
   76
   60
   16
```

7. $\sqrt{1849} = 43$

Guess 40 because 40 × 40 = 1600.

```
   46      46      43
40)1849  + 40    2)86
   160     86
   249
   240
     9
```

8. $\sqrt{3364} = 58$

Guess 60 because 60 × 60 = 3600.

```
   56      56      58
60)3364  + 60    2)116
   300     116
   364
   360
     4
```

9. $\sqrt{4624} = 68$

Guess 70 because 70 × 70 = 4900.

```
   66      66      68
70)4624  + 70    2)136
   420     136
   424
   420
     4
```

10. $\sqrt{6241} = 79$

Guess 80 because 80 × 80 = 6400.

```
   78      78      79
80)6241  + 80    2)158
   560     158
   641
   640
     1
```

11. $\sqrt{7396} = 86$

Guess 90 because 90 × 90 = 8100.

```
   82      82      86
90)7396  + 90    2)172
   720     172
   196
   180
    16
```

12. $\sqrt{9604} = 98$

Guess 100 because 100 × 100 = 10,000.

```
    96      96      98
100)9604  + 100   2)196
    900     196
    604
    600
      4
```

Geometry Exercise 6

pages 240–41

1. 180 ft² $A = lw$
$A = 15 \times 12 = 180$ ft²

2. 196 ft² $A = s^2$
$A = 14^2$
$A = 14 \times 14 = 196$ ft²

3. 57 in² $A = \frac{1}{2}(bh)$
$\frac{19 \times 6}{2} = 57$ in²

4. 28 in² $A = lw$
$A = 8 \times 3\frac{1}{2}$
$\frac{8}{1} \times \frac{7}{2} = 28$ in²

5. 42.25 m² $A = s^2$
$A = 6.5^2$
$A = 6.5 \times 6.5 = 42.25 \text{ m}^2$

6. $25\frac{1}{2}$ in² $A = \frac{1}{2}bh$
$A = \frac{1}{2} \times 8\frac{1}{2} \times 6$
$A = \frac{1}{2} \times \frac{17}{2} \times 6$
$A = \frac{51}{2} = 25\frac{1}{2} \text{ in}^2$

7. 68 cm²
$A = lw$
$A = 12.6 \times 5.4$
$A = 68.04$ to the nearest whole $= 68 \text{ cm}^2$

8. $20\frac{1}{4}$ in²
$A = s^2$
$A = (4\frac{1}{2})^2$
$A = 4\frac{1}{2} \times 4\frac{1}{2}$
$\frac{9}{2} \times \frac{9}{2} = \frac{81}{4} = 20\frac{1}{4} \text{ in}^2$

9. (4) $\frac{1}{2}(12 \times 9)$
$A = \frac{1}{2}(bh)$
$A = \frac{1}{2}(12 \times 9)$

10. 0.009 m²
$A = lw$
$A = 0.15 \times 0.06$
$A = 0.0090 = 0.009 \text{ m}^2$

11. $\frac{9}{64}$ in²
$A = s^2$
$A = (\frac{3}{8})^2$
$\frac{3}{8} \times \frac{3}{8} = \frac{9}{64} \text{ in}^2$

12. $76\frac{1}{2}$ cm²
$A = \frac{(bh)}{2}$
$A = \frac{17 \times 9}{2} = \frac{153}{2} = 76\frac{1}{2} \text{ cm}^2$

13. 314 yd²
$A = \pi r^2$
$A = 3.14 \times 10^2$
$A = 3.14 \times 100 = 314 \text{ yd}^2$

14. 154 ft²
$A = \pi r^2$
$A = \frac{22}{7} \times 7^2$
$A = \frac{22}{7} \times \frac{49}{1} = 154 \text{ ft}^2$

15. a. radius = 0.5 m
b. 0.785 m²
$A = \pi r^2$
$A = 3.14 \times 0.5^2$
$A = 3.14 \times 0.5 \times 0.5$
$A = 3.14 \times 0.25$
$A = 0.7850 = 0.785 \text{ m}^2$

16. 4.5 m²
$A = \pi r^2$
$A = 3.14 \times 1.2^2$
$A = 3.14 \times 1.2 \times 1.2$
$A = 3.14 \times 1.44$
$A = 4.5216$ to the nearest tenth $= 4.5 \text{ m}^2$

Geometry Exercise 7
pages 243–44

1. 108 ft²
$A = lw$
$A = 12 \times 9 = 108 \text{ ft}^2$

2. 9 ft
Area of rectangle: $3 \times 27 = 81 \text{ ft}^2$
Area of square: $A = s^2$
$81 = s^2$
$\sqrt{81} = s$
$9 = s$

3. 80 in²
$A = lw$
$A = 8 \times 10 = 80 \text{ in}^2$

4. 45 tiles
Area of room: $A = lw$
$A = 18 \times 10 = 180 \text{ ft}^2$
Area of one tile: $A = s^2$
$A = 2^2$
$A = 2 \times 2 = 4 \text{ ft}^2$
$180 \div 4 = 45$ tiles

5. 5 gallons
$A = lw$
$A = 25 \times 40 = 1000 \text{ ft}^2$
$1000 \div 200 = 5$ gallons

6. 6 sq yd
$A = \frac{bh}{2}$
$\frac{12 \times 9}{2} = 54 \text{ ft}^2$
$\frac{1 \text{ sq yd}}{9 \text{ sq ft}} = \frac{x}{54}$ $54 \times 1 = 54$ $54 \div 9 = 6$ sq yd

7. 16 yd²
$A = lw$
$A = 6 \times 8 = 48 \text{ ft}^2$
3 windows $= 3 \times 48 = 144 \text{ ft}^2$
$\frac{1 \text{ sq yd}}{9 \text{ sq ft}} = \frac{x}{144}$ $144 \times 1 = 144$ $144 \div 9 = 16 \text{ yd}^2$

8. 78.5 ft²
radius $= \frac{1}{2} \times 10 = 5$ ft
$A = \pi r^2$
$A = 3.14 \times 5^2$
$A = 3.14 \times 25 = 78.5 \text{ ft}^2$

9. 1256 ft²
$A = \pi r^2$
$A = 3.14 \times 20^2$
$A = 3.14 \times 400 = 1256 \text{ ft}^2$

10. 20 ft
One side is the square root of the area.
$\sqrt{400} = 20$ ft

11. 40 ft
Divide the area by 10 feet.
$10\overline{)400}$ = 40 ft

12. (2) 8×12.5

13. (4) $(5.8 \times 6)/2$

Geometry Exercise 8
pages 246–47

1. 1620 in³ $V = lwh$
$V = 12 \times 9 \times 15 = 1620 \text{ in}^3$

2. 512 in³ $V = s^3$
$V = 8^3$
$V = 8 \times 8 \times 8 = 512 \text{ in}^3$

3. 352 m³ $V = \pi r^2 h$
$V = \frac{22}{7} \times 4^2 \times 7$
$V = 22 \times 16$
$V = 352 \text{ m}^3$

4. 49 in³ $V = lwh$
$V = 5\frac{1}{4} \times 2\frac{2}{3} \times 3\frac{1}{2}$
$\frac{21}{4} \times \frac{8}{3} \times \frac{7}{2} = 49 \text{ in}^3$

5. 1.728 cm³ $V = s^3$
$V = 1.2^3$
$V = 1.2 \times 1.2 \times 1.2 = 1.728 \text{ cm}^3$

6. 75 ft³ $V = lwh$
$\frac{30}{1} \times \frac{1}{2} \times \frac{5}{1} = 75 \text{ ft}^3$

7. 10,990 ft³ $V = \pi r^2 h$
$V = 3.14 \times 10^2 \times 35$
$V = 3.14 \times 100 \times 35 = 10{,}990 \text{ ft}^3$

8. $\frac{1}{8}$ in³ $V = s^3$
$V = (\frac{1}{2})^3$
$\frac{1}{2} \times \frac{1}{2} \times \frac{1}{2} = \frac{1}{8} \text{ in}^3$

9. 64 m³ $V = lwh$
$V = 8 \times 4 \times 2 = 64 \text{ m}^3$

10. 1.3 m³
$V = \pi r^2 h$
$V = 3.14 \times 0.4^2 \times 2.5$
$V = 3.14 \times 0.16 \times 2.5$
$V = 1.256$ to the nearest tenth $= 1.3 \text{ m}^3$

11. $1\frac{61}{64}$ in³ $V = s^3$
$V = (1\frac{1}{4})^3$
$\frac{5}{4} \times \frac{5}{4} \times \frac{5}{4} = \frac{125}{64} = 1\frac{61}{64} \text{ in}^3$

12. 113.04 in³ $V = \frac{4}{3}\pi r^3$
$V = (\frac{4}{3}) \times 3.14 \times 3^3$
$\frac{4}{3} \times \frac{3.14}{1} \times \frac{27}{1} = 113.04 \text{ in}^3$

13. 864 sq in
First change 2 feet to inches.
2 feet = 24 in
$V = lwh$
$V = 6 \times 24 \times 6$
$V = 864$ sq in

14. 12 boxes
First find volume of box in inches.
$V = lwh$
$V = 12 \times 6 \times 1$
$V = 72$ sq in
Then divide into volume of carton.
$72\overline{)864}$ = 12

Geometry Exercise 9
pages 249–51

1. 66 in² $A = bh$
$A = 11 \times 6 = 66 \text{ in}^2$

2. 1256 in² $A = 4\pi r^2$
$A = 4 \times 3.14 \times 10^2$
$A = 4 \times 3.14 \times 100 = 1256 \text{ in}^2$

3. 136 ft²
$A = \frac{h(b_1 + b_2)}{2}$
$A = \frac{8(14 + 20)}{2}$
$A = \frac{8 \times 34}{2} = 136 \text{ ft}^2$

4. 42 in
Find the measurement of the horizontal side at the bottom part of the figure. $10 - 7 = 3$ in

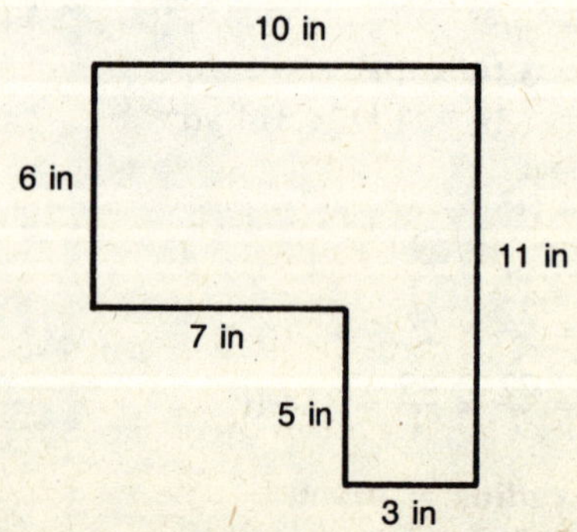

Find the measurement of the vertical side at the right of the figure. $6 + 5 = 11$
Total perimeter:
$P = 10 + 11 + 3 + 5 + 7 + 6 = 42$ in

5. 75 sq in
Area of top: $A = 6 \times 10 = 60$ sq in
Area of bottom: $A = 5 \times 3 = 15$ sq in
Add the areas: $60 + 15 = 75$ sq in

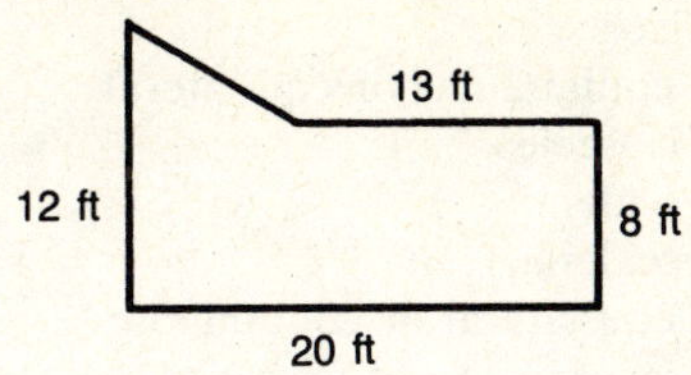

6. 174 ft²
Find the measurements of the base and height of the triangular part of the figure.
$b = 20 - 13 = 7$ ft and $h = 12 - 8 = 4$ ft

Area of triangle: $A = \frac{1}{2}(bh)$
$A = \frac{7 \times 4}{2} = 14 \text{ ft}^2$

Area of rectangle: $A = lw$
$A = 20 \times 8 = 160 \text{ ft}^2$
Add the areas: $160 + 14 = 174 \text{ ft}^2$

7. (5) $12^2 - \pi(4)^2$

8. 32 yd²
Area of dining room: $A = lw$
$A = 9 \times 8 = 72 \text{ ft}^2$
Area of living room: $A = lw$
$A = 18 \times 12 = 216 \text{ ft}^2$
Add the areas: $72 + 216 = 288 \text{ ft}^2$
Change square feet to square yards:
$\frac{1 \text{ sq yd}}{9 \text{ sq ft}} = \frac{x}{288}$ $288 \times 1 = 288$ $288 \div 9 = 32 \text{ yd}^2$

9. 888 ft²
Total area: $A = lw$
$A = 50 \times 30 = 1500 \text{ ft}^2$
Area of pool: $A = bh$
$A = 34 \times 18 = 612 \text{ ft}^2$
Subtract: $1500 \text{ ft}^2 - 612 = 888 \text{ ft}^2$

10. Cylinder D ($r = 3$ inches and $h = 10$ inches). Cylinders A, B, and C are too wide, and cylinder E is too long.

11. (2) The amount of dirt in each truckload.

12. 8 in
Area of rectangle: $A = lw$
$A = 10 \times 6.4 = 64 \text{ in}^2$
Find the square root of 64. $\sqrt{64} = 8$ in

13. 38,130 cm³
Volume of rectangular solid:
$V = lwh$
$V = 40 \times 30 \times 20 = 24{,}000 \text{ cm}^3$

Volume of half cylinder:
(This formula is simply the volume of a cylinder divided by 2. The radius is 10 cm. The height of the cylinder is 40 cm.)

$V = \frac{\pi r^2 h}{2}$

$V = \frac{\pi \cdot (15^2) \cdot 40}{2}$

$V = 14{,}130$
Add the volumes:
$24{,}000 + 14{,}130 = 38{,}130 \text{ cm}^3$

14. 15
Area of rectangle at left: $A = lw$
$A = 12 \times 10 = 120$
Divide 120 by the height of the rectangle at the right.
$120 \div 8 = 15$

15. (5) Not enough information is provided. Find the measurement of the horizontal side at the bottom of the figure. $8 + 5 = 13$. Find the measurement of the vertical side at the right of the figure. $6 + ? = ?$ You do not have enough information to find the area.

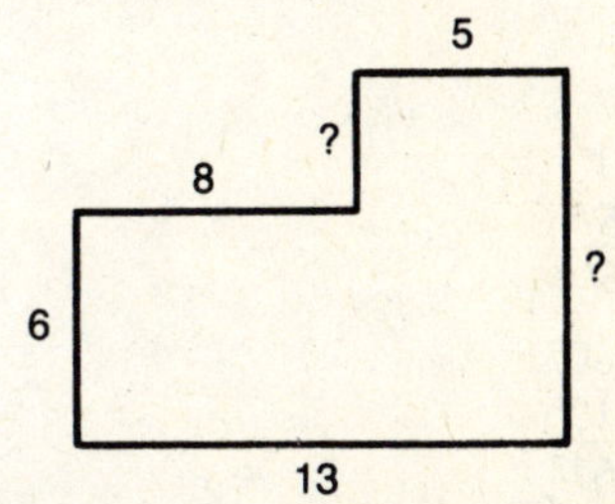

16. (3) 140 Area of rectangle: $A = lw$
$= 12 \times 10$
$= 120$

Area of triangle: $A = \frac{1}{2}bh$
$= \frac{1}{2} \times 4 \times 10$
$= 20$
Total Area: $120 + 20 = 140$

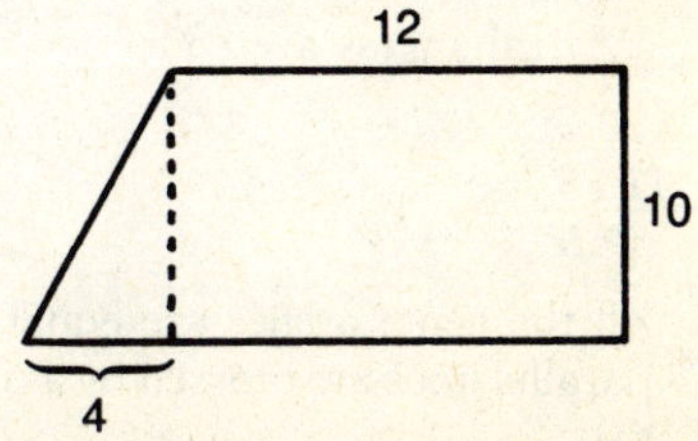

Geometry Exercise 10
page 253

1. acute
2. right
3. obtuse
4. reflex
5. right
6. straight
7. acute
8. obtuse
9. reflex
10. straight
11. right
12. acute
13. acute
14. obtuse
15. straight
16. reflex
17. right
18. acute
19. obtuse
20. reflex

Geometry Exercise 11
pages 255–56

1. $\begin{array}{r} 90^\circ \\ -\ 48^\circ \\ \hline \mathbf{42^\circ} \end{array}$

2. $\begin{array}{r} 180^\circ \\ -\ 48^\circ \\ \hline \mathbf{132^\circ} \end{array}$

3. $\begin{array}{r} 90^\circ \\ -\ 25^\circ \\ \hline \mathbf{65^\circ} \end{array}$

4. $\begin{array}{r} 180^\circ \\ -\ 63^\circ \\ \hline \mathbf{117^\circ} \end{array}$

5. $\angle w$

6. 360°

7. $\begin{array}{r} 180^\circ \\ -\ 75^\circ \\ \hline \mathbf{105^\circ} \end{array}$

8. $\begin{array}{r} 180^\circ \\ -\ 119^\circ \\ \hline \mathbf{61^\circ} \end{array}$

9. $\begin{array}{r} 180^\circ \\ -\ 80^\circ \\ \hline 100^\circ \end{array}$ $\quad \begin{array}{r} \mathbf{50^\circ} \\ 2\overline{)100^\circ} \end{array}$

10. obtuse
The supplement to an angle that measures less than 90° must be more than 90°, and the two angles together must total 180°.

11. $\begin{array}{r} 180^\circ \\ -\ 45^\circ \\ \hline \mathbf{135^\circ} \end{array}$

12. 37° Vertical angles are equal.

13. $\begin{array}{r} 180.0^\circ \\ -\ 110.5^\circ \\ \hline \mathbf{69.5^\circ} \end{array}$

14. 30° All the acute angles are equal when parallel lines are crossed by a transversal.

15. $\begin{array}{r} 180^\circ \\ -\ 120^\circ \\ \hline \mathbf{60^\circ} \end{array}$ The two angles are supplementary.

Geometry Exercise 12
pages 259–60

1. a. scalene
b. right
c. isosceles
d. equiangular or equilateral
e. isosceles
f. scalene
g. scalene
h. equilateral or equiangular
i. isosceles right triangle

2. $\begin{array}{r} 65^\circ \\ +\ 25^\circ \\ \hline 90^\circ \end{array} \quad \begin{array}{r} 180^\circ \\ -\ 90^\circ \\ \hline \mathbf{90^\circ} \end{array}$

3. right

4. hypotenuse

5. $\begin{array}{r} 65^\circ \\ +\ 50^\circ \\ \hline 115^\circ \end{array} \quad \begin{array}{r} 180^\circ \\ -\ 115^\circ \\ \hline \mathbf{65^\circ} \end{array} = \angle Y$

6. isosceles
There are two 65° angles.

7. $\begin{array}{r} 30^\circ \\ +\ 60^\circ \\ \hline 90^\circ \end{array} \quad \begin{array}{r} 180^\circ \\ -\ 90^\circ \\ \hline \mathbf{90^\circ} \end{array} = \angle C$

8. right

9. $\begin{array}{r} 180^\circ \\ -\ 82^\circ \\ \hline 98^\circ \end{array} \quad \begin{array}{r} \mathbf{49^\circ} \\ 2\overline{)98^\circ} \end{array}$

10. $\begin{array}{r} 63^\circ \\ \times\ 2 \\ \hline 126^\circ \end{array} \quad \begin{array}{r} 180^\circ \\ -\ 126^\circ \\ \hline \mathbf{54^\circ} \end{array}$

11. isosceles $\begin{array}{r} 5 \text{ in} \\ +\ 8 \\ \hline 13 \text{ in} \end{array} \quad \begin{array}{r} 21 \text{ in} \\ -\ 13 \\ \hline 8 \text{ in} \end{array}$
Two sides are equal

12. scalene $\begin{array}{r} 25^\circ \\ +\ 35^\circ \\ \hline 60^\circ \end{array} \quad \begin{array}{r} 180^\circ \\ -\ 60^\circ \\ \hline 120^\circ \end{array}$ The three angles are different.

13. scalene $\begin{array}{r} 4 \text{ in} \\ +\ 5 \\ \hline 9 \text{ in} \end{array} \quad \begin{array}{r} 16 \text{ in} \\ -\ 9 \\ \hline 7 \text{ in} \end{array}$ The three sides have different measurements.

14. 90° The right angle measures 90°. The other two angles must add up to 90° to make a total of 180° for all three angles.

15. 110° The two angles given in the triangle total 45° + 65° = 110°. The third angle in the triangle is 180° − 110° = 70°. $\angle X$ is the supplement to 70: 180° − 70° = 110°.

Geometry Exercise 13
pages 263–64

1. No For the larger rectangle, base:height = 12:6 = 2:1. For the smaller rectangle, base:height = 3:1. The sides are not proportional. $\frac{2}{1}$ does not equal $\frac{3}{1}$.

2. Yes For the smaller rectangle, base:height = 10:8 = 5:4. For the larger rectangle, base:height = 15:12 = 5:4. The sides are proportional. $\frac{5}{4} = \frac{5}{4}$.

3. Yes

$$\begin{array}{r} 45^\circ \\ +\ 85^\circ \\ \hline 130^\circ \end{array} \qquad \begin{array}{r} 180^\circ \\ -\ 130^\circ \\ \hline 50^\circ \end{array} = \angle O$$

$$\begin{array}{r} 50^\circ \\ +\ 45^\circ \\ \hline 95^\circ \end{array} \qquad \begin{array}{r} 180^\circ \\ -\ 95^\circ \\ \hline 85^\circ \end{array} = \angle R$$

Each triangle has angles of 45°, 50°, and 85°.

4. No

$$\begin{array}{r} 60^\circ \\ +\ 50^\circ \\ \hline 110^\circ \end{array} \qquad \begin{array}{r} 180^\circ \\ -\ 110^\circ \\ \hline 70^\circ \end{array} = \angle C$$

$$\begin{array}{r} 50^\circ \\ +\ 80^\circ \\ \hline 130^\circ \end{array} \qquad \begin{array}{r} 180^\circ \\ -\ 130^\circ \\ \hline 50^\circ \end{array} = \angle F$$

The angles in these triangles are not the same.

5. Yes Find the ratio of one of the equal sides to the base for each triangle. For the smaller triangle, 6:4 = 3:2. For the larger triangle, 15:10 = 3:2. The sides are proportional. $\frac{3}{2} = \frac{3}{2}$.

6. Yes

$$\begin{array}{r} 90^\circ \\ +\ 50^\circ \\ \hline 140^\circ \end{array} \qquad \begin{array}{r} 180^\circ \\ -\ 140^\circ \\ \hline 40^\circ \end{array} = \angle L$$

$$\begin{array}{r} 90^\circ \\ +\ 40^\circ \\ \hline 130^\circ \end{array} \qquad \begin{array}{r} 180^\circ \\ -\ 130^\circ \\ \hline 50^\circ \end{array} = \angle N$$

Each triangle has angles of 40°, 50°, and 90°.

7. 21 in

$$\frac{\text{short leg}}{\text{long leg}} \quad \frac{8}{12} = \frac{14}{x} \qquad \begin{array}{r} 14 \\ \times\ 12 \\ \hline 28 \\ 14 \\ \hline 168 \end{array} \qquad 8\overline{)168}\ \text{gives}\ 21\ \text{in}$$

8. 78 ft

$$\frac{\text{height}}{\text{shadow}} \quad \frac{6}{5} = \frac{x}{65} \qquad \begin{array}{r} 65 \\ \times\ 6 \\ \hline 390 \end{array} \qquad 5\overline{)390}\ \text{gives}\ 78\ \text{ft}$$

9. Yes $\angle SUT = \angle VUW$ because they are vertical. $\angle S = \angle W$ because they are both right angles. If we subtract the sum of one right angle and one of the vertical angles from 180°, we get the same third angle for each triangle.

10. 40 ft

$$\frac{\text{short leg}}{\text{long leg}} \quad \frac{8}{12} = \frac{x}{60} \qquad \begin{array}{r} 60 \\ \times\ 8 \\ \hline 480 \end{array} \qquad 12\overline{)480}\ \text{gives}\ 40\ \text{ft}$$

11. Yes Each triangle has a right angle. The triangles share $\angle C$. If we subtract the sum of a right angle and $\angle C$ from 180°, we get the same third angle.

12. 10 in

The length of AC = 15 in + 10 in = 25 in

$$\frac{\text{height}}{\text{base}} \quad \frac{6}{15} = \frac{x}{25} \qquad \begin{array}{r} 25 \\ \times\ 6 \\ \hline 150 \end{array} \qquad 15\overline{)150}\ \text{gives}\ 10\ \text{in}$$

Geometry Exercise 14

pages 266–67

1. **Yes** The missing angle in the left triangle is 60°. The missing angle at the right is 75°. The triangles fulfill the *angle, side, angle* requirement.
2. **Yes** The missing side on the triangle at the right is 5 inches. This fulfills the *side, angle, side* requirement.
3. **No** The corresponding sides are not equal.
4. **Yes** The missing side of the triangle on the right is 10 inches. This fulfills the *side, side, side* requirement.
5. **No** The corresponding angles are not equal.
6. **No** The three sides are not the same.
7. **(1)** $AC = DF$ This fulfills the *angle, side, angle* requirement.
8. **(2)** $\angle X = \angle S$ This fulfills the *side, angle, side* requirement.
9. **(1)** $GI = JL$ This fulfills the *side, side, side* requirement.
10. **(5)** B or C: Condition B fulfills the *side, side, side* requirement. Condition C fulfills the *side, angle, side* requirement.

Using the Formulas Page

page 271

(3) 154

The problem asks for area, and the figure shown is a circle, so find the formula for the area of a circle.

$A = \pi r^2$

$A = \pi(7)^2$

$A = \frac{22}{7} \times 49$

$A = 22 \times 7$

$A = 154$

Geometry Exercise 15

pages 271–72

1. $c^2 = a^2 + b^2$
   $c^2 = 30^2 + 40^2$
   $c^2 = 900 + 1600$
   $c^2 = 2500$
   $c = \sqrt{2500}$
   $\mathbf{c = 50}$ **ft**

2. $c^2 = a^2 + b^2$
   $c^2 = 10^2 + 24^2$
   $c^2 = 100 + 576$
   $c^2 = 676$
   $c = \sqrt{676}$
   $\mathbf{c = 26}$ **in**

3. $c^2 = a^2 + b^2$
$c^2 = 12^2 + 5^2$
$c^2 = 144 + 25$
$c^2 = 169$
$c = \sqrt{169}$
$\mathbf{c = 13}$ **in**

4. $c^2 = a^2 + b^2$
$c^2 = 12^2 + 16^2$
$c^2 = 144 + 256$
$c^2 = 400$
$c = \sqrt{400}$
$\mathbf{c = 20}$ **yd**

5. $c^2 = a^2 + b^2$
$c^2 = 60^2 + 45^2$
$c^2 = 3600 + 2025$
$c^2 = 5625$
$c = \sqrt{5625}$
$\mathbf{c = 75}$ **mi**

6. $c^2 = a^2 + b^2$
$30^2 = a^2 + 18^2$
$900 = a^2 + 324$
$576 = a^2$
$\sqrt{576} = a$
$\mathbf{24 = a}$

7. $c^2 = a^2 + b^2$
$34^2 = a^2 + 16^2$
$1156 = a^2 + 256$
$900 = a^2$
$\sqrt{900} = a$
$\mathbf{30 = a}$

8. $c^2 = a^2 + b^2$
$17^2 = a^2 + 8^2$
$289 = a^2 + 64$
$225 = a^2$
$\sqrt{225} = a$
$\mathbf{15 = a}$

9. The diagonal distance is the length of the hypotenuse of a right triangle.
$c^2 = a^2 + b^2$
$c^2 = (2.4)^2 + (1.8)^2$
$c^2 = (5.76)^2 + (3.24)^2$
$c^2 = 9$
$c = \sqrt{9}$
$\mathbf{c = 3}$ **m**

10. $c^2 = a^2 + b^2$
$c^2 = 48^2 + 36^2$
$c^2 = 2304 + 1296$
$c^2 = 3600$
$c = \sqrt{3600}$
$\mathbf{c = 60}$ **mi**

11. $c^2 = a^2 + b^2$
$c^2 = 15^2 + 36^2$
$c^2 = 225 + 1296$
$c^2 = 1521$
$c = \sqrt{1521}$
$\mathbf{c = 39}$ **mi**

Geometry Exercise 16

pages 272–74

1. cube

2. 12.8 m
$P = 4s$
$P = 4 \times 3.2 = 12.8$ m

3. $\mathbf{32\frac{1}{2}}$ **in**
$P = 2l + 2w$
$P = (2 \times 12) + (2 \times 4\frac{1}{4})$
$P = 24 + (\frac{2}{1} \times \frac{17}{4})$
$P = 24 + \frac{17}{2}$
$P = 24 + 8\frac{1}{2} = 32\frac{1}{2}$ in

4. 13 m
$P = a + b + c$
$P = 5.8 + 3.2 + 4 = 13$ m

5. 94 ft
$C = \pi d$
$C = 3.14 \times 30$
$C = 94.2$ to the nearest foot $= 94$ ft

6. $20^3 = 20 \times 20 \times 20 = \mathbf{8000}$

7. .1690
$(.03)^2 = .03 \times .03 = .0009$
$.4^2 = .4 \times .4 = .16$

$$\begin{array}{r} .0009 \\ +\ .16 \\ \hline .1690 \end{array}$$

8. 84
Guess 80 because $80 \times 80 = 6400$.

$$\begin{array}{r} 88 \\ 80\overline{)7056} \\ \underline{640} \\ 656 \\ \underline{640} \\ 16 \end{array} \qquad \begin{array}{r} 88 \\ +\ 80 \\ \hline 168 \end{array} \qquad \begin{array}{r} 84 \\ 2\overline{)168} \end{array} \qquad \begin{array}{r} 84 \\ \times\ 84 \\ \hline 7056 \end{array}$$

9. $\mathbf{30\frac{1}{4}}$ $\mathbf{in^2}$
$A = s^2$
$A = (5\frac{1}{2})^2$
$\frac{11}{2} \times \frac{11}{2} = \frac{121}{4} = 30\frac{1}{4}$ in^2

10. **92.5** $\mathbf{cm^2}$
$A = lw$
$A = 12.5 \times 7.4 = 92.5$ cm^2

11. **6** $\mathbf{cm^2}$
$A = \frac{(bh)}{2}$
$A = \frac{8 \times 1.5}{2} = 6$ cm^2

12. 154 in²

$A = \pi r^2$

$A = \frac{22}{7} \times 7^2$

$A = 154 \text{ in}^2$

13. 525 in³

$V = lwh$

$V = 15 \times 10 \times 3\frac{1}{2}$

$\frac{15}{1} \times \frac{10}{1} \times \frac{7}{2} = 525 \text{ in}^3$

14. (3) (2 × 20) + (2 × 8) − 4

First find the perimeter: $P = 2l + 2w$. Then subtract the width of the walkway, which is 4.

15. (4) 9.6

Area of square:

$A = s^2$

$A = 12^2$

$A = 12 \times 12 = 144 \text{ in}^2$

Divide 144 by the length of the rectangle.

$144.0 \div 15 = 9.6$

16. (3) 216

Find the missing measurements.

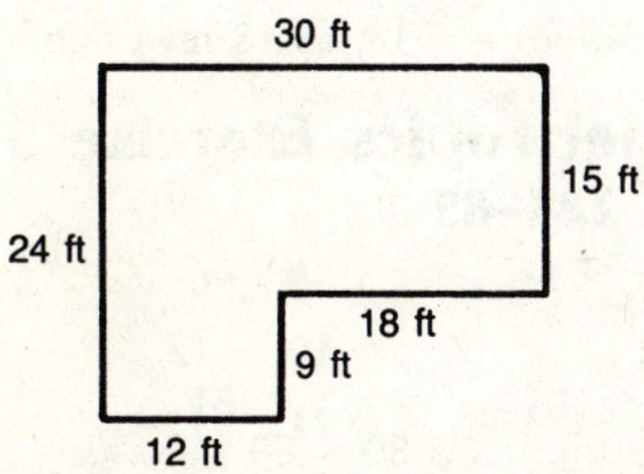

Add the dimensions of each side: $P = 30 + 15 + 18 + 9 + 12 + 24 = 108$ ft. Each edge tile is 6 inches or $\frac{1}{2}$ foot. You need at least two tiles for every foot: $2 \times 108 = 216$ tiles.

17. (2) 62

Area of large rectangle:

$A = lw$

$A = 30 \times 15 = 450 \text{ ft}^2$

Area of small rectangle:

$A = lw$

$A = 9 \times 12 = 108 \text{ ft}^2$

Total: $450 + 108 = 558 \text{ ft}^2$

$\frac{1 \text{ sq yd}}{9 \text{ sq ft}} = \frac{x}{558}$ $558 \times 1 = 558$ $558 \div 9 = 62 \text{ yd}^2$

18. (1) 3348

Volume of large rectangular part:

$V = lwh$

$V = 30 \times 6 \times 15 = 2700 \text{ ft}^3$

Volume of small rectangular part:

$V = lwh$

$V = 12 \times 6 \times 9 = 648 \text{ ft}^3$

Total: $2700 + 648 = 3348 \text{ ft}^3$

19. (3) obtuse

20. (5) 115°

$180° - 65° = 115°$

21. (5) Not enough information is provided.

22. (4) right

$33° + 57° = 90°$ $180° - 90° = 90°$ The triangle has one right angle.

23. (2) 3.6

$\frac{\text{height}}{\text{base}}$ $\frac{12}{10} = \frac{x}{3}$ $12 \times 3 = 36$ $36.0 \div 10 = 3.6$ ft

24. (4) A or B

Either A or B is enough to guarantee that the triangles are congruent. Either condition fulfills the *angle, side, angle* requirement.

25. (1) 15

$c^2 = a^2 + b^2$

$39^2 = a^2 + 36^2$

$1521 = a^2 + 1296$

$225 = a^2$

$\mathbf{15 = a}$

CHAPTER 12

Special Topics Exercise 1

page 277

1. J
2. D
3. B
4. A
5. F
6. G
7. I
8. H
9. C
10. E

11.

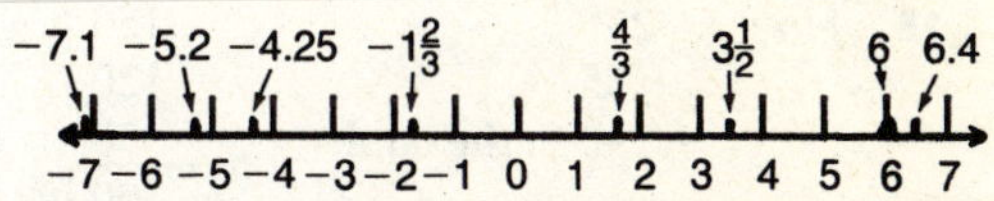

Special Topics Exercise 2

pages 279–80

1. $-9 - 3 = -12$

2. $-8 + 6 = -2$

3. $\begin{array}{r} -20\frac{1}{2} \\ +12 \\ \hline \mathbf{-8\frac{1}{2}} \end{array}$

4. $\begin{array}{r} -20 \\ +12\frac{1}{2} \\ \hline \mathbf{-7\frac{1}{2}} \end{array}$

5. $\begin{array}{r} +5 \\ +9 \\ \hline +14 \end{array}$ $\quad$ $\begin{array}{r} +15 \\ -8 \\ \hline \mathbf{+6} \end{array}$

6. $\begin{array}{r} -6 \\ -5 \\ \hline -11 \end{array}$ $\quad$ $\begin{array}{r} -11 \\ +8 \\ \hline \mathbf{-3} \end{array}$

7. $\begin{array}{r} -18 \\ -2 \\ \hline -20 \end{array}$ $\quad$ $\begin{array}{r} -20 \\ +6 \\ \hline \mathbf{-14} \end{array}$

8. $\begin{array}{r} +8 \\ +11 \\ \hline +19 \end{array}$ $\quad$ $\begin{array}{r} -6 \\ -4 \\ \hline -10 \end{array}$ $\quad$ $\begin{array}{r} +19 \\ -10 \\ \hline \mathbf{+9} \end{array}$

9. $\begin{array}{r} -7 \\ -5 \\ -3 \\ -10 \\ -2 \\ \hline -27 \end{array}$ $\quad$ $\begin{array}{r} +40 \\ -27 \\ \hline \mathbf{+13} \end{array}$

10. $\begin{array}{r} +3 \\ +7 \\ \hline +10 \end{array}$ $\quad$ $\begin{array}{r} +10 \\ -10 \\ \hline \mathbf{0} \end{array}$

11. $\begin{array}{r} -10 \\ -3 \\ -9 \\ \hline -22 \end{array}$ $\quad$ $\begin{array}{r} -22 \\ +5 \\ \hline \mathbf{-17} \end{array}$

12. $\begin{array}{r} +8 \\ +7 \\ \hline +15 \end{array}$ $\quad$ $\begin{array}{r} -9 \\ -4 \\ -2 \\ \hline -15 \end{array}$ $\quad$ $\begin{array}{r} +15 \\ -15 \\ \hline \mathbf{0} \end{array}$

13. $\begin{array}{r} -10^\circ \\ -4^\circ \\ \hline \mathbf{-14^\circ} \end{array}$

14. $\begin{array}{r} +180 \\ +15 \\ \hline +195 \end{array}$ $\quad$ $\begin{array}{r} +195 \\ -37 \\ \hline \mathbf{+158} \end{array}$

15. **No** $\quad$ $\begin{array}{r} +235 \\ +55 \\ \hline +290 \end{array}$ in the bank $\quad$ $\begin{array}{r} -150 \\ -120 \\ -30 \\ \hline -300 \end{array}$ in checks

Special Topics Exercise 3
page 281

1. $(+6) - (+4) = +6 - 4 = \mathbf{+2}$
2. $(-8) - (+3) = -8 - 3 = \mathbf{-11}$
3. $(-9) - (-8) = -9 + 8 = \mathbf{-1}$
4. $(+10) - (-9) = +10 + 9 = \mathbf{+19}$
5. $(+8) - (7) = +8 - 7 = \mathbf{+1}$
6. $(-9) - (-9) = -9 + 9 = \mathbf{0}$
7. $-10 - (12) = -10 - 12 = \mathbf{-22}$
8. $(+6) - (-3) + (-2) =$
   $+6 + 3 - 2 = +9 - 2 = \mathbf{+7}$
9. $(-9) - (+4) - (+10) =$
   $-9 - 4 - 10 = \mathbf{-23}$
10. $(-15) - (20) + (+6) = -15 - 20 + 6 =$
    $-35 + 6 = \mathbf{-29}$
11. $(-8) + (-13) - (+6) =$
    $-8 - 13 - 6 = \mathbf{-27}$
12. $(-3) + (-4) - (-5) - (-6) =$
    $-3 - 4 + 5 + 6 = -7 + 11 = \mathbf{+4}$
13. $(+22) - (-43) = +22 + 43 =$ **65 feet lower**
14. $(+7) - (-17) = +7 + 17 = 24$. The temperature dropped **24 degrees.**

Special Topics Exercise 4
page 283

1. −18
2. +36
3. −45
4. +24
5. −9
6. $-\frac{1}{2}$
7. +96
8. +721
9. −40
10. +5
11. −192
12. −28
13. $(-2)(+6) = -12$
14. $(+3)(-6) = -18$, or 18 fewer children.

Special Topics Exercise 5
pages 284–85

1. +2
2. −2
3. −8
4. $-\frac{2}{3}$
5. −3
6. $-\frac{5}{6}$
7. +8
8. $-\frac{1}{2}$
9. −6
10. +12
11. $+\frac{4}{5}$
12. −13
13. $+\frac{3}{2}$ or $1\frac{1}{2}$
14. $-\frac{9}{5}$ or $-1\frac{4}{5}$
15. $-\frac{3}{4}$
16. $+\frac{1}{10}$

Special Topics Exercise 6
pages 285–86

1. $(15) - (-9) = 15 + 9 = \mathbf{+24}$
2. $(-6)(+20) = \mathbf{-120}$
3. $-4 + 3 - 7 = -11 + 3 = \mathbf{-8}$
4. $\frac{-96}{-8} = \mathbf{+12}$
5. $(-8) - (12) = -8 - 12 = \mathbf{-20}$
6. $\frac{+1{,}000}{-10} = \mathbf{-100}$
7. $(-29) + (-14) = \mathbf{-43}$
8. $(-20) - (-21) = -20 + 21 = \mathbf{+1}$
9. $(20) + (-12) - (+15) = 20 - 12 - 15 =$
   $+20 - 27 = \mathbf{-7}$

10. $(-3)(+\frac{2}{3})(-10) = -3 \cdot \frac{2}{3} \cdot -10 = \mathbf{+20}$
11. $-\frac{14}{21} = \mathbf{-\frac{2}{3}}$
12. $(-1)(+\frac{7}{8}) = \mathbf{-\frac{7}{8}}$
13. $(-3)(5) + \frac{6}{-2} = -15 - 3 = \mathbf{-18}$
14. $(-\frac{8}{4}) + (\frac{15}{-3}) = (-2) + (-5) = \mathbf{-7}$
15. $(-6) + (\frac{30}{5}) = -6 + 6 = \mathbf{0}$
16. $(15)(-\frac{2}{3}) - 8 = -10 - 8 = \mathbf{-18}$
17. $-4 + (-8)(-6) = -4 + 48 = \mathbf{+44}$
18. $(-\frac{36}{9}) - (-10) = -4 + 10 = \mathbf{+6}$
19. $(12)(-2) + (-1)(-15) = -24 + 15 = \mathbf{-9}$
20. $(\frac{10}{-15}) - (\frac{18}{27}) = -\frac{2}{3} - \frac{2}{3} = \mathbf{-\frac{4}{3}}$ **or** $\mathbf{-1\frac{1}{3}}$

Special Topics Exercise 7
page 287

1.
$$\begin{array}{rcr} c - 3 & < & 21 \\ +3 & & +3 \\ \hline \mathbf{c} & < & \mathbf{24} \end{array}$$

2.
$$\begin{array}{rcr} p + 1 & > & 7 \\ -1 & & -1 \\ \hline \mathbf{p} & > & \mathbf{6} \end{array}$$

3. $\frac{9w}{9} \leq \frac{27}{9}$
$\mathbf{w \leq 3}$

4. $\frac{8x}{8} \geq \frac{20}{8}$
$\mathbf{x \geq 2\frac{4}{8} = 2\frac{1}{2}}$

5. $6 \cdot \frac{s}{6} < 4 \cdot 6$
$\mathbf{s < 24}$

6. $\frac{2a}{2} > \frac{-42}{2}$
$\mathbf{a > -21}$

7. **No**
$$\begin{array}{rcr} m - 6 & > & 1 \\ +6 & & +6 \\ \hline m & > & 7 \end{array}$$
Since m is greater than 7, m cannot equal 7.

8. **Yes** $2 \cdot \frac{1}{2}r \leq 4 \cdot 2$
$r \leq 8$
Since r is less than or equal to 8, r can be $\frac{1}{2}$.

9. **Yes**
$$\begin{array}{rcr} d + 7 & \leq & 2 \\ -7 & & -7 \\ \hline d & \leq & -5 \end{array}$$
Since d is less than or equal to -5, d can be -6.

10. **Yes** $\frac{2f}{2} < \frac{12}{2}$
$f < 6$
Since f is less than 6, f can be 4.

Special Topics Exercise 8
page 287

1.
$$\begin{array}{rcr} 5m - 4 & \leq & 26 \\ +4 & & +4 \\ \hline \frac{5m}{5} & \leq & \frac{30}{5} \end{array}$$
$\mathbf{m \leq 6}$

2.
$$\begin{array}{rcr} 3n + 2 & > & 14 \\ -2 & & -2 \\ \hline \frac{3n}{3} & > & \frac{12}{3} \end{array}$$
$\mathbf{n > 4}$

3.
$$\begin{array}{rcr} 4p - 3 & < & 15 \\ +3 & & +3 \\ \hline \frac{4p}{4} & < & \frac{18}{4} \end{array}$$
$\mathbf{p < 4\frac{2}{4} = 4\frac{1}{2}}$

4.
$$\begin{array}{rcr} \frac{x}{3} + 5 & \geq & 7 \\ -5 & & -5 \\ \hline 3 \cdot \frac{x}{3} & \geq & 2 \cdot 3 \end{array}$$
$\mathbf{x \geq 6}$

5.
$$\begin{array}{rcr} 7c - 3 & \leq & 5c + 15 \\ -5c & & -5c \\ \hline 2c - 3 & \leq & 15 \\ +3 & & +3 \\ \hline \frac{2c}{2} & \leq & \frac{18}{2} \end{array}$$
$\mathbf{c \leq 9}$

6.
$$\begin{array}{rcr} 8y + 1 & < & y + 22 \\ -y & & -y \\ \hline 7y + 1 & < & 22 \\ -1 & & -1 \\ \hline \frac{7y}{7} & < & \frac{21}{7} \end{array}$$
$\mathbf{y < 3}$

7.
$$\begin{array}{rcr} 3(s - 2) & \geq & 2s + 10 \\ 3s - 6 & \geq & 2s + 10 \\ -2s & & -2s \\ \hline s - 6 & \geq & 10 \\ +6 & & +6 \\ \hline \end{array}$$
$\mathbf{s \geq 16}$

8.
$$\begin{array}{rcr} 5(a + 3) & < & 2(a - 6) \\ 5a + 15 & < & 2a - 12 \\ -2a & & -2a \\ \hline 3a + 15 & < & -12 \\ -15 & & -15 \\ \hline \frac{3a}{3} & < & \frac{-27}{3} \end{array}$$
$\mathbf{a \leq -9}$

Special Topics Exercise 9
page 289

1. $4x \cdot 3x = \mathbf{12x^2}$
2. $(-4a^2)(-6a^3) = \mathbf{24a^5}$
3. $m^3 \cdot m^4 = \mathbf{m^7}$
4. $a^2 \cdot a = \mathbf{a^3}$
5. $(5b^2)(b^4) = \mathbf{5b^6}$
6. $10x \cdot \frac{1}{2}x = \mathbf{5x^2}$
7. $4x^2y \div xy = \mathbf{4x}$
8. $6x^4y^4 \div 6x^2y = \mathbf{x^2y^3}$
9. $18b^3x^2 \div 2b^2x^2 = \mathbf{9b}$
10. $12t^3v^4 \div 6t^2v^3 = \mathbf{2tv}$

Special Topics Exercise 10
page 290

1. $9 = 3 \times 3$
2. $10 = 2 \times 5$
3. $18 = 2 \times 3 \times 3$
4. $20 = 2 \times 2 \times 5$
5. $24 = 2 \times 2 \times 2 \times 3$
6. $30 = 2 \times 3 \times 5$
7. $40 = 2 \times 2 \times 2 \times 5$
8. $50 = 2 \times 5 \times 5$

Special Topics Exercise 11
page 290

1. $4n + 4 = 4(n + 1)$
2. $3p - 6 = 3(p - 2)$
3. $15a - 10 = 5(3a - 2)$
4. $14c + 35 = 7(2c + 5)$
5. $6b + 8 = 2(3b + 4)$
6. $6f - 30 = 6(f - 5)$
7. $36y - 9 = 9(4y - 1)$
8. $16k + 56 = 8(2k + 7)$

Special Topics Exercise 12
page 291

1. $c^2 + 8c = c(c + 8)$
2. $y^2 - 5y = y(y - 5)$
3. $m^2 + 3m = m(m + 3)$
4. $a^2 - a = a(a - 1)$
5. $d^2 + 4d = d(d + 4)$
6. $n^2 - 8n = n(n - 8)$
7. $p^2 - 2p = p(p - 2)$
8. $s^2 + 9s = s(s + 9)$

Special Topics Exercise 13
page 292

1. $\sqrt{27} = \sqrt{9 \cdot 3} = 3\sqrt{3}$
2. $\sqrt{8} = \sqrt{4 \cdot 2} = 2\sqrt{2}$
3. $\sqrt{20} = \sqrt{4 \cdot 5} = 2\sqrt{5}$
4. $\sqrt{150} = \sqrt{6 \cdot 25} = 5\sqrt{6}$
5. $\sqrt{48} = \sqrt{16 \cdot 3} = 4\sqrt{3}$
6. $\sqrt{50} = \sqrt{25 \cdot 2} = 5\sqrt{2}$
7. $\sqrt{12} = \sqrt{4 \cdot 3} = 2\sqrt{3}$
8. $\sqrt{32} = \sqrt{16 \cdot 2} = 4\sqrt{2}$

Special Topics Exercise 14
page 294

1. **The length is 8 feet.**

$w = \frac{1}{2}x \qquad A = lw$

$l = x \qquad 32 = x \cdot \frac{1}{2}x$

$$32 = \frac{1}{2}x^2$$
$$2 \cdot 32 = \frac{1}{2}x^2 \cdot 2$$
$$64 = x^2$$
$$\sqrt{64} = x$$
$$8 = x$$

2. **The height is 4 inches.**

$h = \frac{x}{2} \qquad A = bh/2$

$b = x \qquad 16 = (x \cdot \frac{x}{2})/2$

$$16 = \frac{x^2}{4}$$
$$4 \cdot 16 = \frac{x^2}{4} \cdot 4$$
$$64 = x^2$$
$$\sqrt{64} = x$$
$$8 = x$$
$$4 = \frac{1}{2}x$$

3. **The length is 12.**

$w = \frac{2}{3}x \qquad A = lw$

$l = x \qquad 96 = x \cdot \frac{2}{3}x$

$$\frac{3}{2} \cdot \frac{96}{1} = \frac{3}{2} \cdot \frac{2}{3}x^2$$
$$144 = x^2$$
$$\sqrt{144} = x$$
$$12 = x$$

4. **The length is 14 meters.**

$w = x \qquad A = lw$

$l = 2x \qquad 98 = 2x \cdot x$

$$98 = 2x^2$$
$$\frac{98}{2} = \frac{2x^2}{2}$$
$$49 = x^2$$
$$\sqrt{49} = x$$
$$7 = x$$
$$14 = 2x$$

5. **The length is 16 feet.**

$w = x \qquad P = 2l + 2w$

$l = 2x \qquad 48 = 2(2x) + 2x$

$$48 = 4x + 2x$$
$$\frac{48}{6} = \frac{6x}{6}$$
$$8 = x$$
$$16 = 2x$$

6. The length is 10 m, and the width is $7\frac{1}{2}$ m.

$w = \frac{3}{4}x$ $\quad P = 2l + 2w$

$l = x$ $\quad 35 = 2(x) + 2(\frac{3}{4}x)$

$35 = 2x + \frac{3}{2}x$

$\frac{2}{7} \cdot 35 = \frac{7}{2}x \cdot \frac{2}{7}$

$10 = x$

$7\frac{1}{2} = \frac{3}{4}x$

7. The base is 12.

$h = x$ $\quad A = bh/2$

$b = 3x$ $\quad 24 = (3x \cdot x)/2$

$24 = \frac{3}{2}x^2$

$\frac{2}{3} \cdot 24 = \frac{2}{3} \cdot \frac{3}{2}x^2$

$16 = x^2$

$\sqrt{16} = x$

$4 = x$

$12 = 3x$

8. The length is 18 inches.

$w = 2x$ $\quad P = 2l + 2w$

$l = 3x$ $\quad 60 = 2(3x) + 2(2x)$

$60 = 6x + 4x$

$\frac{60}{10} = \frac{10x}{10}$

$6 = x$

$18 = 3x$

9. The length is 100 feet, and the width is 60 feet.

$w = 3x$ $\quad A = lw$

$l = 5x$ $\quad 6{,}000 = 5x \cdot 3x$

$6000 = 15x^2$

$\frac{6{,}000}{15} = \frac{15x^2}{15}$

$400 = x^2$

$\sqrt{400} = x$

$20 = x$

$60 = 3x$

$100 = 5x$

10. The length of *AB* is 18.

$h = 3x$ $\quad A = bh/2$

$b = 4x$ $\quad 216 = (4x \cdot 3x)/2$

$216 = 12x^2/2$

$\frac{216}{6} = \frac{6x^2}{6}$

$36 = x^2$

$\sqrt{36} = x$

$6 = x$

$18 = 3x$

11. The length is 30 ft.

$w = 5x$ $\quad P = 2l + 2w$

$l = 6x$ $\quad 110 = 2(6x) + 2(5x)$

$110 = 12x + 10x$

$\frac{110}{22} = \frac{22x}{22}$

$5 = x$

$30 = 6x$

12. *MN* is 7 ft.

$w = x$ $\quad P = 2l + 2w$

$l = x + 3$ $\quad 34 = 2(x + 3) + 2(x)$

$34 = 2x + 6 + 2x$

$34 = 4x + 6$

$-6 \quad\quad -6$

$\frac{28}{4} = \frac{4x}{4}$

$7 = x$

Special Topics Exercise 15

page 296

1. $A = (+9, +5)$ $\quad G = (-7, -2)$
   $B = (+4, +10)$ $\quad H = (-3, -8)$
   $C = (-2, +4)$ $\quad I = (0, -7)$
   $D = (-6, +10)$ $\quad J = (+6, -7)$
   $E = (-11, +3)$ $\quad K = (+10, -3)$
   $F = (-10, 0)$ $\quad L = (+12, 0)$

2.

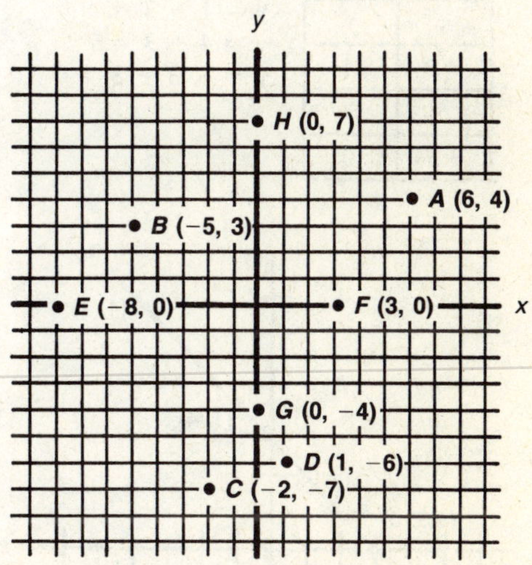

Special Topics Exercise 16

pages 298–300

1. *A* is 3 units above the *x*-axis.
   C is 5 units below the *x*-axis.
   The distance from *A* to *C* is 3 + 5 = 8.
2. *B* is 1 unit to the left of the *y*-axis.
   C is 3 units to the right of the *y*-axis.
   The distance from *B* to *C* is 1 + 3 = 4.

3. E is 2 units to the left of the y-axis.
F is 1 unit to the right of the y-axis.
The distance from E to F is $2 + 1 = 3$.

4. D is 3 units below the x-axis.
F is 1 unit above the x-axis.
The distance from D to F is $1 + 3 = 4$.

5. $d = \sqrt{(2 - 6)^2 + (4 - 1)^2}$
$d = \sqrt{(-4)^2 + (3)^2}$
$d = \sqrt{16 + 9}$
$d = \sqrt{25}$
$d = 5$

6. $d = \sqrt{(6 - 2)^2 + (7 - 2)^2}$
$d = \sqrt{4^2 + 5^2}$
$d = \sqrt{16 + 25}$
$d = \sqrt{41}$
$d =$ approximately 6.4

7. $d = \sqrt{(3 - (-1))^2 + ((-1) - 4)^2}$
$d = \sqrt{4^2 + (-5)^2}$
$d = \sqrt{16 + 25}$
$d = \sqrt{41}$
$d =$ approximately 6.4

8. $d = \sqrt{(+6 - (-10)^2 + (+7) - (-5)^2}$
$d = \sqrt{(+6 + 10)^2 + (+7) + 5^2}$
$d = \sqrt{(16)^2 + (12)^2}$
$d = \sqrt{256 + 144}$
$d = \sqrt{400}$
$d = 20$

Special Topics Exercise 17
page 302

1. $y = x + 3$

x	y
1	4
4	7
−3	0

$y = 1 + 3 = 4$
$y = 4 + 3 = 7$
$y = -3 + 3 = 0$

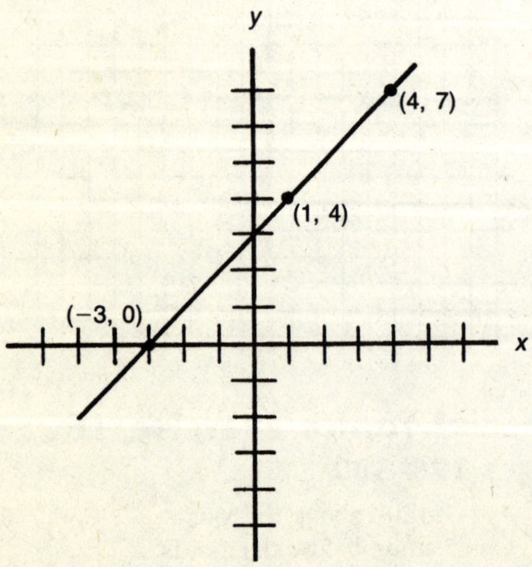

2. $y = 2x - 3$

x	y
5	7
3	3
0	−3

$y = 2(5) - 3 = 10 - 3 = 7$
$y = 2(3) - 3 = 6 - 3 = 3$
$y = 2(0) - 3 = 0 - 3 = -3$

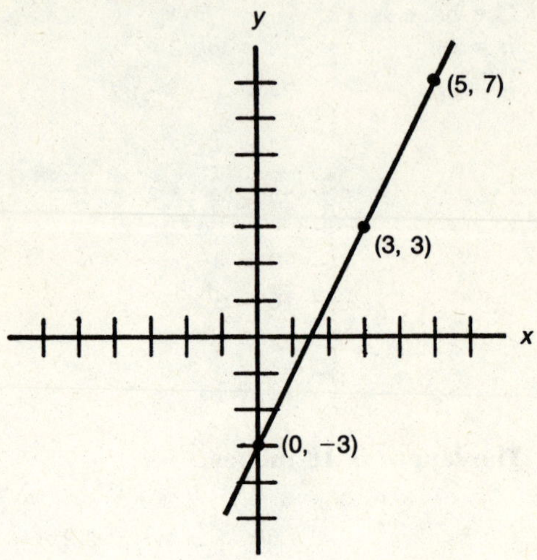

3. $y = \frac{x}{3} + 1$

x	y
6	3
3	2
−3	0

$y = \frac{6}{3} + 1 = 2 - 1 = 3$
$y = \frac{3}{3} + 1 = 1 + 1 = 2$
$y = -\frac{3}{3} + 1 = -1 + 1 = 0$

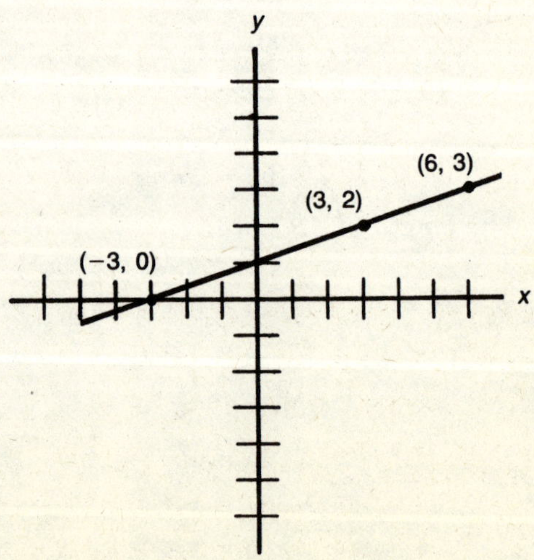

4. $y = -x + 5$

x	y
8	−3
5	0
1	4

$y = -8 + 5 = -3$
$y = -5 + 5 = 0$
$y = -1 + 5 = 4$

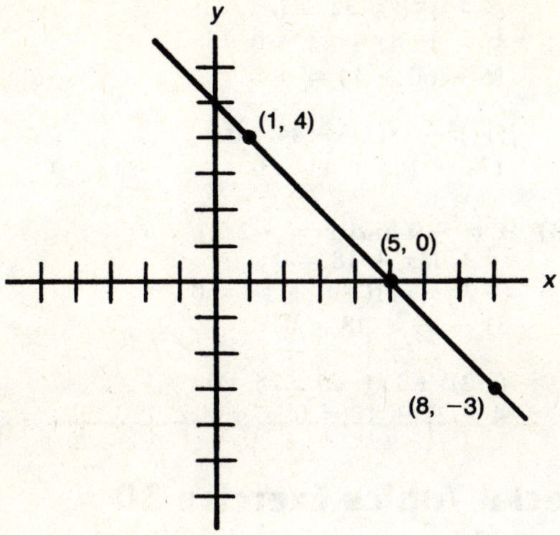

5. Yes $y = 3x - 1$
$y = 3(2) - 1 = 6 - 1 = 5$
(2, 5) is on the graph.

6. No $y = 4x - 5$
$y = 4(3) - 5 = 12 - 5 = 7$
(3, 4) is not on the graph.

7. Yes $y = \frac{x}{2} + 4$
$y = \frac{6}{2} + 4 = 3 + 4 = 7$
(6, 7) is on the graph.

8. No $y = \frac{2}{3}x + 1$
$y = \frac{2}{3}(3) + 1 = 2 + 1 = 3$
(3, 2) is not on the graph.

9. Yes $y = -x + 3$
$y = -8 + 3 = -5$
(8, −5) is on the graph.

10. Yes $y = -\frac{x}{2} + 1$
$y = -\frac{6}{2} + 1 = -3 + 1 = -2$
(6, −2) is on the graph.

Special Topics Exercise 18
pages 304–305

1. a. *u* **b.** *t* **c.** *v* **d.** *s*

2. a. slope $= \frac{8-2}{5-3} = \frac{6}{2} = +3$

b. slope $= \frac{9-5}{10-2} = \frac{4}{8} = +\frac{1}{2}$

c. slope $= \frac{2-6}{5-3} = -\frac{4}{2} = -2$

d. slope $= \frac{10-4}{-7-2} = -\frac{6}{9} = -\frac{2}{3}$

e. slope $= \frac{5-(-3)}{-1-3} = \frac{5+3}{-4} = -\frac{8}{4} = -2$

3. a. x-intercept = (3, 0)
$y = 3x - 9$
$0 = 3x - 9$
$+9 \qquad +9$
$\frac{9}{3} = \frac{3x}{3}$
$\mathbf{3 = x}$
y-intercept = (0, −9)
$y = 3x - 9$
$y = 3(0) - 9$
$y = 0 - 9$
$\mathbf{y = -9}$

b. x-intercept = (6, 0)
$y = x - 6$
$0 = x - 6$
$+6 \qquad +6$
$\mathbf{6 = x}$
y-intercept = (0, −6)
$y = x - 6$
$y = 0 - 6$
$\mathbf{y = -6}$

c. x-intercept =(−10, 0)
$y = \frac{x}{2} + 5$
$0 = \frac{x}{2} + 5$
$-5 \qquad -5$
$2 \cdot -5 = \frac{x}{2} \cdot 2$
$\mathbf{-10 = x}$
y-intercept = (0, 5)
$y = \frac{x}{2} + 5$
$y = \frac{0}{2} + 5$
$y = 0 + 5$
$\mathbf{y = 5}$

d. x-intercept =(−16, 0)
$y = \frac{3}{4}x + 12$
$0 = \frac{3}{4}x + 12$
$-12 \qquad -12$
$\frac{4}{3} \cdot -12 = \frac{3}{4}x \cdot \frac{4}{3}$
$\mathbf{-16 = x}$
y-intercept = (0, 12)
$y = \frac{3}{4}x + 12$
$y = \frac{3}{4} \cdot 0 + 12$
$y = 0 + 12$
$\mathbf{y = +12}$

e. x-intercept $= (\frac{1}{2}, 0)$

$$y = 2x - 1$$
$$0 = 2x - 1$$
$$+1 \quad\quad +1$$
$$\frac{1}{2} = \frac{2x}{2}$$
$$\frac{1}{2} = x$$

y-intercept $= (0, -1)$

$$y = 2x - 1$$
$$y = 2(0) - 1$$
$$y = 0 - 1$$
$$y = -1$$

Special Topics Exercise 19

page 306

1. (3) $x = 6$ **and** $x = 8$

$x^2 - 14x + 48 = 0$
$6^2 - 14(6) + 48 = 0$
$36 - 84 + 48 = 0$

$8^2 - 14(8) + 48 = 0$
$64 - 112 + 48 = 0$

2. (1) $x = 5$ **and** $x = -2$

$x^2 - 3x - 10 = 0$
$5^2 - 3(5) - 10 = 0$
$25 - 15 - 10 = 0$

$(-2)^2 - 3(-2) - 10 = 0$
$4 + 6 - 10 = 0$

3. (4) $x = 4$ **and** $x = 3$

$x^2 - x - 12 = 0$
$4^2 - 4 - 12 = 0$
$16 - 4 - 12 = 0$

$(-3)^2 - (-3) - 12 = 0$
$9 + 3 - 12 = 0$

4. (1) $x = 8$ **and** $x = 1$

$x^2 - 9x + 8 = 0$
$8^2 - 9(8) + 8 = 0$
$64 - 72 + 8 = 0$

$(1)^2 - 9(1) + 8 = 0$
$1 - 9 + 8 = 0$

5. (2) $x = 4$ **and** $x = 11$

$x^2 - 15x + 44 = 0$
$4^2 - 15(4) + 44 = 0$
$16 - 60 + 44 = 0$

$(11)^2 - 15(11) + 44 = 0$
$121 - 165 + 44 = 0$

6. (3) $x = -9$ **and** $x = -2$

$x^2 + 11x + 18 = 0$
$(-9)^2 + 11(-9) + 18 = 0$
$81 - 99 + 18 = 0$

$(-2)^2 + 11(-2) + 18 = 0$
$4 - 22 + 18 = 0$

Special Topics Exercise 20

page 308

1. 7×10^2
2. 4.46×10^{-2}
3. 9.5×10^7
4. 6.6×10^{-4}
5. 1.7×10^4
6. 2740
7. .091
8. 48,820
9. .0000198
10. 5.2

INDEX

A
addition
  facts, 11
  of decimals, 58
  of fractions, 85, 92
  of whole numbers, 14
algebra
  basics of, 202
  expressions, 202
  using algebra in word problems, 216
angle, 224, 251
  acute, 252
  complementary, 254
  obtuse, 252
  reflex, 252
  right, 225, 252
  straight, 252
  supplementary, 254
  vertical, 254
area, 238
associative law, 39
average, 34

B
borrowing
  in fractions, 88
  in whole numbers, 14

C
cancellation, 96
circle, 225
  area of, 239
circumference, 228
common denominator, 89
commutative law, 38
comparing
  decimals, 55
  fractions, 111
  measurements, 165
congruence, 264
corresponding
  angles, 264
  sides, 260
cross product, 82, 121
cube, 226
  volume of, 245
cylinder, 226
  volume of, 245

D
decimals, 50
  adding, 58
  comparing, 55
  dividing, 64
  multiplying, 62
  place value in, 51
  reading, 52
  repeating, 69
  rounding, 68
  writing, 54
  zeros in, 53
denominator, 76
difference, 14
digit, 10
distributive law, 40
dividend, 64
division
  of decimals, 64
  of fractions, 101
  of whole numbers, 16
divisor, 64

E
element, 121
equation, 202, 205
  linear, 300
  longer equations, 210
  one-step, 205
  quadratic, 305
  with multiplication and division, 207
  with parentheses, 213
  with separated unknowns, 215
estimating, 28
exponent, 234

F
factoring, 289
  with variables, 290
formulas, 202, 219, 269
  distance, 220
  formulas page, 270
  interest, 149
  total cost, 221
fractions, 76
  adding and subtracting unlike, 92
  adding like, 85
  borrowing with, 88
  canceling, 96
  changing decimals to fractions, 107
  changing fractions to decimals, 108
  changing improper fractions, 84
  changing mixed numbers, 98
  common fractions and decimals, 108
  comparing, 111
  dividing, 101
  multiplying, 96
  multiplying mixed numbers, 99
  proper and improper, 78
  raising to higher terms, 91
  reducing, 80
  subtracting like, 87
  writing, 83

G
geometry, 224

H
horizontal lines, 225
hypotenuse, 257

I
inequalities, 286
information
  necessary, 22
  not enough given, 152
  what more do you need?, 170
intercept, 303
interest problems, 149
item sets
  introducing, 70
  working with, 148

M
mean, 34
median, 35
meters, 174
monomials, 288
multiplication
  facts, 12
  of decimals, 62
  of fractions, 96
  of whole numbers, 15

N
number
  line, 276
  mixed, 78
  positive and negative, 277
  properties of, 38
  series, 36
  smaller than 1, 47
numerator, 76

O
order of operations, 41, 235, 285

P
parallel lines, 225
parentheses, 39, 213
percent, 128
  common fractions, decimals, and percents, 135
  decimals and, 129
  fractions and, 132
  multistep problems, 142
  proportion and, 135
  rate of increase/decrease, 143
  simple interest, 149
  successive, 146
  word problems, 140
perimeter, 227
perpendicular lines, 225
place value
  in decimals, 51
  in whole numbers, 10
powers and roots, 234
probability, 115
  dependent, 117
product, 15
properties of numbers, 38
proportion, 121
  and percent, 135
  to convert measurement, 159, 163
  to solve word problems, 124
  two-part problems, 126
Pythagorean Theorem, 267

Q
quotient, 16, 64

R
radius, 225
rate
  of increase/decrease, 143
  problems, 220
ratio, 119
rectangle, 225
  area of, 239
  perimeter of, 227
  volume of, 245
rectangular coordinates, 295
  finding distance in, 297
remainder, 16
rounding
  decimals, 68
  whole numbers, 28

S
scales, 174
scientific notation, 307
set-up, 42
signed numbers, 277
  adding, 278
  dividing, 284
  multiplying, 282
  subtracting, 280
slope, 302
sphere, 226
square, 225
  area of, 239
  perimeter of, 227
square root, 234, 236
  simplifying, 291
subtraction
  of decimals, 59
  of fractions, 87, 92
  of whole numbers, 14
sum, 14

T
tables, 197
term, 121
transversal, 255
triangle, 225
  area of, 239
  congruent, 264
  equilateral, 257
  isosceles, 257
  perimeter of, 227
  right, 257
  scalene, 257
  similar, 261

V
variable, 202, 204
vertical lines, 225
volume, 244

W
word problem, 19
  five-step approach to solving, 19
  multistep, 31

Z
zero
  in decimals, 53
  in division, 17

See 147

post test 1-6

7. page 153